CARTOGRAPHIES OF DANGER

Mark Monmonier

CARTOGRAPHIES OF
DANGER

Mapping
Hazards in
America

THE UNIVERSITY OF CHICAGO PRESS
Chicago and London

MARK MONMONIER is professor of geography in the Maxwell School
of Citizenship and Public Affairs at Syracuse University. He is the
author of *Drawing the Line: Tales of Maps and Cartocontroversy* (1995),
as well as of *Mapping It Out: Expository Cartography for the Humanities and Social Sciences* (1993) and *How to Lie with Maps* (2nd edition,
1996), both published by the University of Chicago Press.

The University of Chicago Press, Chicago 60637
The University of Chicago Press, Ltd., London
©1997 by The University of Chicago
All rights reserved. Published 1997
Printed in the United States of America

05 04 03 02 01 00 99 98 97   1 2 3 4 5

ISBN: 0-226-53418-9
ISBN: 0-226-53419-7

Library of Congress Cataloging-in-Publication Data

Monmonier, Mark S.
    Cartographies of danger: mapping hazards in America / Mark
Monmonier.
       p.   cm.
    Includes bibliographical references and index.
ISBN 0-226-53418-9 (cloth: alk. paper).—ISBN 0-226-53419-7
(pbk.: alk. paper)
1. Natural disasters—Maps. 2. Hazardous geographic environments—
Maps. I. Title.
GB5014.M66 1997
363.34`022`33—dc20   96-35082
                                                                 CIP

to
JOHN SNYDER
innovator,
scholar, and
conscience

Contents

Preface

Disasters usually surprise their victims. "Why me?" they ask. "Why now?" "Why here?" These questions often have obvious answers, at least in the sense of relative risk. For example, the campus of the university where I teach abuts a large public park—a nice place to walk, toss Frisbees, or lie in the sun. At dusk, though, Thornden Park becomes a place of danger, especially for female students walking alone. But because we warn students of the risk, tragic surprises here are rare. In large cities, some universities provide "no-go" maps showing areas best avoided at night and even during the day. Crime, of course, is but one of many hazards for which risk varies from place to place. Wouldn't it be helpful to have no-go, no-build, or no-live maps for all kinds of nasty surprises? As this books illustrates, we do—sort of—but what's available is spotty, often unreliable, and still evolving.

How hazard mapping got where it is, and how quickly, is an intriguing story that can increase our ability to view maps critically and analytically as well as introduce new readers to the geographer's perspective. As an educator and researcher specializing in map design and cartographic information, I hope to share my appreciation of maps as

tools for organizing vast amounts of information and for making decisions about land use and emergency preparedness. I hope also to
demonstrate how a geographer thinks about hazards—not just natural hazards like earthquakes and floods but man-made hazards like
chemical accidents and nuclear waste and social ills like crime and disease. The examples I discuss should help the reader understand better
the degree of risk in various parts of the United States and gain insight
on how our individual mental maps lead us to overestimate some dangers and underestimate others. By promoting greater awareness of
hazards of all types, this book can affect decisions we make about
where to live and how involved we become in the environmental concerns of our communities.

This book approaches the geography of environmental hazards
pragmatically—by examining efforts to map them. Although a richly
colored atlas of hazards or a thick listing of place-ratings might be easier to use, in many cases the necessary information simply isn't there.
Data are meager for some hazards and often incompatible from state
to state. And existing maps might be dangerously misleading. California, for example, takes its earthquakes seriously, with a comprehensive
program of seismic mapping, whereas Washington has been comparatively lax about earthquakes in Puget Sound. Even so, California's
detailed seismic-hazard maps reveal the folly of treating metropolitan
areas as homogeneous. In January 1994, for instance, the Northridge
earthquake demonstrated that ostensibly reliable maps readily deceive
anyone who thinks "known faults" are the only faults. Other hazards,
such as radon and chemical releases, are even less accurately invento-
ried than earthquakes, and efforts to map flood hazards sometimes
seem less a step than a stumble in the right direction.

However imperfect, past and current efforts to map hazardous geographical environments are informative: in addition to explaining why
an accurate *Geographical Hazards Almanac* is decades (if not centuries)
in the future, an examination of hazard mapping offers valuable
insights about scientific knowledge, public policy, and popular perceptions. Contrasts among hazards are especially revealing: highly
generalized national maps are useful for some hazards but not others,
several hazards require a variety of maps, and mapping is a promising
strategy for coping with such newly recognized, poorly understood
threats as indoor radon and electromagnetic radiation from power
lines. In addition to discussing the use and limitations of representa-

tive maps, the book identifies sources of additional information important to home owners and local officials.

The chapters here are organized according to "hazardous geographical environments," a Library of Congress subject category with a pertinent juxtaposition of words. The noun *environments* reflects the book's focus on cartographic challenges, rather than physical processes or types of maps. Because environments such as coasts and floodplains are what people occupy and governments must manage, any other approach would be fragmented and cumbersome. Earthquake zones and volcanic areas, after all, are distinct environments, each involving multiple hazards requiring several different types of maps. Although earthquakes and tsunamis (seismic sea waves) result from ruptures in the earth's crust, grouping tsunamis with other coastal hazards makes more sense because earthquake zones and shore environments are very different cartographic challenges. The adjective *geographical* sharpens the book's focus by excluding such personal or workplace hazards as unsafe sex, smoking, and flammable construction materials. Nearly ubiquitous in Western societies, these nongeographical hazards have small (and not always stationary) zones of impact associated with homes or businesses rather than neighborhoods or regions. Finally, the adjective *hazardous* connotes a concern not only with massive, comparatively rare disasters that attract national attention but also with more subtly dangerous, widespread, and persistent phenomena, such as air pollution and hazardous waste.

My selection of hazardous environments is necessarily incomplete. Among the hazards not represented are avalanches, fire storms, mudslides (except when related to volcanoes), pandemics, and transportation accidents not involving hazardous chemicals. Other hazards, such as atomic attacks, landslides, and severe winter weather, are treated only cursorily. While some omitted hazards are either rare, insignificant in their impact on humans, or cartographically uninteresting, others would prove usefully revealing. My choices largely reflect a trade-off between comprehensiveness and representativeness as well as between taxing the reader and telling the story.

Another necessary compromise is a concentration on the United States, the region I know best and the country where most readers are likely to reside. Unless the book became a lifetime work, wider coverage would be spotty in its examples and still reflect an unavoidable Euro-American bias. It seemed better to concentrate on topics I knew or could research efficiently, and let Canadian, European, and other foreign readers view the result as a glimpse of what we Yanks are doing about geographical hazards.

Although maps are its focus, *Cartographies of Danger* is very much concerned with underlying physical processes—good geography never asks *where* without also asking *why*. An insightful evaluation of hazard-zone mapping not only demands a basic understanding of relevant physical processes but also fosters an appreciation of cartographic contributions to scientific knowledge. This interest in understanding the hazards themselves fits well with my emphasis on how maps are used and what they can tell us—although important here and there, map design and cartographic techniques ought not get in the way of the story.

Acknowledgments

Many people contributed to this book by discussing their work, explaining products and processes, providing illustrations and reports, or suggesting additional agencies or people I might consult. I am especially indebted to Ted Algermissen, Earl Brabb, Dan Miller, Jim Reihle, Pam Tatalaski, and Madeline Zirbes, U.S. Geological Survey; Leslie Atkin and K. C. Chartrand, EIS International; Tom Behm and Dan Cotter, Federal Emergency Management Agency; Carolyn Rebecca Block and Daniel Higgins, Illinois Criminal Justice Information Authority; Tom Bowman, Jefferson County (N.Y.) Emergency Management Office; Rodger Brown and J. D. Ziemianski, National Oceanic and Atmospheric Administration; Robert Burdick and Lisa Letteney, Onondaga County (N.Y.) Department of Health; Clay Burns, Niagara Mohawk Power Corporation; Carol Cavalluzzi, Geology Librarian, Syracuse University; Mike Clemetson, Deborah Janes, and Frank Marcinowski, U.S. Environmental Protection Agency; Gordon Clickman, New York State Department of Environmental Conservation; Rich Cobb, Wayne County (N.Y.) Emergency Management Office; Robert Dolan, Department of Environmental Science, University of Virginia; Ute Dymon, Department of Geography, Kent State University; Joe Falge, Onondaga County (N.Y.) Emergency Management Office; Steve Forand, Peter Lauridsen, and Aura Weinstein, New York State Department of Health, Albany; Edwin Fox, U.S. Nuclear Regulatory Commission; Richard Friess and Dan O'Brien, New York State Emergency Management Office; Gerald Galloway, U.S. Military Academy; Warren Horst, Fire Marshall's Office, Suffolk County, N.Y.; Nancy Kraft, *Syracuse Post-Standard*; Allan McDuffie and Gene Stakhiv, U.S. Army Corps of Engineers; Matt McGranaghan, Department of Geography, University of Hawaii; Dennis McNeese, State Farm Fire and Casualty Company; Tony Nero, Lawrence Berkeley National Laboratory; Jeanne Perkins, Association of Bay Area Governments; Linda Pickle, National Center for Health Statistics; Mary Schneider, Vermont Yankee Nuclear Power Corporation; Steve Segal, *Greensburg Tribune-Review*; Dan Wartenberg, Environmental and Occupational Health Sciences Institute; and M. Gordon Wolman, Department of Geography and Environmental Engineering, Johns Hopkins University.

I am also grateful to Syracuse University for a sabbatical leave; to the Maxwell School of Citizenship and Public Affairs for an Appleby-Mosher travel grant; to Mike Kirchoff for assistance in scanning facsimile maps; and to Sona Andrews, Risa Palm, and John Western for constructive comments on the manuscript.

Chapter One

Map Scale, Danger Zones, and Safe Places

A convenient place to begin is with the effect of map scale on the perception and portrayal of danger. The most basic of cartographic concepts, scale determines the detail, size, cost, and reliability of a map and restricts the range of appropriate uses. Although its most important ramifications are qualitative, scale is a numerical concept, which cannot be broached without at least a few numbers. I promise, though, to keep the treatment brief and the arithmetic simple.

Cartographers define scale as the ratio of the length of a symbol on the map to the length of the corresponding feature on the ground. On a typical topographic map with a scale of 1:24,000, a 1-inch line represents 24,000 inches on the ground, or 2,000 feet. The ratio is often written as a fraction, in this instance, 1/24,000. By convention, the numerator is 1, and the same units (feet, meters, miles, or whatever) are implied for both numerator and denominator: 1 inch representing 24,000 inches is the same scale as 1 centimeter representing 24,000 cm. Map scale can also be portrayed graphically with a scale bar representing one or more typical distances or stated verbally, as in "One inch represents two thousand feet."

1

Geographers use the fractional form to distinguish *large-scale* maps from *small-scale* maps. The rule to remember is that large-scale maps have larger fractions, with small denominators, and small-scale maps have tiny fractions, with relatively large denominators. For example, 1/24,000 is markedly larger than a scale of 1/250,000, whereas 1/100,000,000 is much, much smaller. Most geographers regard maps with scales of 1:24,000 or larger as "large-scale" maps and maps with scales of 1:500,000 or smaller as "small-scale," with "intermediate-scale" a fuzzy category somewhere in between. These labels are useful because large-scale maps can cover neighborhoods and other small areas in considerable detail, whereas small-scale maps provide highly generalized portraits of states, countries, or continents. At 1:1,000,000, for instance, a one-inch square on a map might encompass a vast city, whereas at 1:10,000, the same square would cover no more than a few city blocks.

Map scale affects the generalization and usefulness of a map. Because a small-scale map cannot hold as much local detail as a large-scale map, its hazard zones necessarily have smoother, comparatively vague boundaries. Moreover, when a small-scale map is compiled from a set of large-scale maps, not as many roads, streams, and boundaries can be shown, and small occurrences of principal features must be amalgamated with neighbors or dropped altogether. Without further generalization, the map would be hopelessly cluttered and useless. Yet even though details must be sacrificed to clarity, a small-scale map can condense an enormous amount of information into a convenient size. For some hazards, a concise, small-scale overview of regions at greatest risk can enlighten policymakers as well as the general public.

The six highly generalized maps in Figure 1.1 have very small scales. The Electric Power Research Institute's monthly magazine used them to illustrate a story on disaster planning. Scattered throughout the article as elements of an attractive page layout, the original maps appeared at a scale of approximately 1:60,000,000. I made them even smaller (about 1:100,000,000) to fit onto a single page. Although such very small scales preclude state boundaries and place names, the grossly distorted coastlines and international boundaries provide an informative caricature of the United States.

Intended largely as a graphic decoration, the crude earthquake map at the upper left in Figure 1.1 identifies seismic hazard zones not only in California and Puget Sound but also in Utah, South Carolina, and the St. Lawrence Valley. And the dark area near the middle of the map

Earthquakes

Ice and snow

Floods

Tornadoes

Hurricanes

Geomagnetically induced currents

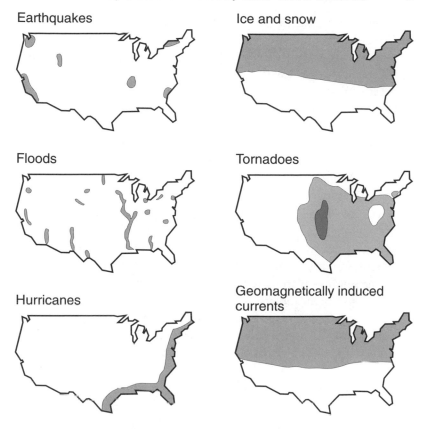

Figure 1.1. Six highly generalized, small-scale hazard maps. Reprinted, with permission, from Bob McGee, "Preparing for Disaster," *EPRI Journal* 17 (September 1992): 22–33.

represents the country's largest known earthquake, a series of four massive shocks centered on New Madrid, Missouri, and felt throughout a large part of the United States during the winter of 1811–12. Other maps in the series show that hurricanes are troublesome along the Atlantic and Gulf coasts, ice and snow affect the northern half of the country more than the southern half, and geomagnetically induced currents (a danger principally to electrical transmission lines) are related to latitude.

Like many cartographic generalizations, the earthquake map lumps together several areas with noteworthy differences. As the next chapter describes, earthquakes in California are more frequent and affect much smaller areas than earthquakes in South Carolina or the central Mississippi Valley. Similarly, the ice-and-snow map does not distinguish among wind-driven blizzards in the Midwest, massive snowfalls east of Lake Ontario, and damaging ice storms farther south.

Another necessity of cartographic generalization, the suppression of small occurrences, is especially obvious on the flood map, which highlights the Mississippi floodplain and fourteen smaller areas experiencing disastrous flooding in recent decades. Many more flood-prone areas exist, but most are too small to include. Moreover, the flood zones on the map must be exaggerated to make them show up. Although flooding is a serious hazard along the upper Mississippi, for instance, the inundated area is usually much narrower than the fifty to one hundred miles suggested by the map.

The tornado map, with two distinct graytones showing different levels of risk, is less generalized than the other five maps. A light gray shading represents a moderate-risk zone covering much of the Midwest and Southeast, while a darker symbol identifies a high-risk belt running southward from Iowa into Texas. However crude this two-level classification and its geographic patterns, the map makes an important distinction between a *hazard*, which threatens life or property, and *risk*, the probability of realizing that threat. All hazard-zone maps are risk maps because they imply a greater threat in the vicinity

Frequency of tornadoes: 1953–1962

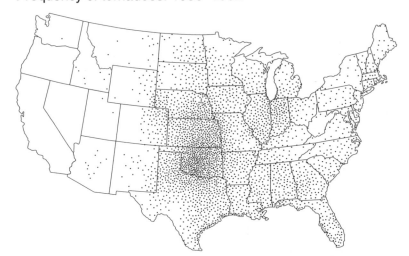

Figure 1.2. Frequency of tornadoes, 1953–1962. Each dot represents the approximate location of two occurrences during the ten-year period. An occurrence is defined as the first point of contact with the ground. Reprinted from U.S. Geological Survey, *National Atlas of the United States* (Washington, D.C., 1970), p. 113.

of the hazard. But in this case, different area symbols express different degrees of risk.

Risk maps are especially important for tornadoes, which concentrate massive amounts of energy on very small areas and can strike anywhere. The typical tornado cuts a narrow swath, less than a thousand feet wide, and, except for flying debris, has little effect on nearby areas. Although its funnel cloud might skip across the surface for a distance of twenty miles or more, touching down briefly here and there, tornadoes seldom leave a trail of destruction more than a few miles long.

Tornadoes are so numerous (perhaps eight hundred a year in the U.S.) that a small-scale map cannot describe individual storms, at least not for more than five years or so. The dot map in Figure 1.2 strikes a visually effective compromise by representing only locations at which tornadoes first touched down, rather than the direction and length of their paths. To avoid graphic clutter in high-risk areas, a single dot represents an average location for two tornadoes. The *National Atlas of the United States* used this map to show tornado frequency for a ten-year period starting in 1953, when widespread use of weather radar and improved record-keeping began to yield reliable data.[1] Closely spaced dots reflect a comparatively high risk in Oklahoma and Kansas, Indiana and Illinois, and parts of the Southeast. These regions are especially hazardous on spring afternoons, when air rising from the heated surface can trigger violent upper-atmospheric encounters between warm, moist air from the Gulf of Mexico and dry, cold air from the north and west. By comparison, tornadoes are rare in the Rocky Mountain and Pacific Coast states, seldom reached by moist tropical air.

An accurate portrait of tornado risk requires more than a single decade of data. Based on twenty-eight years of record, Figure 1.3 offers greater reliability at the price of less precisely reliable symbols.[2] To accommodate additional data as well as illustrate significant regional differences, the mapmaker threaded isolines (similar to the contour lines on topographic maps) through places of approximately equal risk. For places directly on an isoline, decoding seems straightforward: the eye need only follow the risk contour to its label (conveniently stated as an annual average). But because contouring is based on a small number of data points, the positions of the lines and their labeled values can be misleading. For places between isolines, the viewer must accept a range of values or make a calculated guess. At Washington, D.C., for instance, the average yearly tornado inci-

Average annual tornado incidence per 10,000 square miles:
1953–1980

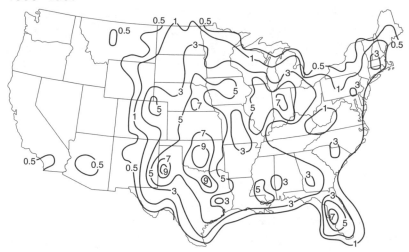

Figure 1.3. Average annual tornado incidence per 10,000 square miles, 1953–1980. Reprinted from National Weather Service, *Tornado Safety: Surviving Nature's Most Violent Storms* (Washington, D.C., 1982), p. 8.

dence is between 1 and 3 tornadoes per 10,000 square miles. That is, in an average year a square 100 miles on a side will experience at least one but no more than three tornadoes. Viewers preferring a single number can note that Washington is slightly closer to a 3.0 contour than to a 1.0 contour and use this information to "interpolate" a rough estimate of 2.2. Though more precise, this single number is less dependable than the range 1 to 3 because interpolation from smoothed contours on highly generalized small-scale maps is notoriously unreliable. Like most small-scale maps, Figure 1.3 is trustworthy only as a geographic overview.

Comparison of the three tornado-hazard maps illustrates another pitfall of map generalization—the possibility that viewers might associate low risk with no risk. Although less common than in Kansas, for instance, twisters occasionally strike New England, upstate New York, and other white areas on the tornado map in Figure 1.1. I know from experience, as well as from the other two maps, that the white is misleading. A few years ago, on a Sunday drive through Fulton, New York, we saw the destructive signature of a small tornado that had skipped across town ten minutes earlier. Although the storm damaged several roofs, cars, and power lines, and started a few minor fires, no one was killed. But had the tornado lingered longer or struck a more populated area, it could have caused a major disaster. Fortunately for area residents who might have been outdoors at the time, National Weather Service personnel responsible for severe storm warnings do not ignore the possibility of tornadoes within the map's white, low-risk zones.

Map scale is no less important than the map author's fears, experiences, and group values. White middle-class Americans, for example, differ markedly from impoverished African Americans in the places they avoid, and, not surprisingly, both groups view danger differently from the scientists and government agencies responsible for most hazard maps. These differences are readily apparent in the *Places Rated Almanac* and similar guides pitched at comparatively affluent readers who either are free to move or enjoy thinking about relocating. A brief examination of the popular quality-of-life genre holds useful insights.

Place-ratings guides adopt one of two designs: the travel narrative, in which individual chapters describe each well-rated place, or the systematic inventory, in which individual chapters address specific locational factors, such as climate, crime, and employment. The prototypical place-ratings travelogue is *Safe Places*, published in 1972 and substantially revised in 1984 as *Safe Places for the 80s*.[3] The earlier edition describes fifty towns and small cities in thirty-five states, which journalists David and Holly Franke visited on a 105-day, 20,000-mile trip that began and ended in Manhattan. Although a map of the Frankes' circuitous route decorates the endpapers, they grouped their chapters alphabetically by state and shared no experiences enroute from one town to the next. But by integrating interviews and personal observations with facts and figures, they avoided much of the tedium of fifty chapters of listy prose constrained by a standard outline. I've had my copy for more than two decades—a gift from Marge, who knows I like to daydream about places—but have never read the 932-page volume cover to cover. Most readers, I suspect, use the book as I do: as an informal geographic encyclopedia.

As its title implies, *Safe Places* is about avoiding crime, urban riots, and other big-city unpleasantness that made "law and order" an effective political slogan for conservative and centrist Republicans. (Husband David had worked on the *National Review* and wrote a book about William F. Buckley, Jr., and wife Holly had been a staff assistant for two Connecticut congressmen.) Although the Frankes also considered affordable housing, low property taxes, and other quality-of-life factors, they selected their fifty places largely because of low crime rates. Not surprisingly, none of their recommended places was a large city, and only one (Green Bay, Wisconsin) was a medium-size city (population between 50,000 and 250,000). Their introductory chapter underscores this point with an anecdote: throughout the trip, luggage left in the back seat was safe at night, but several weeks after the Frankes returned home to New York, someone stole their car.[4]

Although the Frankes later moved to the village of Ridgefield, Connecticut, a safe sixty miles away, crime remained the overriding concern. The introduction to *Safe Places for the 80s* describes their simple three-step screening process:[5]

1. Determine the overall crime rate.
2. Choose a top limit on the overall crime rate.
3. Compile the rates for each category of crime.

The result was an expanded list of 105 low-crime communities in thirty-nine states—only 13 carried over from the earlier edition. To justify their narrow view of "safe," the authors brushed aside two other common hazards:[6] "Each year almost one out of three households in the United States is hit by crime. Crime is more a threat to you than the possibility of an automobile accident or being stricken by cancer or heart disease." In addition to ignoring the more frequently fatal consequences of cancer and cardiovascular disease, they also dismissed the toxic threats of Three Mile Island, Love Canal, and Mount St. Helens.

Safe Places contains no hazard maps, not even a map of crime rates. But for each town or city a small, arty regional map shows principal highways, physical features, and a selection of landmarks and neighboring cities. For example, the map for Belvedere, California (in Marin County, north of San Francisco), verifies that San Francisco Bay isolates the village from "hippies and street people" in Haight-Ashbury and Berkeley. The map also shows the nearby village of Tiburon as well as Sausalito, Oakland, and the direction to Point Reyes National Seashore—but not the Hayward and San Andreas faults.[7] In their accompanying text the Frankes mention the San Andreas fault fleetingly, but only as the site of the few communities on the Peninsula without high rates of property crime. Nowhere do they tell or show the reader that parts of Belvedere are susceptible to violent shaking from ruptures along either of the area's major faults.[8]

The genre's other exemplar is the *Places Rated Almanac*, subtitled

Your Guide to Finding the Best Places to Live in North America.[9] As its name implies, *Places Rated* scores the nation's 323 metropolitan areas (plus another 20 in Canada) on ten factors of interest to anyone thinking of locating a business, relocating a family, or retiring. These factors cover a broad range of concerns: cost of living, jobs, housing, transportation, education, health care, crime, the arts, recreation, and climate. Authors David Savageau (a "personal relocation consultant") and Richard Boyer (an award-winning mystery writer) use statistical data, mostly from government agencies, to rate and rank all major cities separately for each factor. In the final chapter, they add up the ratings and construct a single, composite ranking from "best" to "worst." These rankings and the reactions of local officials in the nation's "best" and "worst" cities are widely reported by the news media every four years, when the book is revised. Boosters who are unabashedly enthusiastic or predictably angry have helped make *Places Rated* a persistent best-seller since its first edition in 1981.

Despite a few chapters that mention environmental threats, *Places Rated* is less aware of hazards now than in the early 1980s. Crime rates computed by the FBI continue to influence the ratings, but no other factors play a direct role. In 1985, for instance, a joint health-care/environment index not only included demerits for each pollutant that exceeded the EPA standard but levied additional penalties against cities with dangerously high ozone levels. Moreover, although earlier editions treated nuclear power as an environmental hazard, the cost-of-living chapter now mentions only the added burden on electric customers of local utilities owning a nuclear plant. The climate chapter, which continues to rate places only for climatic mildness, retains a short discussion of natural hazards (hurricanes, tornadoes, and earthquakes) but dropped an earlier section on water quality.

Even though hazards have little direct effect on its ratings, *Places Rated* includes a few informative small-scale hazard maps. In the climate section, for instance, a U.S. map shows high- and moderate-risk areas for tornadoes and hurricanes, and a similar map describes geographic variation in both the severity and the frequency of earthquakes. And another map identifies places mentioned in the chapter's discussion of weather extremes. These maps are part of the rich supplementary information overlooked by journalists and other readers interested primarily in the almanac's famous (or infamous) composite rankings.

Although *Places Rated* dominates the genre in sales, several other

books serve important niches. In 1993, for instance, Norman Crampton based *The 100 Best Small Towns in America* on a screening of government statistics and more than five hundred telephone interviews
with local officials and other boosters.[10] By focusing on communities
with populations between 5,000 and 15,000, he appealed to readers
seeking a quality of life very different from *Places Rated*'s metropolitan
areas, defined by a central city with at least 50,000 residents. Although
crime rate is only one of fourteen selection factors, Crampton's
approach and organization is similar to *Safe Places*. (In 1974, David
Franke's *America's 50 Safest Cities* offered Conservative Book Club
members and other readers a somewhat similar, crime-only treatment
of cities with populations *over* 50,000.[11]) Crampton ignores other hazards anecdotally as well as systematically. For example, the chapters
for Moses Lake, Washington, which received ash from the Mount St.
Helens eruption in 1980, and St. Helens, Oregon, only fifty miles away,
reveal nothing about volcanic hazards. And his book has only one map,
a double-page spread showing state boundaries and identifying the
towns by name.

Published a decade earlier, *The Best Towns in America: A Where-to-
Go Guide to a Better Life* demonstrates that older is not always less
informative.[12] In selecting fifty places with populations between
25,000 and 100,000, author Hugh Bayless consulted maps as well as
numerical data. In addition to crime, taxes, accessibility to a major
city, and various regulations on land use, gun ownership, and pollution, Bayless considered a variety of natural hazards. His second chapter, "Best Towns Are Safe," presents small-scale versions of several cartographic sources, including maps of tornadoes, hurricanes, potential
nuclear targets, potential pollution (nuclear power plants, chemical
waste dumps), and earthquakes. Less reliant on statistical data than
other writers, Bayless was circumspect as well as confident in his selec-
tions: "While nowhere is completely safe, there are many places where
safety is great and hazards are distant. All of our best towns are reasonably safe—much more so than the big cities, and appreciably more
so than other towns."[13]

Concerned more with removing hazards than merely avoiding
them, environmental scientist Benjamin Goldman published a guide
similar to *Places Rated* in its organization and use of numerical data but
narrowly focused on technological hazards.[14] His title and subtitle *The
Truth about Where You Live: An Atlas for Action on Toxins and Mortality*
aptly describe his approach. With more than a hundred maps, the book

clearly qualifies as an atlas. Almost half of its maps address mortality directly, by showing death rates for the 3,073 counties of the forty-eight contiguous states. Nearly as many maps provide equally detailed county-level portraits for a variety of industrial toxins and other pollutants, and additional maps address several relevant demographic, socioeconomic, education, and health services factors.

Although "where you live" exaggerates the precision of county data, Goldman offers a comprehensive treatment of both effects and causes. Too coarse to show individual neighborhoods and the effects of minor polluters, county data nonetheless provide a practicable compromise between state and metropolitan area data, which lack geographic detail, and city and town data, for which mortality rates can fluctuate wildly from year to year—particularly for less common causes of death. Rates for small places are intrinsically unstable because an extra death during the year can boost the death rate by 10 percent or more, whereas one fewer death could cause an equally prominent drop. To provide greater reliability, Goldman computed average rates for a fifteen-year period and also adjusted for age differences among counties. Without age adjustment, high death rates would typically occur—unsurprisingly and with little meaning—in counties with comparatively older populations. To identify influences of race and gender, he prepared separate maps for white males, white females, minority males, and minority females for ten environmentally sensitive mortality factors, including infant mortality, birth defects, and lung cancer.

Goldman's typical treatment is a two-page spread with a map on the left and a table on the right. While the table lists rates and names for both the most and the least hazardous 2 percent of the nation's counties, the map emphasizes only the worst places. Solid black area symbols point out the 61 most hazardous counties in the top 2 percent, while progressively lighter graytone symbols represent counties in the next lower 3, 5, and 15 percentiles. Although white (blank) symbols for the remaining 2,305 counties might (perhaps quite wrongly) imply little or no risk in the lowest 75 percentile, the contrast between this larger white area and the black and gray counties appropriately highlights high-risk regions, where need for action is greatest. Nonactivist readers can, of course, use Goldman's identifications of best and worst areas as an environmental supplement to the *Places Rated Almanac*.

My final example is only an honorary member of the place-ratings genre: *Risks and Hazards: A State by State Guide* does not rate places and was never sold commercially.[15] I obtained my copy by writing to the Federal Emergency Management Agency (FEMA) and asking for publication no. 196, a 130-page booklet distributed free to promote disaster preparedness. This colorful and informative guide to seven hazards (earthquakes, volcanoes, hurricanes, tornadoes, snow and extreme cold, nuclear attack, and nuclear power plants) includes checklists of preparations and responses as well as brief treatments of tsunamis, floods, hazardous dams, and radioactive fallout. Our taxes paid for it, of course, and as collections of small-scale maps go, it's a good buy. My principal complaint is its state-by-state organization, which accords exactly two pages of maps to each state. The U.S. Senate might approve, but does tiny Rhode Island need—or deserve—as much space as complex California?

Each state's double-page treatment consists of a single "Nuclear Attack" map on the left-hand page facing several smaller hazard maps

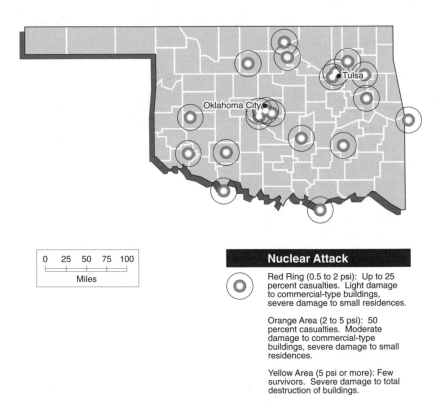

Figure 1.4. Nuclear attack map for Oklahoma. Reprinted from Federal Emergency Management Agency, *Risks and Hazards: A State by State Guide,* FEMA publication no. 196 (Washington, D.C., 1990), p. 86.

on the right-hand page. Although the nuclear-attack map reflects a threat less troubling now than during the cold war, its detailed depiction of likely targets is a compelling argument for any defense, including Star Wars and disaster preparedness, against the madness of "mutually assured destruction." As Figure 1.4 illustrates in black and white for Oklahoma, vivid blast-like symbols mark major cities, military bases, large dams, power plants, and other important industrial sites. Each three-part symbol uses yellow to identify an inner area with "few survivors," orange to show a larger surrounding zone with "50 percent casualties" and significant structural damage, and a thin red ring to bound an even larger outer zone with "up to 25 percent casualties" and lighter damage. (Densely packed, overlapping symbols in parts of Colorado, the Dakotas, Missouri, Montana, Nebraska, and Wyoming leave little doubt about the locations of missile silos.) Other maps suggesting the vulnerability of metropolitan populations inspire awe and respect, if not fear and anger, in the face of the nuclear standoff. In dramatizing the consequences of a nuclear war, these fifty dismal maps argue not only for civil defense and fallout protection (a FEMA responsibility) but also for peace negotiations and nuclear disarmament.

Each opposing page carries two to six maps of other prominent threats, the number of which varies from six for California to only two for Idaho, Nevada, North Dakota, and Utah. Smaller and less detailed than the nuclear-attack map, these maps vary from hazard to hazard. Each map of nuclear power plants, for instance, identifies the state's plants by name and location. (Commercial nuclear plants are hazardous because an accidental release of radiation, however unlikely, is nonetheless possible, and evacuation plans are required for ten- and fifty-mile "emergency planning zones" around each plant.) For each Pacific state a map with a red band along the coast points out areas "historically subject to tsunami." For states with active volcanoes, a hazard map identifies each volcano by name and uses a light, medium, or dark point symbol (little triangles) to show whether eruptions occur every two hundred, one thousand, or ten thousand years. Similar intensity codes describe three levels of risk for hazard zones on the earthquake, hurricane, and tornado maps.

Figure 1.5 illustrates in black and white two of the three risk maps provided in color for Oklahoma. Most prominent on the red-brown earthquake map is the absence of the solid red area symbol representing "high-hazard" areas; although a noteworthy threat, earthquakes

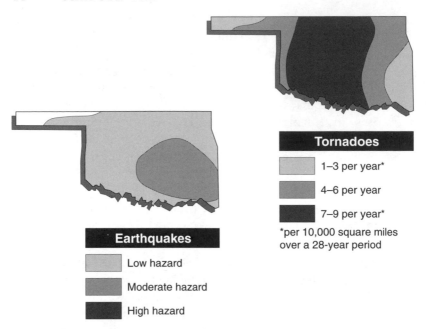

Figure 1.5. Earthquake and tornado maps for Oklahoma. Reprinted from Federal Emergency Management Agency, *Risks and Hazards: A State by State Guide*, FEMA publication no. 196 (Washington, D.C., 1990), p. 87.

are less hazardous here than in California or Alaska. In contrast, on the greenish tornado map, a dark-green area symbol (representing an average annual tornado frequency of 7–9 per 10,000 square miles) indicates that residents of central Oklahoma must prepare diligently for destructive twisters. A third map (not shown) reveals that extreme cold and freezing (but not heavy snowfall) is a significant hazard only in the northwestern quarter of the state.

Although its maps are highly generalized, *Risks and Hazards* is a useful source of hazards information not available in conventional place-ratings handbooks. But don't expect precision or easy answers. Careful page-by-page inspection of *Risks and Hazards* reveals that even for this limited subset of seven hazards, no part of the country is risk-free. And researchers tempted to construct a composite ratings map by piecing together the various maps and assigning weights to risk categories ought not ignore FEMA's warnings that flooding and fallout radiation are potential hazards in all areas. As the following chapters indicate, a thorough search for "relatively safe" places demands a careful and critical examination of more detailed sources.

Chapter Two

Shaky Preparations

E arthquakes are big news—at least those that kill several thousand people or occur near a media center such as Los Angeles or San Francisco. Much of our knowledge of seismic disasters comes from television, which reported the 1989 Loma Prieta and 1994 Northridge earthquakes as live, breaking news, much like the Persian Gulf War. In contrast, tragic earthquakes in Asia or Latin America—any place too remote for live, on-the-scene reports by well-known network personalities—are noted briefly, if at all. Newspaper accounts are a bit more comprehensive; a locator map and a few pungent quotes from Cal Tech seismologists catch the eye and make good copy. But the story is off page one the following day and forgotten within a week. Ten thousand Armenians dying amid crumbled masonry is nothing for the local paper's news editor to get worked up about.

Earthquakes make me cynical about the news media and our elected officials. Journalists rarely follow up on how governments, corporations, and individuals are preparing for the next major quake, and politicians readily forget promises to restrict land use, tighten building codes, and promote disaster planning—a pity that Gaia cannot be cast as the Evil Empire. But in chastising our leaders for not leading, we cannot ignore public aversion to regulations and taxes. As an old friend at the U.S. Geological Survey (USGS) once remarked, "The people get pretty much what they deserve."

Halfhearted attempts at seismic preparedness reflect an awkward mix of awareness and denial, evident in a distaste for long-range planning and a misperception of the consequences of a massive earthquake in a heavily populated area. Although long-range forecasts are scientifically valid, voters who have trouble balancing a checkbook have little interest in arcane claims that an earthquake of magnitude 6.5 or greater somewhere along a one hundred-mile fault is 10 percent likely within the next fifty years. Preoccupied with more immediate needs, most citizens recognize earthquakes as a threat but ignore the cost of rebuilding their metropolises. And even scientists savvy about probabilities and chaos theory seldom, if ever, ponder the jolt to the world economy when the long-awaited Big One hits downtown Los Angeles, San Francisco, or Seattle. No wonder, then, that seismic preparedness often means little more than a slow, back-burner commitment to retrofitting.

Despite the glacial pace of replanning and rebuilding, science has made substantial progress in understanding seismic activity and estimating its effects. As this chapter reveals, these advances are readily apparent in the maps of geologists and urban planners. Maps of field and laboratory data account for most of what we've learned about earthquakes, and mapping plays a key role in how we are using this knowledge to reconstruct cities. I should say "roles," though, because the maps operate at two levels: analytical and rhetorical. In providing dramatic, geographically detailed descriptions of the consequences of inaction, seismic maps are becoming a powerful tool of public policy.

What most people know about earthquakes is a naïve mix of fact and myth. Although perpetuated by maps, these misconceptions result mostly from misinterpretation of maps, not from factual errors. The most widespread popular fallacy is the half-truth that earthquakes are most hazardous along fault lines. It's true, of course, that ruptures in the earth's crust produce linear fractures geologists call faults. And it's also true that movement along faults often recurs, and that buildings and bridges on surface faults are indeed vulnerable. But it's folly to think that places several miles from the fault line are necessarily less hazardous than places directly on the fault—among other factors, risk depends on the stability of slopes and the susceptibility of poorly consolidated soil to shaking. In 1989, for instance, the Loma Prieta earthquake severely damaged buildings in San Francisco's Marina district, where eighty years earlier the city had filled a lagoon with sand and debris.[1] Ground shaking was intense and destructive even though the center of the earthquake was more than thirty miles away.

Another popular delusion is the belief that earthquakes in the United States are largely a California problem, with occasional occurrences along the Pacific coast in Oregon, Washington, and Alaska. That's what we learned in high school earth science, where the geologically impressive "Pacific Ring of Fire" described the association of earthquakes and volcanoes in a sweeping band extending up the west coast of South America and Mexico, northward past British Columbia, westward along the Alaskan Peninsula, southward through Japan and the Philippines into the East Indies, and then eastward around Australia and south toward New Zealand. To reinforce this notion, the circum-Pacific seismic belt not only contains most of the world's active volcanoes but is clearly evident on world maps of major earthquakes (Figure 2.1). Additional testimony is the world pattern of *tsunamis*, powerful sea waves caused by submarine earthquakes, which threaten coasts throughout the Pacific basin, including Hawaii and other midocean islands. When highlighted in red or accented with pictograms of erupting volcanoes—artists who illustrate high school texts and educational films seem to have a rich arsenal of disaster clichés—the

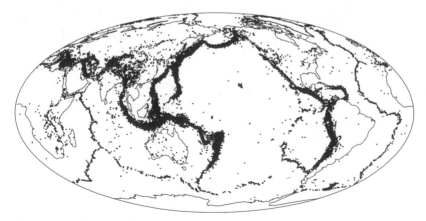

Figure 2.1. Epicenters of earthquakes of magnitude 4.0 or greater, 1980–1990. Courtesy of the National Earthquake Information Center, U.S. Geological Survey.

"ring" presents a vivid, memorable image that readily overshadows seismic dangers elsewhere on the map.

The intriguing theory of *plate tectonics* explains not only the circum-Pacific earthquake belt but prominent outliers in the Mediterranean and central Asia. Better known as *continental drift*, this grand geological narrative treats the earth's crust as a set of slowly moving slabs, or *plates*. Figure 2.2 shows the names and boundaries of plates in the circum-Pacific region. Fissure-like lines represent a gradual separation of plates through *sea-floor spreading* away from the East Pacific Rift, a deep trench bordered on both sides by long, parallel ridges. The Pacific plate is moving westward, away from the rift, at about four inches a year, while the Nazca plate is moving eastward at about two inches a year. Warm, fluid rock rising slowly to the surface from deep within the mantle replenishes the diverging plates and prevents the rift from becoming a vast canyon. The effect is similar to a

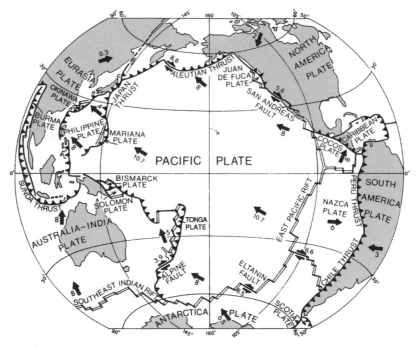

Figure 2.2. Major tectonic plates of the circum-Pacific region. Numbers represent annual rate of movement, in centimeters. Reprinted from Millington Lockwood and others, "Explanatory Notes for the Natural Hazards Map of the Circum-Pacific Region Pacific Basin Sheet," to accompany map CP-35 (Reston, Va.: U.S. Geological Survey, 1990), p. 22.

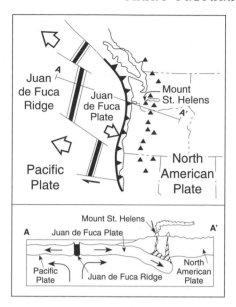

Figure 2.3. Subduction of the Juan de Fuca plate under the North American plate accounts for volcanic and seismic activity in the Pacific Northwest. Reprinted from Robert I. Tilling, *Volcanoes* (Reston, Va.: U.S. Geological Survey, 1992), p. 36.

diverging pair of huge conveyor belts. To complete the analogy, the plates collide along their opposite boundaries, typically with a more rapidly moving plate *subducting* (diving) beneath a relatively resistant neighbor. Line symbols with barbs showing direction of movement mark subduction zones that correlate nicely with the linear pattern of earthquake centers (Figure 2.1).

A map and corresponding cross section (Figure 2.3) describe subduction in greater detail for the Pacific Northwest. Named after the Greek mariner who assumed a Spanish surname and claimed to have explored the Pacific coast in 1592, the Juan de Fuca plate is a subsection of the Pacific plate moving eastward away from the Juan de Fuca Ridge, an impressive feature on maps of the sea floor. A chain of volcanoes extending from northern California into Canada marks the plate's subduction beneath the North American plate. The earth's interior is hundreds of degrees hotter than the surface, and as the plate descends, some of it melts into magma to feed volcanoes, such as Mount St. Helens. Friction between the moving plates produces numerous small earthquakes, particularly prominent during volcanic eruptions. The region has also experienced large, destructive earthquakes. Between 1,000 and 1,100 years ago, for instance, a massive rupture in the North American plate near Seattle abruptly raised the land as much as twenty-three feet in some places but dropped the surface elsewhere, creating numerous new lakes.[2]

Not all plate boundaries are spreading ridges or subduction zones. On the tectonic map of the Pacific (Figure 2.2), for instance, the plate boundary marked by the famous San Andreas fault lacks barbs indicating subduction. This omission is intentional because in this area the Pacific plate is mostly sliding northwestward past the North American plate. Geologists recognize the San Andreas as a *strike-slip fault*, in which blocks of land on opposite sides of the fault move horizontally and vertical displacement is less significant. Slippage is slow but relentless, with Los Angeles, west of the fault line, approaching San Francisco, slightly east of the fault, at an average rate of two inches per year. Stress energy, which accumulates continually, is released abruptly by periodic ruptures along short sections of the fault.[3] Although roads crossing the fault were offset nearly twenty-one feet by the infamous 1906 San Francisco earthquake, big earthquakes with a large offset are less typical than small earthquakes with little or no offset. Each year thousands of tiny slippages are evident only to highly sensitive seismographs.

More than eight hundred miles long and at least ten miles deep, the San Andreas fault is a spectacular feature, readily visible from the air and easily mapped. But treating this portion of the plate boundary as a single, well-marked fault is a dangerous generalization readily refuted by more detailed geologic maps. Ruptures in the crust occur not along a single fault line but within an elongated zone in which the main fault not only bifurcates into long offshoots but is accompanied by numerous shorter faults with roughly the same trend. Many of these parallel faults are only a few miles away, but a few lie fifty miles or more to the east. The possibility of a break along any of several fault lines is an aggravation for planners, who could more easily restrict land use along a single, well-identified fault line. Dispersed faulting is also a challenge for geologists, who must look carefully for old but deadly faults hidden by centuries of erosion and landslides.

However obvious the link between faults and earthquakes, California lawmakers largely ignored the connection until the 1971 San Fernando earthquake demonstrated a need for stronger building codes and tighter land-use controls. Enacted the following year, the

Alquist-Priolo Special Studies Zones Act outlawed home building on active faults. In a two-pronged effort to reduce damage, the legislature ordered the State Geologist to delineate *special studies zones*, now called *earthquake fault zones*, and encouraged cities and counties to regulate land use along faults. The mandate was pointed but flexible:

> to assist cities and counties in their planning, zoning, and building-reg-ulation functions, the State Geologist shall . . . [identify] appropriately wide special studies zones to encompass all potentially and recently active traces of the San Andreas, Calaveras, Hayward, and San Jacinto faults, and such other faults, or segments thereof, as he deems suffi-ciently active and well-defined to constitute a potential hazard to struc-tures from surface faulting or fault creep.[4]

Relevant state and local officials must review drafts of the maps before the Division of Mines and Geology (the State Geologist's staff) dis-tributes "official copies" to cities, counties, and state agencies.

California saved time and money by compiling its earthquake fault zone (EFZ) maps on large-scale (1:24,000) USGS topographic maps, which combine contours representing the land surface with symbols and labels for roads, rivers, boundaries, place names, and other refer-ence features. As Figure 2.4 illustrates, the EFZ maps reproduce these topographic symbols—originally printed in five or six colors—as a faint, single-color background for bold lines and labels marking active faults and zone boundaries. The heavy lines show faults where a sur-face rupture within the last eleven thousand years suggests the possi-bility of recurrence. (In geologic time, an eleven-century-old fault is sufficiently recent to be considered "active.") A solid line represents an "accurately located" fault, which geologists can observe and trace with confidence, whereas lines consisting of long dashes, short dashes, or dots express increasing uncertainty for faults that are approximately located, inferred, or concealed. Annotations include a question mark to show "additional uncertainty," the year of any recent offset with a known date, and the letter *C* to identify *creep*, the cumulative effect of slow, repeated displacement in small increments.

The map's lighter solid lines are official fault-zone boundaries, delineated as short straight-line segments joined at turning points marked by circles. To help planners and surveyors determine whether a parcel is within a fault zone, mapmakers position turning points on streams, roads, and other readily locatable features. Reflecting its offi-cial policy on relative risk, the Division of Mines and Geology usually sets zone boundaries 500 feet away from "major active faults" but only

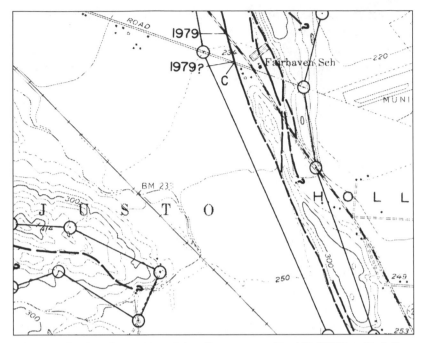

Figure 2.4. Portion of a typical special studies zones map. From Earl W. Hart, *Fault-Rupture Hazard Zones in California*, Special Publication no. 42 (Sacramento: California Division of Mines and Geology, 1992), p. 6.

200 to 300 feet away from well-defined, minor faults. These distances are not recommended setbacks but belts within which geologists and engineers should look carefully—through "special studies"—for active branches of the fault and other hazards. Zones can be markedly wider in localities with complex faulting or nonvertical faults. For example, a north-south fault dipping at a low angle to the east would require a wider corridor on the east side of the fault line, where the fault is not far below the surface.

The Alquist-Priolo act requires a continuing review of new data as well as the drafting of new and revised maps. Because of further study and refined procedures, the Division of Mines and Geology revised nearly a quarter of the 534 maps issued between 1973 and 1992, and extended map coverage as additional active faults were identified and officially zoned.[5] Revisions reflect newly discovered faults as well as zone-boundary changes resulting from more detailed local observation. Landowners or developers eager to convince officials the risk is tolerable or the map is wrong submit data from test drilling, exploratory excavation, or electronic geophysical surveys. Applications to develop residential subdivisions and other projects within an earthquake fault zone must include a geologic report by a registered geologist.[6] Although the State Geologist can waive the required report if the local government and its own geologist concur, developers requesting a waiver must present a detailed, well-documented justification. Waivers, reports, and appeals as well as information collected in the state's own evaluation occasionally lead to revisions.

Local zoning boards, with their own restrictions on land use, interpret the Alquist-Priolo maps in different ways. Most boards prohibit habitable buildings within 50 feet of a surface fault—well within the zones delineated on the state's maps—unless a detailed geologist's report demonstrates the absence of active branches.[7] But some jurisdictions are more restrictive. Portola Valley, for instance, accepts the standard 50-foot setback for known faults but requires a 100-foot setback where the fault's location is merely inferred (indicated on the maps by short-dash lines). But these restrictions apply only to single-family wood-frame buildings and other relatively resistant structures. For more vulnerable buildings, Portola Valley mandates greater setbacks, 125 and 175 feet respectively, for known and inferred fault lines.

Earthquake fault zone maps are easy to find if you live in the area. City and county planning departments usually display the maps covering their own jurisdiction, and some planning offices also sell inexpensive blue-line copies, similar to architectural drawings, made from reproducible masters provided by the state. But finding EFZ maps for other places can be difficult. The Division of Mines and Geology, which makes the maps, refers customers to the Blue Print Service Company in San Francisco.[8] To find out which areas are covered and the names of map sheets, the public can consult index maps in the Alquist-Priolo Earthquake Fault Zones guide the agency sells for three dollars.[9] As Figure 2.5 illustrates, the index maps identify only quadrangles with fault-rupture hazard zones mapped by the State Geologist. (Quadrangle names refer to the 1:24,000 USGS topographic maps used as base maps. Each 7.5-minute quadrangle covers 1/8 of a degree of latitude and 1/8 of a degree of longitude.) For the San Fernando quadrangle, near the top right of the excerpt, "R79" indicates that a revised map was issued in 1979.

Another way to see EFZ maps is to visit a real estate agent. Although California lawmakers were reluctant to ban the sale of existing homes on fault lines, they did insist that sellers (or their agents) inform prospective buyers that real property is within an earthquake fault zone.[10] Not messing around, the Alquist-Priolo act requires a formal disclosure, in writing, whenever a buyer offers to purchase a house or building lot. Wary of penalties and eager to respond knowledgeably to concerned customers, real estate firms usually keep copies of the maps. To help members comply with the law as well as to alleviate the fears of concerned home buyers, the Santa Clara County Board of Realtors and similar trade groups in other counties have even published their own street maps showing earthquake fault zones.[11]

Risa Palm, a geographer who studied Californians' perception of earthquake hazards for two decades, is skeptical about the real estate industry's use of the Alquist-Priolo maps. The disclosure requirement, she found, has little effect on the behavior of home buyers, who often don't discover their dream home is next to a fault line until they've

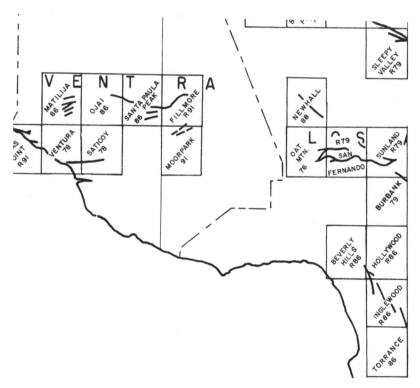

Figure 2.5. Portion of the Special Studies Zones index map covering parts of Ventura and Los Angeles counties. From Earl W. Hart, *Fault-Rupture Hazard Zones in California*, Special Publication no. 42 (Sacramento: California Division of Mines and Geology, 1992), p. 14.

already decided to make a purchase offer.[12] For example, in a survey of well-educated, middle-class Bay Area home buyers who recalled receiving an earthquake fault zone disclosure, only 24 percent had purchased earthquake insurance and a mere 9 percent had made structural improvements. Palm is particularly critical of the disclosure form's vague, one-sentence declaration that the house is "in an Alquist-Priolo zone"—hardly a revealing admission about earthquake hazards. A more effective strategy, she argues, is to require disclosure when the agent first introduces the purchaser to the property.

Whether Californians have learned much from two decades of fault-zone mapping is questionable. Although interest in geology, hazard insurance, and mitigation typically rises immediately after a significant earthquake, knowledge of local faults remains astonishing vague. In Santa Clara County, for instance, less than a year after the Loma Prieta earthquake, Palm and coworker Mike Hodgson found that 77 percent of home owners overestimated the distance to the nearest fault—in other words, a known fault was much closer to home than most residents thought.[13] Moreover, many could not correctly name the closest active fault. Allowed to choose among the San Andreas, Hayward, and Calaveras faults, only 21 percent of those living closest to the Hayward fault identified it correctly—a remarkable 57 percent incorrectly perceived the better-known San Andreas fault as closer. Perhaps too well aware of the recent rupture on the San Andreas fault, residents tended to ignore less charismatic faults more deadly because of accumulated stresses.

S urface faulting can be destructive, but fault lines are not the most hazardous locations. The 1994 earthquake that killed fifty-six people in and around Northridge and caused more than $15 billion in damage to buildings, highways, and personal property demonstrates the folly of hazard-mitigation planning focused largely on surface faults.[14] For a convincing argument, look at the Alquist-Priolo index map in Figure 2.5 and find the Canoga Park quadrangle, which includes Northridge. You can't—the Canoga Park map, just south of the Oat Mountain quadrangle, has no official earthquake fault zones. The earthquake that devastated Northridge originated far below the surface on an unknown fault.

Geologists readily identified the break as a *blind thrust fault*. "Blind" is obvious: the fifteen-mile-long rupture occurred at depth, about eight miles below Northridge.[15] A thrust fault is an inclined fracture in which one section of the crust is thrust upward over another section. Caused by a buildup of stress, the break shortens the crust, hurls upward an enormous amount of energy, and sometimes noticeably elevates the land surface.[16] Thrust faults are common in the Northridge area: the 1971 San Fernando earthquake raised the San Gabriel Mountains about six feet, and the more recent 1991 Sierra Madre quake injured more than thirty people and caused more than $33 million in damage.[17]

When a thrust fault doesn't break the surface, maps typically record only its *epicenter*, the point at the surface directly above the *hypocenter*, the point below the surface where the rupture began. Although the epicenter gives the earthquake a location and a name, the location is only an estimate. Two weeks after the Northridge quake, more precise calculations placed the epicenter a quarter mile south of Northridge, in Reseda.[18] But images of the disaster were welded to Northridge, and the name stuck.

Except as rhetorical warnings to complacent citizens, epicenters are of little value to planners. Damage is not necessarily most severe at the epicenter, after all, nor is a map of epicenters any more useful in forecasting damage than a map of fault lines. To plan effectively for earthquakes, governments must consider not only where a temblor might originate but also the vulnerability of the land surface to violent shaking, landslides, and *liquefaction*, a phenomenon in which sandy soil suddenly acts like a liquid. Also important, of course, are populations and structures at risk because of their locations.

Efficient integration of information about geologic hazards requires a new kind of interactive map called a *geographic information system*.[19] Better known by its acronym, a GIS is a computerized system for storing, retrieving, analyzing, and displaying geographic data. Figure 2.6 describes the most widely used GIS operation, the electronic overlay of individual *coverages*, each representing a specific risk factor or feature category. By registering several coverages to a common grid or coordinate system, a GIS can generate meaningful maps based on a combination of factors. For example, a planner could construct a three-factor map highlighting residential areas with either ten feet or more of unconsolidated fill or a slope angle exceeding fifteen degrees: populated areas vulnerable to shaking or landslides. To examine the impact of an earthquake along a particular fault, she might overlay a fourth coverage showing all areas within four hundred feet of a fault

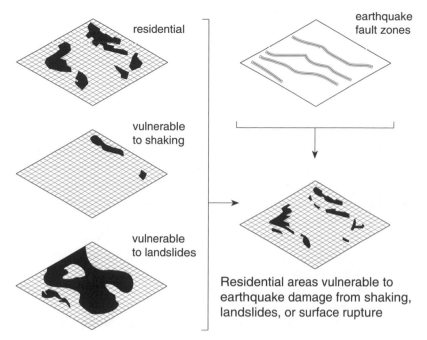

residential

earthquake
fault zones

vulnerable
to shaking

vulnerable
to landslides

Residential areas vulnerable to
earthquake damage from shaking,
landslides, or surface rupture

Figure 2.6. A geographic information system constructs multifactor hazard-zone maps by overlaying and combining single-factor maps, or coverages.

line. Although these operations can be carried out manually, using geologic and topographic maps and large sheets of tracing paper, the GIS not only calculates slopes and buffer zones but allows the analyst to experiment with different thresholds and combinations of factors. Without computers, this analysis—if attempted at all—would be less thorough.

GIS lets researchers simulate hypothetical earthquakes.[20] Figure 2.7 describes the use of a fine-grained grid to predict geographic variations in ground shaking. A geologist first constructs a map of surface vulnerability to seismic pressure waves; after examining data on soils and geology, she assigns each cell to a shaking susceptibility category. In an eight-level classification, for example, a rating of 1 indicates very low intensity while an 8 represents extremely high susceptibility. She then designs an ideal, plausible earthquake, with a given magnitude at a particular location along a specific fault. Geophysical theory lets her calculate and map the pattern at the surface of energy transmitted through bedrock from the hypothetical rupture.[21] She enters this "energy-attenuation" map into the GIS, which estimates the seismic energy level for each grid cell.[22] The system then multiplies the shaking-susceptibility and seismic energy coverages to yield a single map of ground-shaking intensity. Although earthquake modeling involves complex mathematics, this simplified description illustrates the need for a GIS to manage massive amounts of data for tens of thousands of small cells.

Jeanne Perkins helped develop hazard-zone mapping based on earthquake scenarios.[23] An information specialist trained as a geologist, she works for the Association of Bay Area Governments (ABAG). Although concerned with a broad range of hazards, Perkins focuses on ground shaking, a greater threat to the San Francisco region than landslides or surface faulting. The U.S. Geological Survey funds her work, and USGS seismologists provided the energy-attenuation maps for hypothetical ruptures along thirty-one active faults.[24] Each scenario represents a large earthquake occurring at a given depth and position along the fault. In consultation with other geologists and the USGS seismologists, she specifies length of the break and the magnitude of the earthquake.

When Perkins started modeling earthquakes in the late 1970s, the

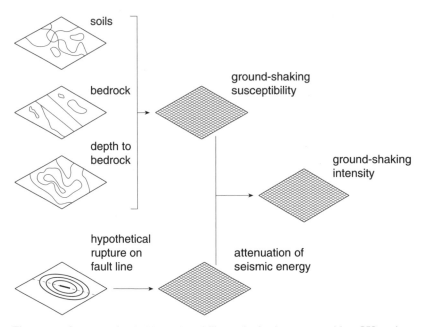

Figure 2.7. Coverages for shaking vulnerability and seismic energy enable a GIS to simulate the intensity of ground shaking for a hypothetical earthquake.

seismic energy map was a pattern of concentric circles, with the contour at the center representing the greatest energy and the outermost
contour representing the least energy.[25] To provide more realistic simulations, seismologists later modified the pattern to represent a rupture that starts in the middle and moves in both directions to opposite
ends of the break. Perkins calls the resulting pattern a "hotdog," in
contrast to the "donut" for the earlier scenarios. A more recent refinement is a pair of "teardrop" patterns, representing hypothetical ruptures starting at opposite ends of the break. As the rupture gathers
force, energy increases in the direction of movement to produce a very
lopsided pattern of shaking and damage. Each scenario now needs
three seismic energy maps: a hotdog model for a rupture starting in
the middle and twin teardrop models for ruptures starting at opposite
ends.

Perkins uses a GIS to integrate the work of seismologists, computer experts, cartographers, and planners. Her electronic map is an enormous grid of one-hectare (2.47 acre) square cells measuring 100
meters by 100 meters (328 feet by 328 feet). Known by the catchy
acronym BASIS, for Bay Area Spatial Information System, her GIS
generates *intensity map files*, some of which have been plotted and distributed to local governments. To explain these maps to local officials
and the public, Perkins prepared a booklet entitled *The San Francisco
Bay Area—On Shaky Ground*. Foldouts with excerpts for the central
San Francisco Bay Region highlight a selection of risk maps. *Shaky
Ground* also serves as an introduction to a varied set of twenty intermediate-scale (1:125,000, with one inch representing two miles) hazard maps covering ABAG's nine-county area.

Figure 2.8, an enlargement of a small portion of one foldout map,
illustrates the overprinting of area symbols representing ground-shaking intensity on a base map showing streets, freeways, and shorelines.
I chose this example because the intensity symbols, printed in red on
the original, don't overpower the base map, printed in black. (Red
turns black on black-and-white facsimiles.) The scenario is a magnitude 6.9 earthquake on the Hayward fault, about twelve miles east, on
the other side of the Bay—but not so large or close as to subject San
Francisco's waterfront to "violent" or "very violent" shaking. (On
other maps, though, the dense red symbols for these categories would
completely obscure other features on a black-and-white facsimile.)
Even so, diagonal-line symbols signifying "very strong" shaking point
out the vulnerability of filled land between Fort Point and North

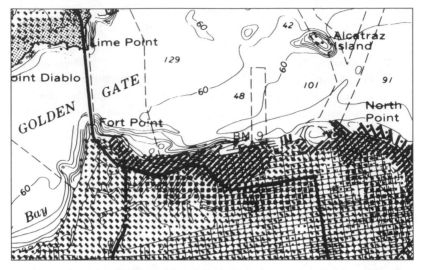

Figure 2.8. Enlarged excerpt of map showing ground-shaking intensity for a 6.9 earthquake on the Hayward fault. From Jeanne B. Perkins, *The San Francisco Bay Area—On Shaky Ground* (Oakland, Calif.: Association of Bay Area Governments, 1987), p. 7.

Point—the Marina district, devastated by the 1989 Loma Prieta earthquake, a magnitude 7.1 rupture on a fault thirty miles away.

Perkins's maps are persuasive and easy to read. Shading symbols printed in red connote danger, and progressively denser area shadings provide a logical light-to-dark sequence for the six categories of shaking intensity. The symbols range from nearly solid red to a sparsely textured pattern of fine red dots. Names and letter codes describe the corresponding levels of intensity:

A—Very Violent
B—Violent
C—Very Strong
D—Strong
E—Weak
<E—Negligible

These category names reflect the intensity scale developed by the pioneer earthquake cartographer Harry Wood for the 1908 *Carnegie Report* on the 1906 San Francisco earthquake.[26] "I'm not happy with the abrupt jump from 'strong' to 'weak,' and am thinking of changing 'weak' to 'moderate,'" Perkins told me in her office in Oakland.[27] "When would shaking from a massive rupture on the nearby Hayward fault ever be weak?"

An exemplar of cartographic risk communication aimed at lay readers, *Shaky Ground* explains unfamiliar or poorly understood terms. A chart entitled "How Big is Big?—Measuring Earthquakes" makes the important distinction between magnitude and intensity. *Magnitude* refers to the energy released by a rupture and is commonly measured by the widely familiar Richter scale. Decimal places are important in magnitude ratings—because the Richter scale is logarithmic, a 7.0 earthquake has roughly thirty-one times the destructive power of a 6.0 earthquake. In contrast, *intensity* measures the local impact of that energy—because surface materials and distance from the break vary geographically, intensity can be mapped for a particular temblor, actual or hypothetical. Most maps of earthquake intensity are based on the Modified Mercalli (MM) scale, which uses roman numerals to describe twelve typical levels of damage. Three categories on ABAG's "San Francisco scale" roughly match three MM levels, while the two most destructive shaking categories fall between pairs of MM levels:[28]

	XII—Massive Destruction
Very Violent—A	
	XI—Utilities Destroyed
	X—Most Small Structures Destroyed
Violent—B	
	IX—Heavy Damage
Very Strong—C	VIII—Moderate to Heavy Damage
Strong—D	VII—General Non-structural Damage
Weak—E	VI—Felt By All People

Although less detailed than the twelve-level MM scale, ABAG's six-category classification presents a less ambiguous picture of shaking intensity. Simple labels are straightforward, and fewer categories make the map's intensity symbols easier to print.

Shaky Ground includes shaking-intensity maps for only two of thirty-five earthquake scenarios—a magnitude 8.4 rupture on the San Andreas fault and a 6.9 break on the Hayward fault. The maps look different and demand different mitigation measures. For example, the more massive San Andreas earthquake causes very violent shaking on the fault itself, south of San Francisco, as well as violent shaking on bay-fill waterfront land in San Francisco and Oakland. But the smaller Hayward quake, which would inflict violent shaking on many parts of Berkeley, Oakland, and other East Bay locations, puts much more red on the map. These and numerous other differences raise questions about which map is more important. Should planners respond mostly to the map for the bigger earthquake, or should they try to accommodate both maps? And what should they do with maps for the thirty-three other earthquake scenarios? Simulation models easily overwhelm decision makers with huge amounts of conflicting advice.

To cope with thirty-five different scenarios, Perkins produced several composite maps. For a map of "Maximum Ground Shaking Intensity," she used the GIS to assign each cell the highest intensity attained for any scenario. Although useful in identifying critical areas for emergency-response planning, this map presents an inaccurate portrait of relative risk—each scenario might reflect its fault's history and accumulated stresses, but all thirty-five simulated earthquakes are not equally likely. The 8.4 San Andreas earthquake, for instance, has a "recurrence interval" estimated at one thousand years, while the 6.9 Hayward scenario describes a rupture occurring, on average, once every two hundred years. To accord the smaller, more likely Hayward rupture greater clout than the larger but less likely San Andreas quake, Perkins added together weighted maps for all scenarios.

The process is complicated but reasonable. The inverse of the recurrence interval (1/200 for Hayward and 1/1,000 for San Andreas scenario) provides relative weights that give the Hayward scenario (0.005) five times the impact of its San Andreas counterpart (0.001). But these weights cannot be applied directly to categories representing levels of shaking—what, for instance, would

$$(\text{"very strong"} \times 0.005) + (\text{"violent"} \times 0.001)$$

mean? To convert shaking-intensity levels to numbers that can be weighted and added together, Perkins used earthquake-engineering data that relate Modified Mercalli intensity to a damage ratio (the cost of repairing a building as a percentage of the cost of replacing the building).[29] With unreinforced masonry, for example, damage ratios of 25% for very strong shaking and 46% for violent shaking provide proportions that can be weighted and combined:

$$(0.25 \times 0.005) + (0.46 \times 0.001) = 0.00125$$
$$+ 0.00046 = 0.00171.$$

In this simplified, two-scenario example, dividing the weighted sum (0.00171) by the sum of the weights (0.006) yields a weighted average damage ratio of 28.5%, a readily interpreted, mappable risk estimate—but only for unreinforced masonry. Other types of construction have their own coefficients for converting shaking-intensity levels to damage ratios, and each construction type yields a different map. With this approach, Perkins reduced thirty-five earthquake scenarios to only three separate risk maps representing the area's most common types of construction.

To help developers and public officials choose among various types of construction, Perkins modified the risk maps to show probable loss as a percentage of present value. For example, a mapped value of 5 percent means that a building worth $100 million at that location might experience earthquake damage its owner could cover by putting $5 million into an interest-bearing account. Known as *discounting*, this modification is trickier than converting shaking intensities into damage ratios; the analyst must also estimate repair costs of damage incurred when construction costs are higher. However problematic the underlying (and unrevealed) assumptions about interest rates and inflation, discounted damage is more useful than unmodified damage ratios for assessing costs and benefits.

Although discounted damage is a complex concept, the maps have a straightforward key showing eight levels of risk described by interpretative labels:

6.1+%	Extremely high cumulative damage potential
5.1–6.0%	Very high
4.1–5.0%	High
3.1–4.0%	Moderately high
2.1–3.0%	Moderate
1.1–2.0%	Moderately low
0.3–1.0%	Low
0.0–0.2%	Very low

Symbols increasing in intensity from light red to nearly solid red represent the range from "low" to "extremely high" damage. Leaving the lowest category blank provides an unambiguous light-to-dark sequence of shading symbols and also helps point out "very low" damage on ridges with little or no soil.[30] These easily understood labels compensate for the maps' comparatively involved subtitle ("Cumulative Damage Potential Expressed as Expected Damage Discounted to Present Value").

Shaky Ground includes risk maps for three types of construction: tilt-up concrete buildings, concrete and steel buildings, and wood-frame dwellings. Excerpts in Figure 2.9 compare their predictions for San Francisco's North Bay waterfront. Because some construction types are relatively resistant to shaking, not every map includes sites in

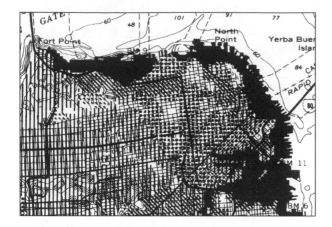

Tilt-up concrete
buildings

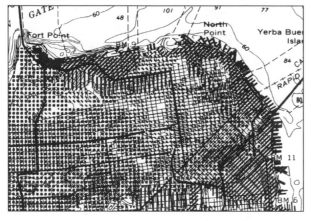

Concrete and steel
buildings

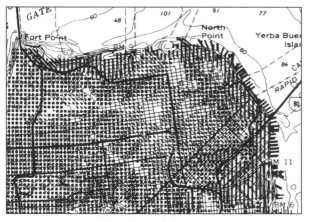

Wood-frame dwellings

Figure 2.9. Excerpts from composite risk maps showing cumulative expected damage
for three types of construction. On these 1:125,000-scale maps one inch represents two
miles. From Jeanne B. Perkins, *The San Francisco Bay Area—On Shaky Ground* (Oakland, Calif.: Association of Bay Area Governments, 1987), pp. 12, 14, and 16.

all eight categories. In fact, for wood-frame dwellings, which resist shaking better than most buildings, nowhere in the Central Bay region does the map show the highest three categories. By contrast, for tilt-up concrete buildings, which often fall apart during severe shaking, filled-in waterfront areas are especially hazardous. To help readers understand differences in construction, the page facing each map contains pictures of a typical building before and after a damaging earthquake.

Perkins encourages readers to compare the risk maps with her other shaking-intensity maps. Careful comparison, she points out, will demonstrate convincingly that a Bay-side site typically is more hazardous than a fault-line site, where severe damage usually requires an earthquake on the fault itself. She also cautions viewers not to misread the maps. A particularly pointed caveat, highlighted in bold type, warns: "These risk maps are based on data related to faults, geology, frequency of earthquakes, and statistical damage patterns. They do NOT include any information on EXISTING land use or EXISTING building stock. Therefore, they cannot be used alone in making estimates of current property at risk."[31] Like most authors of risk maps, Perkins is wary of hasty interpretations by lawyers, realtors, and bankers.

Mountainous areas, with "very low" cumulative damage potential, stand out in white on ABAG's risk maps, but when an earthquake triggers landslides, these hillslopes can be as fatal as bay-fill land. A significant seismic hazard, landslides are less deadly than ground shaking in California, where steep slopes are sparsely inhabited. Part of the reason is money: higher elevations with better views attract comparatively affluent residents, who prefer low-density neighborhoods and can build relatively resistant houses. But because of the region's growth and prosperity, landslides are an increasing threat, as real estate developers seek to colonize attractive uplands. If zoning boards in the Bay Area didn't restrain development, landslides, debris flows, and mudflows would be more hazardous than earthquakes.

Just south of San Francisco, in San Mateo County, Earl Brabb uses maps to fight development on landslide-prone slopes. Brabb lives here and works at the U.S. Geological Survey's western regional offices in Menlo Park. He is a geologist known worldwide for his maps of earthquake and landslide hazards. And he collaborates frequently with Jeanne Perkins and other planners and geologists committed to mapping hazards and averting disasters. Several of his maps have been written into county ordinances.[32]

Brabb gave me a walking tour of Building 7, where dozens of maps, some quite large, enliven two floors of narrow, otherwise banal hallways. The experience was impressive visually as well as scientifically—a gallery of abstract portraits with vivid colors to stimulate the brain's right hemisphere while an eclectic treatment of earthquake hazards fascinates the left. Secondary earthquake hazards prominent in this collection include debris flows and groundwater contamination—fractured aquifers invaded by underground plumes of hazardous chemicals from Superfund sites. Many of the maps describe San Mateo County in considerable detail, at 1:62,500 (one inch representing one mile), and most have more then one author. The Geological Survey encourages collaborative research.

The San Mateo County maps reflect the efforts of fourteen USGS scientists and the cooperation of county officials.[33] Led by Brabb and geographer Leonard Gaydos, the team employed a geographic information system based on a grid of 30-meter cells (squares about 99 feet on a side). In addition to modeling earthquake intensity, Brabb's group studied the effects of slope and soil moisture. Landslides and debris flows are common on steep slopes, where otherwise insignificant earth tremors can initiate movement. But the soil's ability to hold moisture as well as recent amounts of rainfall in the area are also important. One of the team's goals is a "real-time" method for warning residents of imminent landslides during periods of heavy rain.[34]

In using a GIS to model the effects of slope, geologists turn to the *digital elevation model* (DEM), consisting of elevations measured at the center of each cell in the grid. A recent item in the USGS list of cartographic products, DEMs allow the GIS to compute slope maps and identify dangerously steep terrain. DEMs also help earth scientists recognize two types of hilly terrain, hazardous in distinctly different ways. As the close contour lines in Figure 2.10 illustrate, "hard terrain" consists of steep slopes, on which landslides, debris flows, and avalanches occur and move rapidly. In contrast, "soft terrain" has gentler, more rounded slopes, on which slumping and slow-moving landslides are more common. The GIS also computes azimuth (slope direction), which geologists use to estimate the direction, speed, and cumulative magnitude of landslides, mudflows, and other forms of *mass movement*. Although results with 30 m resolution are promising, higher spatial resolution could more reliably point out areas susceptible to landslides.

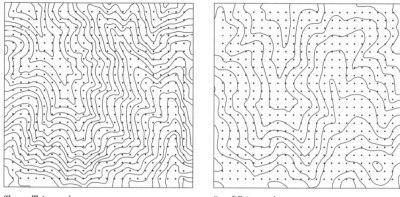

"hard" terrain   "soft" terrain

Figure 2.10. Contrasting examples of "hard" (left) and "soft" (right) terrain. Lines are elevation contours (lines of equal elevation above sea level), with a contour interval of 40 feet. Dots representing the DEM show grid-cell centers spaced 30 m (approximately 100 feet) apart. Reprinted from Earl E. Brabb, "Analyzing and Portraying Geologic and Cartographic Information for Land-Use Planning, Emergency Response, and Decision-making in San Mateo County, California," *GIS '87—San Francisco: Second Annual International Conference, Exhibits and Workshops on Geographic Information Systems* (Falls Church, Va.: American Society for Photogrammetry and Remote Sensing, and American Congress on Surveying and Mapping, 1987), pp. 362–74, fig. 2.

Would greater detail justify the added cost of more exact inventories and better DEMs? Not necessarily: mapping takes time as well as money, after all, and scientists must ask whether more accurate hazard maps would warrant mapping new areas at a markedly slower pace. To assess these trade-offs, Earl Brabb asked geologists and geographers throughout the world to rank nine options, which included various mapping projects, research studies, and workshops.[35] Topping the priorities list was landslide susceptibility mapping, followed by an intermediate-scale (1:20,000 to 1:80,000) inventory of landslides. Susceptibility maps are a "translated product" based on scientific analysis but presented in a form understandable to decision makers who control land use and plan for disasters. These two high-priority strategies go hand in hand, though, because intermediate-scale inventories of landslides compiled from air photos provide a base for susceptibility mapping. Next to the bottom was a large-scale (1:10,000) inventory of landslides—a useful trove of scientific data perhaps, but a less rewarding public investment.

Brabb makes good use of San Mateo County's intermediate-scale landslide inventory but is hardly optimistic that traditional methods—

and a strained public purse—can identify and map the entire country's twenty million landslides. But he is willing, if not eager, to accept almost-as-good susceptibility maps based on less precise methods, including satellite data, DEMs, and a GIS. What's more, he argues, maps covering the whole world could be produced in a decade for a mere $30 million per year.[36] Although the total cost of $300 million might seem a bit much, it's small compared to the proposed project's enormous benefit—Brabb estimates that landslide susceptibility mapping could reduce the annual worldwide loss of $25 billion to a mere $2 billion. Even if these estimates are somewhat inflated, the idea is too intriguing to ignore.

Mitigation and response are the "before and after" of emergency management, and both phases require maps. When a disastrous earthquake occurs, emergency-response maps help rescuers get in and victims get out. Seismic planners must pay particular attention to transport routes, energy supply lines, and communications links— "lifelines" vulnerable to surface faults, severe shaking, and secondary hazards, such as landslides. In addition to identifying where lifelines might be severed and for how long, emergency-response maps must also highlight dams, fuel-storage areas, and other potentially catastrophic technological hazards.

Disaster preparedness depends on drills to test emergency communications and educate public safety personnel. Successful drills require realistic mock emergencies, based on anticipated earthquakes. To promote effective emergency-response planning, California's Division of Mines and Geology developed earthquake-planning scenarios for five massive yet credible earthquakes: two near San Francisco, two near Los Angeles, and one in San Diego.[37] Each scenario is presented as a report consisting of several small-scale foldout maps and a detailed discussion of each map's principal features and intended use. The maps not only point out places requiring special attention but encourage local governments and public utilities to examine collateral effects such as fire, flooding, and the release of hazardous chemicals.

The first report, published in 1982, describes the damage to lifelines of a magnitude 8.3 earthquake—as big as the 1906 quake—on the San Andreas fault near San Francisco.[38] Of the report's ten maps, only one shows ground shaking, surface faulting, and ground failure. The other nine represent specific kinds of earthquake damage—three maps address transportation and communications (highways and airports, railroads and marine facilities, telecommunications), one focuses on water-supply and waste-disposal systems, and three deal with energy (electric power, natural gas, and petroleum fuels).

Facsimiles of map keys from two of them (Figure 2.11) demonstrate that earthquake scenarios are translated products, intended for non-scientists. The Highways and Airports map shows not only where principal routes are likely to be closed but also the severity of damage and duration of each closing. The Communications map divides the area into four regions, each with a distinct pattern of recovery represented by a small but straightforward graph. Wary of misinterpretation, the report's authors introduced each map with an identical caveat, printed on the preceding page in large, bold type on light red paper and warning readers that the map reflects a hypothetical chain of events and is not a substitute for detailed geotechnical evaluations of specific facilities. Concerned that maps can be photocopied or removed from the report, the authors repeated a smaller version of the warning on each map. However visually striking and abundant, the complexly worded admonition could use a translation as well as a reminder that what map viewers don't see can hurt them.

Warnings about omissions are essential—a map based on a single hypothetical earthquake cannot predict every plausible structural failure. But equally serious are deadly hazards mentioned only fleetingly in detailed notations about vulnerable locations. Recall, for example, the Cypress Street Viaduct—the double-deck freeway in Oakland that collapsed when the magnitude 7.1 Loma Prieta earthquake interrupted the World Series on 17 October 1989. I watched TV the entire evening, and remember the tragedy well. Intense shaking caused concrete supports to disintegrate, and the upper deck collapsed onto the lower roadway and killed forty-two people.[39] This freeway was a segment of Interstate 880 (formerly part of California Route 17), identified on the 1982 Highways and Airports scenario as section H13 and deemed likely to be "Closed for up to 36 hours." Buried in the ten-line, six-sentence notation for section H13 is a single veiled hint of disaster: "The elevated section through downtown Oakland is expected to be extensively damaged."[40] Whatever "extensively damaged" might mean, I-880 was closed much longer than thirty-six hours.

To say the scenario was fatally flawed, literally, would unfairly condemn an essential disaster-response tool. Regionwide earthquake-planning scenarios are intended for "emergency planning purposes only"—the scale (about 1:400,000) is too small and the budget too

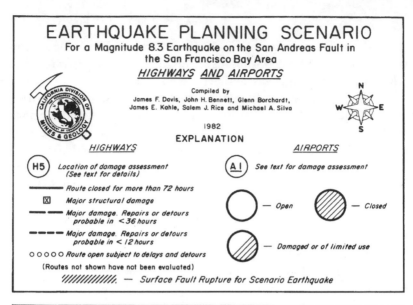

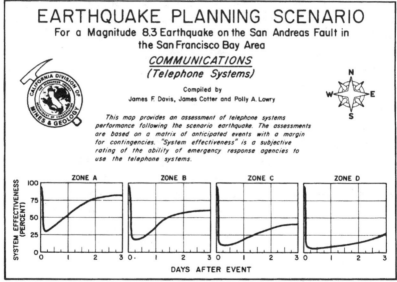

Figure 2.11. Keys from two maps in an earthquake-planning scenario for the San Francisco Bay area. From James F. Davis and others, *Earthquake Planning Scenario for a Magnitude 8.3 Earthquake on the San Andreas Fault in the San Francisco Bay Area*, Special Publication no. 61 (Sacramento: California Division of Mines and Geology, 1982), pp. 66 and 92.

limited for a more comprehensive assessment of risk. Moreover, if a magnitude 8.3 earthquake had struck the Bay Area, a collapsed freeway would account for only a small fraction of thousands of deaths and hundreds of thousands of injuries.[41] Planning scenarios have an important role in seismic preparedness, and that role includes warning would-be users not to misinterpret their message. All hazard-zone maps share this concern.

The Loma Prieta earthquake might have been less deadly if the designers of Oakland's Cypress Street Viaduct, built in 1971, had had available a new kind of seismic risk map, developed in the 1970s for architects and structural engineers.[42] These new maps show peak ground acceleration (the maximum horizontal stress on a bridge, foundation, or other structure) as a percentage of gravity. Maximum acceleration tells structural engineers how flexible to make a bridge or high-rise building—structures designed to bend or move when stressed but return to normal after a big earthquake. Ground-acceleration maps also tell highway designers how stiff to make a freeway so that its sections do not shake apart. Maximum acceleration is a convenient measure for architects, who can compute the horizontal force on the whole structure as well as its individual components. In some parts of California, buildings and bridges anchored in bedrock must resist accelerations nearly as large as gravity.

To be useful to designers, maps of ground acceleration must represent a broad range of possible earthquakes centered in many different places and varying widely in magnitude and likelihood. Research for these maps is a daunting task, entailing the systematic collection, evaluation, and integration of massive amounts of seismic and geologic information. To combine the unique contributions of many credible earthquakes on a single map, geologists establish a level of certainty (such as 90 percent) and an appropriate time span (such as 50 years). The result is a map of maximum ground acceleration with, in this example, a 90 percent probability of not being exceeded in 50 years. Stated differently, there is only a 10 percent chance the acceleration values on the map will be exceeded in 50 years.

Because buildings and bridges vary substantially in their design life, it is not surprising that geophysicist Ted Algermissen, who directed ground-acceleration mapping for the U.S. Geological Survey, produced two national maps showing maximum horizontal acceleration with a 90 percent likelihood for 50 and 250 years.[43] According to probability theory, earthquakes this large have average return periods of 475 and 2,375 years respectively. Probabilistic risk maps also help designers recognize that the largest earthquake to hit an area thus far is not the largest earthquake possible.

Engineers must treat ground-acceleration maps as empirically based generalizations subject to revision. Figure 2.12 illustrates for a portion of the Northeast the effects of more complete data, more reliable methods, and more solid theory.[44] Both panels show peak ground acceleration for a large earthquake with a 2,375-year return period. The upper panel, an excerpt from Algermissen's earlier maps, published in 1982, shows a maximum acceleration in western New York (between Buffalo and Rochester) as high as 46 percent of gravity. By contrast, in the lower panel, a revised map published in 1990 estimates the maximum acceleration in the same area as only 35 percent of gravity.[45] Viewers must not infer, though, that ground-acceleration estimates for the earlier map were uniformly more conservative—careful comparison of the two panels reveals discernible increases, for example, from 15 to 19 percent in eastern New York and from 32 to 34 percent in eastern Massachusetts. Designers of skyscrapers, I hope, have adopted safety factors that can accommodate further refinements. I hope, too, they have not ignored vertical ground acceleration, which can exceed the force of gravity.[46]

Although Algermissen's recent work merits the respect of architects and civil engineers, the three independent groups that write most of the country's building codes rely largely on an older, less exact ground-acceleration map he developed in 1976.[47] Figure 2.13 illustrates a typical adaptation: the seismic risk map for the *Uniform Building Code*, used widely in the western United States.[48] Smooth contours divide the country among five seismic-risk zones, identified by the integers from 0 through 4.[49] These zones reflect peak accelerations with a 90 percent certainty over a 50-year period—the typical design life of ordinary buildings. Zone 4, where risk is greatest, departs from Algermissen's map by treating as equal all peak accelerations greater than 40 percent of gravity—important distinctions, of course, but too detailed for a small, one-page map.[50] The *Code* itself is a small (5.5-by-8-inch) book, and a simple, highly generalized map is consistent with its authors' goal of a conveniently enforced minimum standard. This simplification is politically expedient as well: the International Conference of Building Officials avoided dissent within its membership by including in zone 4 some higher-risk areas a more detailed map might

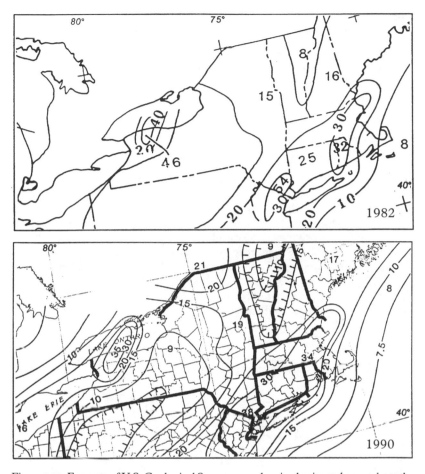

Figure 2.12. Excerpts of U.S. Geological Survey maps showing horizontal ground-accelerations with a 90 percent probability of not being exceeded in 250 years. Original map published in 1982 (upper panel) was revised in 1990 (lower panel). From Ensearch Environmental Corporation, *Generic Site Characteristics Report* (Troy, N.Y.: New York State Low-Level Radioactive Waste Siting Commission, April 1994 draft), appendix figs. B-1 and B-2.

have portrayed more reliably as zones 5 and 6. Many of these higher-risk areas are too narrow to stand out on a small map, and incorporating neighboring areas to make them visible would misclassify relatively safe sites and needlessly increase construction costs. This stratagem is not as reckless as it seems because a vulnerable structure proposed for zone 4 would need a special seismic study.

Seismic maps in building codes can inflame long-standing tensions between theory and practice. In 1993, for instance, the California Seismic Safety Commission ordered its representative to the Building Seismic Safety Council to vote against new ground-shaking maps proposed by the U.S. Geological Survey. At issue were a pair of probabilistic "spectral response" maps: one for tall, relatively flexible buildings with a 1.0-second period of vibration and the other for shorter, stiffer buildings estimated to complete a cycle of shaking in 0.3 second.[51] A refinement of Algermissen's earlier work, the maps addressed the diverse design requirements of tall and short structures. But as the commission's executive director noted, the new maps represented

> potentially sweeping and drastic changes to seismic design practice for both new buildings and the retrofit of existing buildings throughout the nation and particularly in California. Requirements would be raised in some areas and lowered in many more areas. For example, over 60 years ago, California prohibited the use of unreinforced masonry in new construction. The USGS maps may form the basis for the reintroduction of this type of poorly-performing construction for new buildings in California in low-seismic regions. . . . In contrast, high seismic values near the San Andreas fault may imply the need for added restrictions beyond those required in our building codes.[52]

Seismic zones of the Uniform Building Code

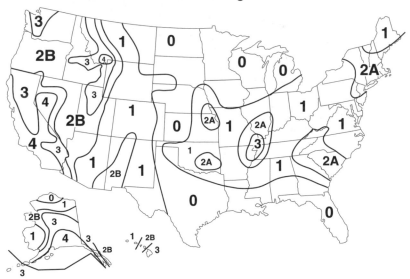

Figure 2.13. Seismic hazard zones for the 1991 Uniform Building Code. Redrawn from International Conference of Building Officials, *Uniform Building Code, 1991 Edition* (Whittier, Calif., 1991), fig. 23-2.

Although theoretically valid, the maps were too radical for safety officials as well as builders. As the commission official wryly observed, "Building codes are not purely scientific."[53]

Precision doesn't always surrender to simplicity, though. Zone boundaries on building-code maps once followed county boundaries, thereby simplifying the work of county-level planners and code-enforcement officers.[54] But because several California counties are too seismically diverse for a single countywide risk zone, code officials adopted a contour map, which allowed appropriate intracounty variations in design standards.[55] Even so, some codes give counties the option of adopting a single zone—a useful strategy for avoiding lawsuits over whether a property is "inside" or "outside" a rigidly interpreted, cartographically reified boundary.

Conspicuous patches of zones 2 and 3 east of the Rocky Mountains (Figure 2.13) warn that earthquakes are not just a West Coast problem. History reinforces this alert: the largest earthquake to strike North America since the arrival of European explorers was centered over New Madrid, a small river town in southeastern Missouri.[56] Between 16 December 1811 and 7 February 1812 a series of large earthquakes destroyed New Madrid, altered the course of the Mississippi River, and toppled chimneys in Richmond, Virginia. Felt as far away as Boston, the largest shock, with an estimated Richter magnitude of 8.7, packed two and a half times the punch of the 1906 San Francisco earthquake.[57] Massive earthquakes have also occurred in South Carolina, the St. Lawrence Valley, and eastern New England, and seismic evidence suggests neighboring areas are equally vulnerable.

Eastern earthquakes differ markedly from the West Coast's comparatively shallow fault-line temblors.[58] Although less frequent, eastern earthquakes originate much deeper in the crust—so deep that intense shocks from a single rupture can reach cities as much as a thousand miles apart. Because of huge damage zones and poor earthquake preparedness, a big eastern earthquake can be enormously destructive.[59] To plan for a disaster geologists believe is inevitable, if not imminent, the U.S. Geological Survey and the Federal Emergency Management Agency (FEMA) have encouraged earthquake planning in all eastern states. In Massachusetts, for instance, the Office of Emergency Preparedness has used ground-shaking maps to argue for more strict building codes, especially on unconsolidated soils surrounding Boston harbor.[60] In New York, the State Geologist and the State

Emergency Management Office are collaborating on a GIS to model ground shaking and other soil-related earthquake hazards.[61] The Central U.S. Earthquake Consortium (CUSEC), a cooperative multistate initiative based in Tennessee, is promoting seismic safety through public awareness, hazards mapping, and stronger building codes.[62]

Essential to disaster preparedness, risk maps occasionally contribute to public hysteria. In 1990, for instance, misinterpreted risk maps added to the credibility of Iben Browning, an inventor, business consultant, and self-promoting genius who "projected" a massive earthquake in the vicinity of New Madrid between December 1 and 5.[63] Browning claimed that a rare alignment of the earth, the moon, and the sun would produce a combined solar and lunar tide large enough to trigger a huge earthquake on the New Madrid fault. He first announced his prediction in late 1989, in his $225/year business newsletter and during his $2,500 talks on the corporate lecture circuit. Because of his numerous patents and his success as a highly paid consultant, the media took Browning seriously enough to disseminate his prediction—in American journalism's notion of fairness, the angry protests of earth scientists were merely the other side of a lively controversy.

Throughout the summer and fall of 1990, newspapers and wire services embellished the story with seismic risk maps from official sources—maps that suggested Browning had merely pinpointed in time what science had already forecast in space. On September 26 a magnitude 4.7 earthquake near Cape Girardeau, Missouri, added to the excitement, as did the NBC miniseries "The Big One: The Great Los Angeles Earthquake (a soon to be true story)," televised in mid-November. School districts in several states canceled classes for Monday, December 3. When "Quake Day" arrived, nine TV crews with mobile satellite uplinks were transmitting live broadcasts from New Madrid, and a team of "observers" from the U.S. Geological Survey were answering (and dignifying) the media's questions. By Monday afternoon, though, reporters and technicians started to pack, and by Tuesday most had left. Much to everyone's relief or amusement, Browning's earthquake had missed its appointment.

Scientists were appalled by the enthusiasm of journalists and educators for Browning's prediction, even after professional geologists had pronounced his claims unscientific and highly unlikely. To understand and document the geologic hoax of the century, the U.S. Geological

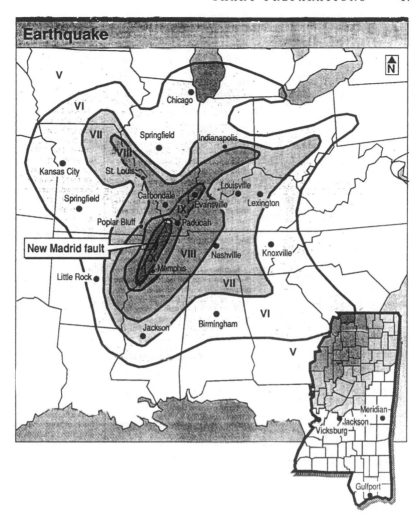

Figure 2.14. Used to represent a single earthquake, this map of maximum estimated intensity exaggerates the areal impact of a magnitude 7.5 earthquake on the New Madrid fault. Reprinted from William Spense and others, *Responses to Iben Browning's Prediction of a 1990 New Madrid, Missouri, Earthquake,* U.S. Geological Survey Circular no. 1083 (Washington, D.C., 1993), p. 69; map appeared originally in the Jackson, Mississippi, *Clarion-Ledger,* 15 July 1990.

Survey commissioned a study. Its 248-page report, richly illustrated with facsimiles of news stories, lists seventeen specific reasons Browning was taken seriously. None of these items is explicitly cartographic, but maps embellish many of the news accounts. And heading the appendix of media facsimiles is the map in Figure 2.14, which uses thick contours, somber shading, and arcane roman numerals to describe maximum Modified Mercalli intensities for a magnitude 7.5 earthquake on the New Madrid fault. Many newspapers misused this map to describe the predicted earthquake's geographic impact. But their interpretation was wrong: instead of describing a single earthquake, the map portrays the geographic extent of maximum intensity for an earthquake *anywhere* along the 120-mile fault. Because no single earthquake would have so wide an impact, the map not only made Browning's prediction credible but exaggerated its scope.

Despite Iben Browning's ridiculous specificity, short-run seismic forecasting seems an attainable goal. The National Seismic System, an ambitious plan to connect regional seismological networks and better understand crustal processes, hopes to provide very short-term earthquake warnings—"less than several minutes" but long enough for most people to seek safe shelter.[64] Although precise long-range forecasts remain elusive, encouraging theories about "seismic gaps" promise educated speculation about the likely locations of major earthquakes within the next few decades.[65]

Of what value, though, is a forecast that anticipates a massive seismic disaster by a few minutes, or even a year? A short-run warning might save countless lives, but the Big One, when it strikes, will inflict enormous losses because of corrective measures not undertaken years or decades earlier. However vague as forecasts, seismic risk maps are timely warnings telling governments to restrict land use, reinforce bridges and buildings, and prepare the public for "low-probability events" with enormous consequences. If Iben Browning were right, wouldn't a somewhat vague warning twelve years in advance, say, be more useful than a more temporally precise prediction too late for effective mitigation?

Media coverage of Iben Browning's prediction missed an important point: seismic forecasting is less about time than about location. In this sense, Browning's prognosis was no more newsworthy than the predictions of Jeanne Perkins, Earl Brabb, and Ted Algermissen—geographic predictions, to be sure, but backed by better evidence and clearer reasoning. What's more, these and other seismic cartographers now have the ear, the eye, and the confidence of state and local officials.

Chapter Three

Lavas and Other Strangers

The popular image of an erupting volcano belching black smoke and bleeding bright orange lava is colorful but misleading. Photogenic Hawaiian volcanoes with relatively frequent, comparatively gentle eruptions of sluggish lava, easily outwalked, perpetuate a geologic stereotype of molten rock advancing slowly down the slope, frying or burying everything in its path. Spectacular, of course, but an inaccurate picture of volcanic threats to human life: despite enormous damage to agriculture and buildings, lava flows account for only 0.1 percent of the estimated 76,000 eruption-related deaths worldwide between 1900 and 1986—far less than a dozen other volcanic hazards that are more rapid, more widespread, or more poisonous.[1] Volcanic risk involves a diversity of hazards and eruptive styles, including explosive lateral blasts; deadly gases; various falls, surges, and flows of ash and other *pyroclastic material* (hot fragments of rock); and secondary effects such as tsunamis and flooding from dams collapsed by volcanic mudflows. Fortunately for emergency planners, seldom does a single volcano merit risk maps for more than four or five specific hazards.

Although volcanologists and seismologists must consider multiple hazards, hazard assessments for volcanoes rely less on computer simulations and more on field mapping, air photos, and laboratory analyses. Maps of volcanic deposits are directly useful—without simulation—because volcanoes have a narrower range of locations: eruptions usually reactivate an existing volcano or open a new vent in a well-defined volcano field, whereas earthquakes can occur anywhere along known faults as well as on new or buried faults. For *lahars* (mudflows) and other volcanic hazards steered by gravity and terrain, areas invaded decades or centuries ago are demonstrably vulnerable. Moreover, extending the hazard zone downhill to accommodate a somewhat bigger eruption is straightforward. And for ashfalls, which vary with wind direction, geographic forecasting depends on cartographic evidence of the frequency, magnitude, and areal extent of past eruptions.

Like seismologists, volcanologists worry about very-low-frequency events with enormous consequences. Human history records several immense volcanic disasters. In 1883, for instance, the eruption of Krakatoa, a small volcanic island between Java and Sumatra, triggered a 130-foot tsunami that killed 36,000 people. But none of these geologically "recent" tragedies compares with the volcanic catastrophes once blamed for massive worldwide species extinctions.[2] We know about ancient volcanic calamities because of their immense imprints on the landscape, such as the Long Valley caldera in California's Mammoth Lakes area, east of Yosemite Park. This 170-square-mile elongated volcanic depression was formed 700,000 years ago by an eruption that released 150 cubic miles of ash and pumice—perhaps a thousand times what Mount St. Helens belched out in 1980—and covered areas 300 miles east to a depth of six inches.[3] Our prehistoric ancestors were no doubt depressed and chilled by the "volcanic winter" in which dark skies blocked sunlight and reduced temperatures worldwide.[4] However unthinkable the impact of massive crop failures on today's huge human populations, geologists consider an eruption of this magnitude possible but highly unlikely. Volcanic catastrophes have occurred in the Long Valley–Mono Lakes volcano field more or less regularly every 200,000 years. Although geologic history suggests the next calamity is a safe 100,000 years in the future, the area has experienced comparatively minor explosive eruptions more frequently, one as recent as 500 years ago.

To assess the need for emergency planning, geologists commonly classify volcanoes as *active, dormant,* or *extinct*—everyday terms that beg the questions What does "active" mean? and How dormant is "dormant"? Turning to radiocarbon dating for answers, Roy Bailey and his colleagues at the U.S. Geological Survey developed a nation-wide volcano-hazards map that replaced "active" and "dormant" with three risk groups based on carefully reconstructed eruption histories:[5]

1. Volcanoes that have short-term eruption periodicities (100–200 years or less) and/or have erupted in the past 200–300 years;
2. Volcanoes that appear to have eruption periodicities of 1,000 years or greater and last erupted 1,000 or more years ago; and
3. Volcanoes that last erupted more than 10,000 years ago, but beneath which exist large, shallow bodies of magma (molten rock) that are capable of producing exceedingly destructive eruptions.

Despite the numbers, their new typology is far from objective: experts must resolve the fuzzy boundary between the first and second groups as well as decide which older volcanoes belong in the third group, requiring monitoring because of low-probability eruptions with cata-strophic consequences. Very old volcanoes not qualifying for inclusion in one of these three groups are deemed extinct.

Developed in the early 1980s, Bailey's three groups influenced the volcano-hazard maps in the Federal Emergency Management Agency's 1990 atlas *Risks and Hazards: A State by State Guide*.[6] But FEMA simplified the map key by assigning each group a single fre-quency

    1 eruption per 200 yrs.
    1 eruption per 1,000 yrs.
    1 eruption per 10,000 yrs.

using tiny triangles with increasingly warmer yellow, orange, and red interiors to represent the progressively greater destructive potential of lower-frequency eruptions. However volcanologically imprecise, these generalized labels and color-coded intensity symbols offer nontechni-cal readers a straightforward and balanced translation of relative risk. Another modification, needed only for Alaska, Hawaii, and Washing-ton, is a thin red border to highlight "volcanoes that have erupted since 1950"—a pointed reminder that the map's 200-year volcanoes are a more pressing concern for emergency managers.

Volcano-hazard maps have two distinct yet complementary styles. At the less-detailed continental or national level, volcanologists identify active volcanoes and other areas requiring monitoring, and at the more-detailed regional or local level, they describe specific hazards around active volcanoes. Figure 3.1, typical of the less-detailed view, is a broad-brush portrait of volcanic risk in the western U.S. Point symbols varying in shape and prominence reflect a modified classification presented in a 1992 report describing the U.S. Geological Survey's Volcano Hazards Program.[7] A single threshold of 2,000 years distinguishes active volcanoes from "potentially active volcanoes"—a change that not only simplifies the map key for lay readers but puts more volcanoes in the highest-risk group. (In hazard-zone mapping, erring on the side of caution often yields useful propaganda.) Equally persuasive are the map's solid, geologically suggestive triangles representing active volcanoes, which contrast sharply with open, visually weaker circles connoting lower-risk volcanoes. In addition, pound signs—rarely used as map symbols—signify the distinctly different risk of *volcano fields*, where geysers, geophysical evidence, and geologically "recent" volcanic activity suggest the possibility of new volcanoes and small lava flows.

Framing the map to show only western states is significant: no remotely active volcanoes occur east of the Rocky Mountains. Moreover, active volcanoes in the West largely occur along plate boundaries in the "Pacific Ring of Fire" (Figure 2.2). An inset for Alaska reinforces the relevance of plate tectonics and continental drift: the state's forty-three active volcanoes are mostly on the Alaskan peninsula and in the Aleutian Islands. An inset for Hawaii shows five active and two potentially active volcanoes, but none on Oahu, the most densely populated island. In the middle of the Pacific plate, Hawaii's volcanoes are atypical because almost 95 percent of volcanoes occur where plates converge or diverge.[8] Figure 3.1 also reveals several volcanoes within the North American plate, including the Long Valley caldera, the Yellowstone caldera, and the Bandera volcano field in New Mexico. Less common and generally less active, intraplate volcanoes nonetheless require monitoring and emergency planning.

Small-scale volcano-hazard maps typically show populations at risk. On the original, unabbreviated version of the map described in Figure 3.1, red triangles, open circles, and pound signs representing volcanic threats stand out from light blue circles portraying urban centers. Four sizes of circle show four groups of cities, the smallest with 40,000 to 100,000 residents and the largest with populations of over a million. Although the conurbations of Los Angeles and San Francisco are safely distant from volcanic threats, a few sizable cities in the Pacific Northwest seem uncomfortably close. Bellingham, Washington,

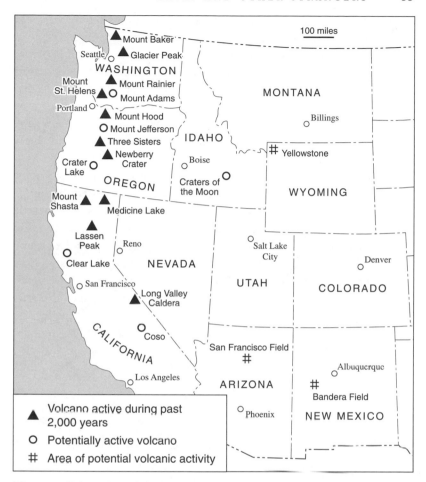

Figure 3.1. Volcano hazards in the western United States. Reprinted from Thomas L. Wright and Thomas C. Pierson, *Living with Volcanoes: The U.S. Geological Survey's Volcano Hazards Program*, U.S. Geological Survey Circular no. 1073 (Washington, D.C., 1992), p. vii.

near the Canadian border, is less than forty miles due west of Mount Baker—a comparatively safe direction only if high-altitude winds maintain their prevailing west-to-east pattern during an eruption rich in ash. Two larger cities are exposed to two threats each: Portland, Oregon, is less than sixty miles from Mount St. Helens and Mount Hood, while Seattle is within seventy miles of Glacier Peak and Mount Rainier. Although neither Portland nor Seattle is considered downwind, detailed maps are needed to assess the vulnerability of their outer suburbs to specific hazards.

When Mount St. Helens exploded in May 1980, U.S. Geological Survey scientists were excited but not surprised. Over a decade earlier, USGS director William Pecora had warned that the "lofty snow covered cone . . . may blow its top at any time."[9] And five years before the eruption, in a lengthy article in *Science*, three USGS geologists cautioned that "an eruption is likely within the next hundred years, possibly before the end of this century."[10] Concerned with the hazard's geographic dimensions, USGS geologists Rocky Crandell and Don Mullineaux prepared two hazard-zone maps, published in 1978 as a USGS bulletin provocatively entitled *Potential Hazards from Future Eruptions of Mount St. Helens Volcano*.[11] A small, highly generalized map (about 1:4,100,000 in the original report) described the ashfall hazard, while a larger, more detailed map (1:250,000) outlined hazard zones for lava flows, pyroclastic flows, mudflows, and floods. To stress the urgent need for emergency planning, Crandell and Mullineaux asserted that "future eruptions of Mount St. Helens are a near certainty." Two years later an explosive eruption confirmed their cartographic scenario and thrilled their bosses and colleagues—even as they mourned the loss of sixty-five lives, including one of their own volcanologists, U.S. Geological Survey staff were euphoric over the forecast's accuracy.

Based on years of scouting a wide region around the volcano and

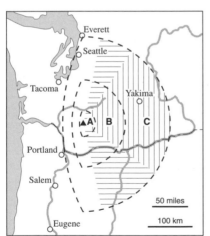
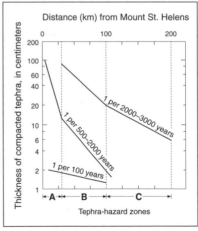

Figure 3.2. Ash-hazard zones for Mount St. Helens (left) and the relationship between distance downwind and thickness of ash deposit for three eruption periodicities (right). Reprinted from Dwight R. Crandell and Donal R. Mullineaux, *Potential Hazards from Future Eruptions of Mount St. Helens Volcano*, U.S. Geological Survey Bulletin no. 1383-C (Washington, D.C.: U.S. Government Printing Office, 1978), pp. C18–9.

mapping ash deposits from past eruptions, the ash-hazard map (Figure 3.2) is simple in appearance but hardly straightforward. Three concentric zones extend downwind 30, 100, and 200 km (approximately 19, 62, and 124 miles, respectively) to the east. Zone A, closest to the volcano, has the greatest potential thickness of *tephra* (ash particles), which declines progressively through zones B and C. Lopsided zones reflect the dominant east-to-west air flow, and rounded dashed-line boundaries imply uncertainty about distance and wind direction. To make their forecast more precise, Crandell and Mullineaux divided the zones into sectors based on climatological data for strong, high-altitude winds: a vertical-line pattern describes a 90-degree sector toward which the wind blows 50 percent of the time, and the addition of horizontal-line patterns on both sides marks a wider, 157-degree sector toward which the wind blows 80 percent of the time.

Because volcanologists cannot reliably forecast the force and geographic impact of an eruption, hazard-zone maps must accommodate a range of large, medium, and small eruptions. Crandell and Mullineaux addressed this uncertainty with a map key relating thickness of ash deposits (scaled along the vertical axis in the right-hand panel of Figure 3.2) to distance from the volcano (scaled along the horizontal axis). Three lines represent thickness-distance relationships estimated from field data for eruptions with different frequencies. (For the largest of these eruptions, with a periodicity of 2,000–3,000 years, the outer boundary of zone C represents an ashfall 4.7 inches [12 cm] deep—twice the 6 cm thickness indicated by the graph because erosion and compaction can reduce a fresh ash deposit to half its original thickness.)

Had the weather behaved normally, the ashfall forecast might have been more impressive. When Mount St. Helens exploded at 8:32 A.M. on May 18, 1980, and erupted vigorously for the next nine hours, southwest winds carried the ash plume north and east of the vent through the center of Washington state rather than directly east through both Washington and Oregon.[12] In Figure 3.3, *isopachs* (thickness contours) describe an elongated pattern of ash deposits, which extend through Idaho into eastern Montana. Although the thickest ash at the outer boundary of zone C is less than the 124 mm depth for the largest eruption in the map key, the May 18 ashfall had a strong linear pattern, with depths over 30 mm in Moses Lake and Ritzville, well beyond zone C. However unlike the forecast map's arc-like boundaries, this linearity was not unexpected; Crandell and Mullineaux had warned that "only a small part of any tephra-hazard zone

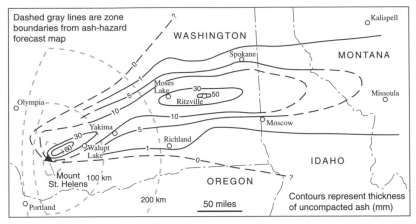

Figure 3.3. Uncompacted ash deposits from the May 18, 1980, eruption of Mount St. Helens. Isopachs (thickness contours) are in millimeters. Dashed lines are hazard-zone boundaries from the forecast map in Figure 3.2. Reprinted from Diane M. McKnight, Gerald L. Feder, and Eric A. Stiles, "Effects on a Blue-Green Alga of Leachates of Ash from the May 18 Eruption," in *The 1980 Eruptions of Mount St. Helens, Washington*, U.S. Geological Survey Professional Paper no. 1250 (Washington, D.C., 1981), pp. 733–41; isopachs from map on p. 734.

is likely to be affected by any one eruption."[13] Moreover, "If winds are strong and uniform, the same volume of material would form thicker deposits along a narrow band leading away from the volcano. If the material erupted is fine grained, or if winds at levels reached by the material are strong and uniform in direction, thick deposits could reach relatively far from the volcano."[14] The visual mismatch of hazard zones and ash deposition underscores a fundamental caveat of risk mapping: low risk doesn't mean no risk.

If the criterion for accuracy is the degree to which hazard zones incorporate all areas with notable damage, the comparatively detailed flowage-hazard map was no more accurate than the ash-hazard map. Figure 3.4 compares the extent of potential hazards delineated by Crandell and Mullineaux with the extent of damage from the eruption.[15] The geographic forecast in the upper panel shows a more or less symmetrical hazards area around the summit as well as several narrow hazard zones extending westward toward the Columbia River. The slopes of the volcano are susceptible to lava flows, pyroclastic flows, debris flows, and floods, and nearby areas are vulnerable to ash clouds associated with pyroclastic flows. In addition, rivers draining to the west provide corridors for debris flows and floods. By contrast, the lower panel of Figure 3.4 reveals a highly asymmetric damage area with immense devastation north of the newly opened caldera but relatively little impact to the south. Although the hazards map forecast the extent of pyroclastic flows and debris flows, it anticipated neither the enormous debris avalanche just north of the summit nor the devastating lateral blast immediately following. The blast leveled trees as far as fifteen miles north of the new caldera and singed trees beyond the blowdown zone for another mile or so. Despite the unanticipated force of the directed blast, the hazards map accurately forecast debris flows that reached the Columbia River as well as shorter debris flows on the volcano's south slope. Created when hot debris rapidly melted snow and

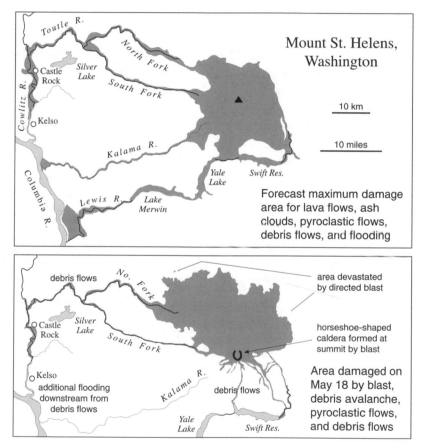

Figure 3.4. Combined hazard area from the 1978 flowage-hazard map (above) compared with areas damaged by the May 18, 1980, eruption (below). Generalized from Thomas L. Wright and Thomas C. Pierson, *Living with Volcanoes: The U.S. Geological Survey's Volcano Hazards Program*, U.S. Geological Survey Circular no. 1073 (Washington, D.C., 1992), map on p. 22.

ice, posteruption debris flows, mudflows, and floodwaters devastated valley bottoms well beyond the volcano.

A revised map released on May 26 underscored the need to consider a range of eruptive scenarios. Geologists added a wedge-shaped area facing north from the crater formed on May 18 but left intact the hazard zones on the south slope because the geologic history of Mount St. Helens suggested the possibility of an even more powerful eruption.[16] Volcanoes are notoriously unpredictable, and the U.S. Geological Survey was taking no chances.

After watching the volcano erupt on the evening news, an alarmed public demanded increased government support for volcanological research and monitoring. From around $1 million per year in the 1970s, the annual budget for volcano-hazards assessment and related work rose to over $15 million a year in the 1990s.[17] Despite impressive progress during the 1980s, the USGS's Volcano Hazards Program faces numerous challenges. Under the heading "Tasks for the 1990s," a 1992 report listed five major elements and twenty-seven different "priority activities," many of which cannot be hurried to completion by more money.[18] Tasks range from intensive field mapping, electronic monitoring, and satellite remote sensing to computer analysis, modeling, and public education. The final item on a list of thirteen goals for the year 2000 is a highly interactive geographic information system (GIS) to help local public officials control land use and plan evacuations—a long-range goal because the necessary data are largely incomplete or lacking. Of the thirteen volcanoes in the Cascades, three had no hazards assessment, one had a partial assessment, three were "in progress," five had preliminary assessments, and one (Mount St. Helens) required updating.[19]

Although detailed mapping is incomplete, a recent evaluation of potential sites for nuclear power plants provides a broad-brush portrait of volcanic hazards in the Pacific Northwest.[20] The left panel in Figure 3.5 reveals a nearly continuous hazard zone stretching over 600 miles from the Canadian border into northern California. Circles with a 50 km (31-mile) radius estimate the maximum reach of directed blasts, debris avalanches, and pyroclastic flows and surges around thirteen active volcanoes. Finger-like protrusions beyond these "proximal hazard zones" show the added threat to valley bottoms of mudflows and flooding, and the diagonal-line pattern marks a large secondary hazard zone, where nonexplosive eruptions of slow-moving lava might

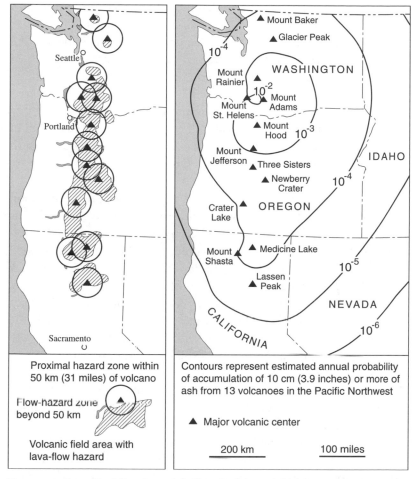

Figure 3.5. Generalized flow-hazard (left) and ash-hazard (right) maps for the Pacific Northwest. Compiled from C. Dan Miller, "Volcanic Hazards in the Pacific Northwest," *Geoscience Canada* 17 (1990): 183–7; maps on pp. 185 and 186.

affect small areas. Although Portland is at the edge of a lava-flow area, no major city is threatened by the far more sudden hazards represented by the 50 km circles and their "distal fingers."

The right panel of Figure 3.5 estimates the combined hazard of volcanic ash from all thirteen volcanoes. Isolines represent the annual probability of 10 cm (4 inches) or more of ash—quite enough to interfere with the operation of a nuclear plant.[21] Based on distance, eruption frequency, and estimated volume of ash, the map's probabilities are highest between Mount St. Helens and Mount Adams—relatively close to each other—where a probability above 10⁻² (0.01) reflects a chance greater than 1 in 100 during a single year of a highly significant ashfall. Because the map's contour values decline exponentially, the risk at Portland, inside the 10⁻³ (0.001) line, is much closer to 1 in 1,000. By contrast, the estimated risk on the 10⁻⁶ (0.000001) line in central Nevada is a nearly negligible 1 in 1,000,000. Although subsequent analysis might alter the shapes and positions of the probability contours, the map describes an ashfall hazard that threatens much of the Pacific Northwest, especially areas east of active volcanoes in south central Washington and north central Oregon.

A 1989 report illustrates the detailed volcanic-hazards information the U.S. Geological Survey hopes to provide to local officials throughout the Pacific Northwest.[22] *Potential Hazards from Future Volcanic Eruptions in California* is a seventeen-page booklet with a folded "plate" inserted in a pocket (envelope) glued to the inside back cover—a typical format for USGS bulletins and research reports. Unfolding the awkwardly large, 49-by-39-inch plate reveals six maps showing more than seventy-five volcanic vents that have erupted in the past 10,000 years, and many more vents active between 10,000 and 100,000 years ago. A large map key explains color symbols for six vent categories and eleven hazard categories. Because the areas are remote, the maps' 1:500,000 scale (one inch represents almost eight miles) accommodates all paved highways and many streams, mountain peaks, and other landmarks.

Dan Miller, who compiled the report from his own work and more than a dozen other studies, wanted to make land-use planners and other California officials aware of "dispersed hazards" involving many different kinds of volcanism.[23] Although Long Valley and Mount Shasta are among the three most hazardous volcanic areas in the continental U.S.—the third, Mount St. Helens, is in Washington—other dangerous locations in the state ought not be overlooked. Useful in restricting access and organizing evacuations during a crisis, the maps' greatest payoff is in long-range planning. "Volcanoes are attractive areas," Miller told me, "but governments, which own most of the land, should behave wisely and not attract permanent investment and sizable populations."

Miller's largest map covers the Mount Shasta, Medicine Lake, and Lassen Peak area in northern California. Around each volcano a red line describes a circular pyroclastic flow-hazard zone and separate blue lines describe tephra-hazard zones likely to receive at least 20 cm (8 inches) and 5 cm (2 inches) of compacted ash. Based on extensive field mapping, these hazard zones estimate the extent of the largest eruption or ashfall within the past 10,000 years. As on Crandell and Mullineaux's hazards map for Mount St. Helens, narrow debris-flow zones extend beyond a larger, irregularly shaped "combined flowage-hazard zone." Along a typical stream directly below a debris-flow zone lies a flood-hazard zone. The zones around Mount Shasta are based on earlier maps Miller plotted at 1:62,500 (one inch represents one mile), the scale of USGS base maps for the area.[24] Learning enough about Mount Shasta to make a reliable hazards map took five years of fieldwork.

To inform residents and visitors about the hazards of Mount Shasta, geologists Rocky Crandell and Don Nichols prepared a short booklet illustrated with four page-size hazards maps.[25] Written for lay readers, *Volcanic Hazards at Mount Shasta, California*, uses photographs of the Mount St. Helens eruption to dramatize the destructiveness of ash clouds, pyroclastic flows, and mudflows. Air photos of volcanic landforms in the Mount Shasta area describe the massive size of debris landslides, lava flows, and Black Butte, one of several cone-shaped volcanic domes formed at the base of Mount Shasta. The first half of this twenty-one-page booklet discusses volcanic threats to the region, and the second half presents the hazard-zone maps.

Concentric zones on three of the maps illustrate the dangers of volcanic ash, pyroclastic flow and lateral blast, and lava flow. Roughly the same size, the maps vary in scale to reflect the extent of the hazard. The outermost of the ashfall map's five zones has a radius of 40 miles, the distance at which the largest ash eruption in the past 10,000 years deposited an inch of ash. (As a concession to nonscientists, all measurements are in inches and miles, rather than metric units.) In contrast, the broadest of three zones threatened by pyroclastic flow and lateral blast reaches only 19 miles from the volcano, and the least hazardous of three lava-flow zones has a radius of about 11.3 miles. A fourth map at a slightly larger scale uses three irregularly shaped zones to show areas threatened by mudflows.

Zone boundaries on the ash-hazard map are full circles, rather than the lopsided arcs used for Mount St. Helens (Figure 3.2). Although a caption warns that areas east of the volcano are particularly vulnerable because of prevailing west-to-east high-altitude winds, past eruptions of Mount Shasta did not produce large amounts of ash. Because ashfalls from low-volume eruptions are buffeted by less predictable surface winds, areas west of the volcano are clearly at risk. Moreover, minor eruptions of Mount St. Helens on May 25 and June 12, 1980, demonstrated that low-altitude winds in the Cascades can direct ash-falls to the northwest or southwest.[26]

Figure 3.6, a slightly smaller black-and-white replica of the mud-flow-hazard map, is typical of the booklet's straightforward cartography. Three major highways, four settlements, three volcanic land-marks, and several locally important lakes and streams provide a sparse yet efficient geographic frame of reference, and area symbols in three intensities—printed in a threatening burnt orange—present a lucid overview of relative risk. A concise caption explains the symbols and interprets their pattern:

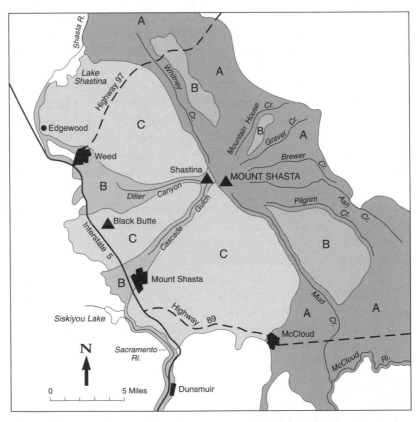

Figure 3.6. Mudflow-hazard map for Mount Shasta. Reprinted from Dwight R. Crandell and Donald R. Nichols, *Volcanic Hazards at Mount Shasta, California*, a U.S. Geological Survey General Interest Publication (Washington, D.C., 1993), p. 18.

Zones designated by letters show relative likelihood of being affected by future mudflows. Zone A is most likely and zone C is least likely to be affected. No mudflow hazard exists on high areas within or beyond the zones. Hazard decreases everywhere within the zones with greater height above stream channels and greater distance from Mount Shasta.[27]

Toward the end of the booklet, in a section on "How to Prepare for and Cope with an Eruption," the last of ten suggestions refers indirectly to all four maps:

Know what hazards are predicted in the general area where you live; if you are in an area that has a relatively high likelihood of being affected if an eruption occurs, think about what kinds of actions you and your family would take if volcanic activity was predicted or actually began.[28]

An exemplar of risk communication in its straightforward translation of volcano hazards into specific suggestions for the general public, *Volcanic Hazards at Mount Shasta* was published in 1987, seven years after Dan Miller presented his detailed hazard maps to state and local officials in a USGS bulletin and two years after geographers Tom Saarinen and Jim Sell called attention to the USGS's traditional reluctance to communicate with the public. In a study of government handling of the Mount St. Helens eruption, Saarinen and Sell praised USGS geologists for a warning that saved countless lives.[29] (The U.S. Forest Service, which heeded the warning, banned all activity near the summit a month before the eruption.) But the geographers called the failure to prepare residents of the ashfall area a "major oversight." Their interviews revealed, for instance, that Crandell and Mullineaux's USGS bulletin never reached local officials in the eastern Cascades. Although state officials were also derelict, Saarinen and Sell placed most of the blame on the U.S. Geological Survey:

By relying too much on others to follow through . . . the USGS contributed to the failure. They were most aware of the potential problems and they were charged by the federal government with the responsibility for warning the public about geologic hazards. To properly give warnings, the USGS must learn from this experience and do more than merely provide the hazard information. They must also deliver it to those who should use it, in a form the latter can understand.[30]

Harsh words, perhaps, because federal volcanologists have no jurisdiction over state and local officials or anyone else.

USGS staff maintain that the Mount Shasta booklet was their own idea, not a response to the agency's critics. And they make a good case.

In his 1980 bulletin on Mount Shasta, for instance, Dan Miller's advice on planning for the "next eruption" included

> Preparation of a pamphlet or other means of describing official plans and procedures for dealing with various aspects of a volcanic eruption. Such a pamphlet should be distributed to the populace around Mount Shasta if an eruption seems imminent and should include information on the kinds of events that might occur, their probable ranges in severity, the expected effects of those events, and what people should do if an eruption does occur.[31]

This turned out to be a concise summary of the volcano-hazards booklet. According to Miller, the idea was Crandell and Nichols's, who obtained the advice and support of two California agencies, the Division of Mines and Geology and the Office of Emergency Services. Like other federal agencies, the U.S. Geological Survey is keenly aware of the authority and sensitivity of state agencies upon whose cooperation it depends.

Local and state officials, who helped the USGS introduce the booklet at a public workshop, were important partners in its distribution.[32] Several thousand copies were placed in post offices and other public places for anyone who wanted them. An equal number were held in reserve, to be mailed directly to residents near and downwind of Mount Shasta when the volcano became restless—officials feared an immediate mass mailing would send a false alarm, creating needless anxiety as well as mistrust in future warnings. Officials were also wary of local resistance to information that might diminish property values or undermine development plans. Someone—perhaps local realtors—took huge quantities that were never seen again.

Risk communication is a complex process, especially in the rural Pacific Northwest, where residents don't like being told what to do, particularly by the federal government. (Remember Harry Truman, the folk hero of the Mount St. Helens eruption who perished with his cats after officials failed to convince him to evacuate?) Jim Riehle, who directs the USGS volcano-hazards effort, is well aware that "the Geological Survey has no authority to tell anyone to do anything."[33] Instead of edicts, volcanologists must rely on careful planning and shrewd cajoling—strategies that account for the relatively few fatalities resulting from the 1991 eruption of Mount Pinatubo in the Philippines. USGS advisors approached mayors in the hazard area individually and used videos of actual eruptions to illustrate what they thought would occur. But one mayor, Riehle reports, "said he didn't want any gringos coming in telling him what to do—and he's dead."

Chapter Four

Uncertain Shores

S everal years ago I spent a delightful week in Hawaii, lecturing at the university and talking with faculty and students. Awakening far too early my first morning there, I fought jet lag by paging through the local phone book. Telephone directories, I've found, are a rich source of information about an area's economy, ethnicity, religious heritage, tolerance of vice, and geographic hazards. But this time, my first visit to the islands, I was looking for something specific—a tsunami (pronounced *soo-NAH-mee*) evacuation map, pointing out low-lying areas whose residents should flee to higher ground or an upper story if a seismic sea wave were imminent. And I quickly found the map where I thought I would, in the public-service section of the Oahu white pages.

Because almost all homes, offices, and hotel rooms get a new one every year, phone books are effective in disseminating evacuation maps in a variety of areas. Coastal communities threatened by Atlantic or Gulf hurricanes often have evacuation maps in their telephone directories, as do some areas with nuclear power plants. The maps typically include an explanation of the local warning system as well as suggested preparations and advice for persons living outside a hazard zone—unnecessary evacuation could clog roads and overwhelm shelters.

Evacuation maps are valuable to Hawaiians because a worldwide seismic monitoring network can detect the large submarine earthquakes associated with most tsunamis.[1] Underwater faults as far away as Alaska, Asia, or South America can attack the islands with a wall of water fifty feet high, but the four to fifteen hours before the arrival of a distant tsunami allows a thorough evacuation of threatened areas.

In the Oahu phone book nineteen tsunami evacuation maps cover 177 miles of shoreline at scales ranging from 1:16,800 to 1:48,000, depending on degree of development.[2] Arranged in sequence, two to a page, each strip map overlaps slightly the area covered by the preceding map. Figure 4.1, the right half of the sectional map for Waikiki Beach, illustrates the landmarks and principal streets used as reference features. Parallel-line shading highlights evacuation areas, and triangle-in-circle civil defense symbols identify public shelters. When sirens sound, residents should "evacuate all shaded areas."

Like many efforts to map hazards, Hawaii's tsunami evacuation maps are rooted in tragedy. In 1946, four and a half hours after a mag-

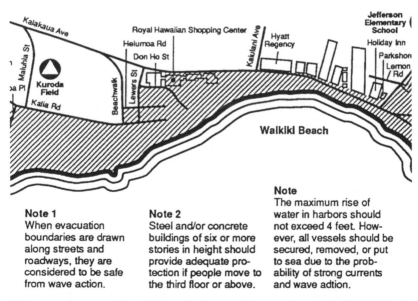

Note 1
When evacuation boundaries are drawn along streets and roadways, they are considered to be safe from wave action.

Note 2
Steel and/or concrete buildings of six or more stories in height should provide adequate protection if people move to the third floor or above.

Note
The maximum rise of water in harbors should not exceed 4 feet. However, all vessels should be secured, removed, or put to sea due to the probability of strong currents and wave adtion.

Figure 4.1. Center portion of the sectional tsunami-evacuation map for Waikiki Beach. From *The Everything Pages: Oahu, March 1994–1995* (Honolulu: GTE Hawaiian Telephone Company, 1994), p. 48. Courtesy of the Office of the Director of Civil Defense, State of Hawaii.

nitude 7.3 earthquake in the Aleutian Islands, a tsunami struck the islands, killing 159 people and causing $26 million in property damage.[3] In 1960 a warning system developed after the 1946 disaster successfully predicted the arrival of a destructive tsunami fifteen hours after a magnitude 8.6 earthquake off the coast of Chile.[4] But many residents ignored or misunderstood the warning, and 61 people died. To provide clearer guidance for future warnings, Hawaiian officials delineated hazard zones based on a worst-case fifty-foot wave and placed generalized evacuation maps in local phone books.[5] In 1986 a difficult evacuation of new neighborhoods revealed the need to update the maps. Two years later, researchers at the University of Hawaii convinced the legislature to fund refined estimates of the inundation area based on procedures endorsed by the Army Corps of Engineers. Introduced in Hawaiian telephone directories around 1991, the new maps describe smaller, more accurate hazard zones with clear boundaries based on roads and other physical features.

A Japanese word meaning "harbor wave," *tsunami* refers to a train of waves set in motion by vertical faulting beneath the sea, an underwater landslide, or a volcanic explosion near the coast. Traveling as fast as 500 miles an hour, six to fifteen wave crests five to ninety minutes apart might slowly raise and then lower the water only a few feet over the open ocean—imperceptible to ships at sea. Approaching land, the waves slow down but increase in height. The first wave arrives quietly, typically as a gradual rise in water level above the high-tide mark. Minutes after flooding low-lying areas, the water begins to retreat with an audible sucking sound. The withdrawing wave exposes a wide expanse of beach below the low-tide mark and leaves ships in the harbor high if not altogether dry. The second assault is noisy and vicious, with the wave moving up the beach in a steep front and crashing into buildings, dunes, and anything else in its path. Wave attacks can last eight hours or longer.

A tsunami's magnitude is measured by its *runup height*—the elevation above sea level reached by water running up the beachfront.[6] During the 1946 tsunami, some Hawaiian coasts experienced runups greater than thirty feet. Tsunamis striking Japan have registered runups of nearly one hundred feet. Even so, tsunamis rarely wet areas more than a half mile inland, and damage typically is confined to a narrow belt along the shore.[7]

Like volcanoes, tsunamis are almost exclusively a problem for Alas-

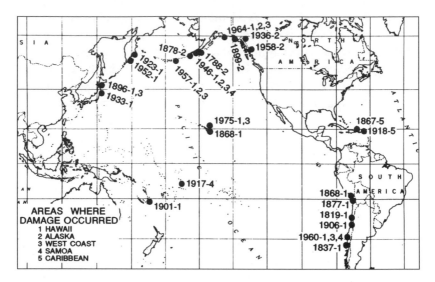

Figure 4.2. Epicenters of earthquakes associated with destructive tsunamis in the United States and its possessions. Reprinted from James F. Lander and Patricia A. Lockridge, *United States Tsunamis, 1690–1988* (Boulder, Colo.: National Geophysical Data Center, 1989), p. 11.

ka, Hawaii, and our other Pacific states. *United States Tsunamis, 1690–1988*, a comprehensive catalog compiled by the National Oceanic and Atmospheric Administration (NOAA), lists only eleven tsunamis between Maine and Texas.[8] No deaths occurred, and except for a 1926 "tidal wave" that struck Bass Harbor, Maine, damages were negligible. Earthquakes along the Mid-Atlantic Ridge produce very little vertical movement, and even so, the continental shelf would absorb much of the energy of a long-distance Atlantic tsunami. In contrast, the NOAA catalog includes eleven tsunamis with known fatalities for Hawaii, six for Alaska, and three for the West Coast.

Tsunamis are closely related to the Pacific Ring of Fire (Figure 2.1), responsible for most of the world's earthquakes and volcanoes. Figure 4.2, a NOAA map of epicenters of earthquakes producing tsunamis destructive to the United States and its possessions, reflects the distinctive circum-Pacific pattern as well as significant clusters along the Alaskan and Chilean coasts. A single-digit code following the year indicates the part of the U.S. damaged by each tsunami. A preponderance of *1*s, the code for Hawaii, attests to the islands' vulnerability from a variety of directions. And multiple damage-location codes indicate that distant tsunamis can strike widely separated shores. In 1960, for example, the Chilean tsunami that killed 61 people and caused $24 million in damage in Hawaii also inflicted more than a half million dollars' damage on California. And the 1964 tsunami that killed 107 Alaskans produced runups as high as fifteen feet along the West Coast and killed 11 people in Crescent City, California.[9] Although infrequent, tsunamis can be enormously destructive.

Storm surge is an equally fearsome threat to the Atlantic and Gulf coasts. Although strong winds and flooding damage both inland and coastal locations, a hurricane's most lethal attack is a wall of water that carries the destructive attack of breaking waves up a gently sloping beach and over a mile inland in the most extreme cases. Surge is computed by subtracting the predicted height of astronomic tides from the water level measured with a tide gauge. (Calculated tides published in almanacs are accurate only when seas are calm.) Typical surge heights of three to ten feet can be especially devastating when a hurricane strikes at high tide. Blamed for 90 percent of coastal fatalities, storm surge is especially deadly to persons who ignore orders to evacuate.[10]

Surge is largely the product of two forces.[11] Intense low pressure sucks water toward the hurricane's center and elevates the water surface about a foot for an atmospheric pressure drop of an inch of mercury. That might account for a two-foot rise. More important, water piles up against the shore as wind-driven waves force more water onto the land than drains back to the sea. Pileup raises the water level another two to ten feet.

Figure 4.3 describes the storm surge created by hurricane Carol, which attacked the New England coast in August 1954.[12] The map traces the erratic advance of the hurricane's eye. After moving slowly westward toward the Georgia coast for several days, Carol turned sharply to the north and advanced rapidly to landfalls on eastern Long Island and southern Connecticut, where the water rose nine feet above expected tides. Surge charts describe increased water levels over a three-to-five-day period. Letters link each surge chart to its location on the map, and a dashed line marking the closest approach of the eye traces the storm's movement northward. An intense low-pressure cell, Carol created a noticeable storm surge over 100 miles from its path.[13] Although surge elevations were generally higher along the storm track, the most pronounced surge occurred at Woods Hole, Massachusetts, on the southern shore of Cape Cod, where 120 mph winds attacked the coast. (Because of counterclockwise circulation around the eye,

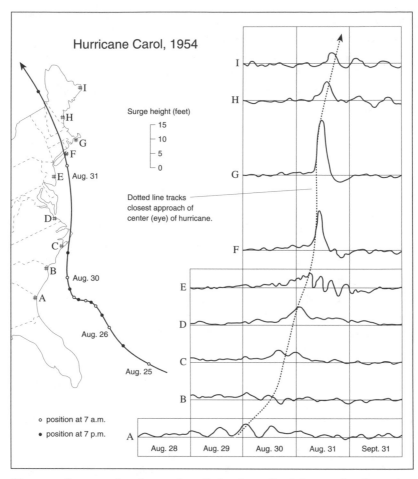

Figure 4.3. Storm track and surge charts for hurricane Carol, August 28 to September 1, 1954. Compiled from N. Arthur Pore, "The Storm Surge," *Mariners Weather Log* 5 (1961): 151–6; map and graphs on p. 152.

wind speed and surge height are greater to the right of the storm track.) Large surge heights in southern New England also reflect the storm's comparatively rapid movement just before landfall as well as an increased pileup on the shallow continental shelf south of New England. Dense coastal development contributed to an overall loss of 60 lives and nearly $500 million in damage.[14]

Not all coastal areas are equally hazardous. Figure 4.4, a hurricane-probability map in the Federal Emergency Management Agency's *Coastal Construction Manual*, shows that severe tropical storms are especially troublesome on the Gulf coast, in south Florida, and along the Outer Banks of North Carolina.[15] Based on eighty-five years of experience, 1886 through 1970, the map divides the coast into 50-mile sections and presents two series of annual percentage probabilities.[16] The inner band of numbers reports the likelihood of a landfall by any large tropical storm with winds stronger than 73 mph—the meteorologist's cut-point between hurricanes and lesser tropical storms—and the outer band shows the probability of a "great hurricane," with sustained winds over 125 mph. According to the map, the Miami area, where Marge and I encountered Hurricane Betsy during our honey-

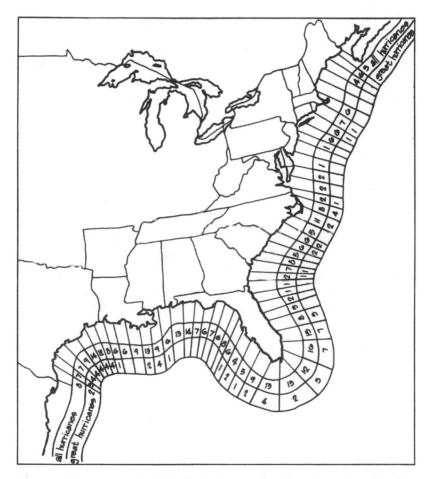

Figure 4.4. Percentage probabilities for hurricanes along fifty-mile portions of the East Coast. Reprinted from Federal Emergency Management Agency, *Coastal Construction Manual,* FEMA publication no. 55 (Washington, D.C., 1986), p. 2-4.

moon, is the most hurricane-prone section of the East Coast, with a
16-in-100 chance of a hurricane in any one year and a 7-in-100 likelihood of a great hurricane. Had we known the odds, we still would have
gone—like owners of coastal property, the young take chances.

A long the northern New England coast, *northeasters* are more
destructive than hurricanes.[17] Large, slow-moving storms with
high winds and heavy, prolonged rain, "nor'easters" occur largely
between November and March at a rate of three or four a month.[18]
Meteorologists call them *extratropical cyclones* because, unlike hurricanes, which originate in the Atlantic Ocean south of the Tropic of
Cancer, a typical northeaster starts out as a low-pressure cell over the
south central states and moves east and then north along the coast.[19]
Cyclonic (counterclockwise) circulation around the storm's low-pressure core occasionally produces intense onshore winds ahead and to
the right of the storm. Although usually not of hurricane velocity, these
northeast-to-southwest winds pick up enormous amounts of moisture
over the ocean and aggravate flooding around tidal estuaries. Persistent
direct winds can create a storm surge of four feet or more, and a slowly advancing northeaster often intensifies the surge by catching the
high tide.

Although most northeasters are neither fatal nor notably destructive, their vast size, slow pace, and frequent occurrence accounts for
considerable beach erosion. And on occasion a northeaster is deadly
powerful. On March 5–9, 1962, for instance, the famous "Ash Wednesday Storm" killed at least 33 people from North Carolina to New Jersey and caused nearly $200 million in damage between Florida and
Maine.[20] In the winter months, other large, slow-moving northeasters
have crippled parts of the New England and Middle Atlantic regions
with deep, wet snow.

B ecause of varied forces at work, a concise nationwide summary of
tsunamis, hurricanes, and northeasters seems a daunting task.
But the coastal hazards sheet added to the U.S. Geological Survey's
National Atlas map series in 1985 offers a comprehensive overview.
Compiled by geographer Robert Dolan and U.S. Army Corps of Engineers researchers Fred Anders and Suzette Kimball, the 1:7,500,000
map uses colored bands to represent nine different factors. Figure 4.5,
a black-and-white replica of the map's 1:5,000,000 inset for the Gulf
coast south of New Orleans, illustrates the sheet's intricate treatment
of relative hazardousness.

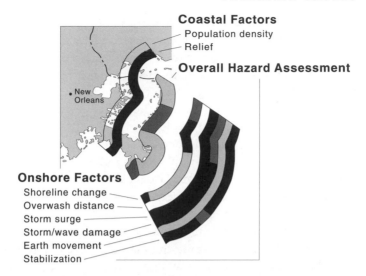

Figure 4.5. The U.S. Geological Survey's coastal hazards map uses intensity symbols arranged in bands to represent onshore and coastal factors. A thicker band along the shore provides an overall assessment of hazards. A slightly enlarged inset (1:5,000,000) describes the contorted shoreline along the Gulf coast below New Orleans. From Fred Anders, Suzette Kimball, and Robert Dolan, *Coastal Hazards*, National Atlas 1:7,500,000 map (Reston, Va.: U.S. Geological Survey, 1985).

Differentiated by hue as well as position, three sets of bands represent greater threats with darker colors. In shades of purple, the two inland bands describe a pair of onshore factors: *population* and *relief*. The light innermost band in the excerpt represents a sparsely settled coastline, with relatively few people at risk, whereas the darker adjacent band depicts the vulnerability to storm surge of elevations rarely above 3 meters (10 feet). Elsewhere on the full map, a dark innermost band represents a densely populated coast, while a light second band might indicate seaside cliffs.

In contrast, the outer bands use four intensity symbols ranging from dark brown for most hazardous to light brown for least hazardous to evaluate six coastal factors: *shoreline change* (innermost of the six bands) contrasts accreting (growing), stable, and eroding coasts; *overwash penetration* describes the likelihood of water and sand breaching the dune line; *storm surge* reflects the surge height of a hypothetical storm with a ten-year recurrence interval; *storm/wave damage* indicates the frequency and severity of tropical storms, extratropical storms, or tsunamis at least 1 meter (3.3 feet) high; *earth movement* describes susceptibility to earthquakes and landslides; and *stabilization* measures the density of jetties, seawalls, and other structures for retarding shore-line erosion. A cream-color symbol along the second band in Figure 4.5 indicates "no data" for dune-line penetration, whereas dark brown along the third band reflects the high storm-surge hazard of exposed headlands with gently sloping underwater terrain, and dark and medi-um brown along the outer band describes the success—thus far—of Corps of Engineers efforts to retard erosion.

Between the onshore and coastal bands, a single, thick band pre-sents an overall assessment of all coastal factors except stabilization. The authors weighted these five hazards equally but gave "greater emphasis to extreme events which generally constitute the greatest hazard."[21] Seven intensity levels rely on a red/green traffic-light metaphor: four symbols ranging from bright red for "very high risk" to light pink for "moderate risk" describe the nation's more hazardous shores, while three symbols ranging from light green for "moderate to low risk" to bright green for "very low risk" describe less hazardous coasts. Despite a significant storm-surge hazard southeast of New Orleans, less than half the shoreline in Figure 4.5 is rated high risk, and the remainder is considered moderate risk. In contrast, a bright red band along the Alaska peninsula and the Aleutian Islands warns of very high risk because of tsunamis, while for Hawaii slightly lighter

Figure 4.6. A simplified version of the National Atlas map's overall hazard assessment. Compiled from Fred Anders, Suzette Kimball, and Robert Dolan, *Coastal Hazards*, National Atlas 1:7,500,000 map (Reston, Va.: U.S. Geological Survey, 1985).

shades of red mark levels of risk between high and moderate. In combining the effects of hurricanes and tsunamis, the map underscores the
threat of submarine earthquakes to Alaska and Hawaii.

Despite praise from coastal scientists, the National Atlas coastal
hazards map never caught the attention of the public, the media, or the
Congress. "It was just too complicated," Bob Dolan lamented recently.[22] "Most people said it was like looking at a bucket of snakes." When
I asked whether anyone had ever prepared a smaller version focusing
on the overall hazard assessment, Dolan said no, not to his knowledge.

To understand how a simplified cartographic summary might be
useful, I compiled the map in Figure 4.6 by transferring to a smaller,
more generalized base map the four highest risk categories of the overall hazard assessment. Two hours' work on a Macintosh was effective,
I think, in providing a concise summary of coastal hazards. But is
this—or any—summary adequate? Will "moderate risk" be read—
incorrectly—as "minor risk"? Will viewers overlook the seriousness of
"moderate risk" along densely settled coasts? Can a title adequately
emphasize the map's portrayal of only the "most dangerous" coasts?
Because of possible misinterpretation, my dumbed-down summary
seems useful only as a cartographic teaser, to call attention to the original, more complete version. But what better way to advertise a richly
informative map?

However useful in understanding coastal forces and regional differences, small-scale cartographic overviews like the National
Atlas coastal hazards map say little to planning boards, developers, and
home buyers who need to understand the effects of changing sea level
on communities and individual homesites. While some shorelines have
been stable for hundreds of years and should remain so—barring an
accelerated worldwide rise in sea level—other coastlines are certain to
change, this year or next perhaps, but certainly within twenty or thirty years.[23] Forecasting that change, or at least describing its direction
and pace, can be an important goal of detailed coastal maps.

A case in point is the most fragile-looking part of our Atlantic shoreline, the Outer Banks. It's a fascinating feature: a chain of long, narrow
barrier islands stretches nearly two hundred miles from Cape Henry,
Virginia, to Cape Lookout, North Carolina, in two streamlined arcs
intersecting at Cape Hatteras, nearly twenty-five miles from the mainland. On small-scale maps the Outer Banks appear to enclose Pamli-
co Sound, but more detailed views reveal a number of inlets. Although

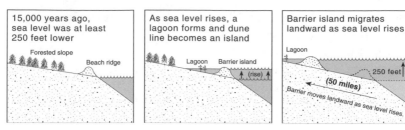

Figure 4.7. Formation and inland advance of the Outer Banks reflect rising seas and a gently sloping continental shelf. Redrawn from Robert Dolan and Harry Lins, *The Outer Banks of North Carolina*, U.S. Geological Survey Professional Paper no. 1177-B (Washington, D.C., 1985), p. 12, fig. 21.

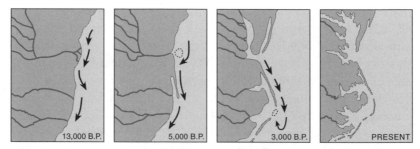

Figure 4.8. Longshore drift and rising seas account for the evolution and shape of the Outer Banks. Redrawn from Robert Dolan and Harry Lins, *The Outer Banks of North Carolina*, U.S. Geological Survey Professional Paper no. 1177-B (Washington, D.C., 1985), p. 12, fig. 21.

contemporary maps show a single inlet between Cape Henry and Cape Hatteras, geologists have identified twelve different inlets since Walter Raleigh established the ill-fated Lost Colony on Roanoke Island in 1585—in Raleigh's time there were six inlets.[24] The shorter barrier arc between Cape Hatteras and Cape Lookout has six inlets now, but only one has been open continuously since the sixteenth century. Triggered by storm surge and high waves during severe storms, formation and closure of inlets is nature's principal means of transferring sediment from ocean to sound. The Corps of Engineers can slow the process—at great cost to taxpayers—but they can't stop it.

To coastal scientists, barrier islands reflect rising seas. Along the North Carolina shore, the water has been rising more or less consistently for the past fifteen thousand years—a lot longer than recent fears about global warming and melting polar ice caps. Figure 4.7 explains the effects of the estimated 250- to 300-foot rise in sea level. The left-to-right sequence begins with a line of coastal dunes fifty to seventy-five miles seaward of the present barrier islands. As sea level rose, a lagoon formed behind the dune ridge. Periodic overwash and the formation of inlets moved sediment from the ocean side to the lagoon side, and as the sea continued to rise, the dune line advanced inland along the gently sloping continental shelf. Radiocarbon dating suggests the rise in sea level slowed down about four thousand years ago, but old maps and aerial photographs show that for the past hundred years the Outer Banks have moved toward the mainland at an average rate of 3 to 5 feet per year.[25]

A sequence of four maps (Figure 4.8) explains the barrier islands' origin and unique shape. Arrows represent *longshore drift*, whereby the beach acts as a sluggish yet powerful river of sand. Along this stretch of coast the drift is southward: because winds and waves are generally from the northeast, a sand grain washed up the beach comes to rest, on average, south of where it started. The left panel shows longshore drift about thirteen thousand years ago—*B.P.* on the diagram is geology-speak for "Before Present"—when the coastline was comparatively smooth and simple. By 5000 B.P. the longshore current had formed a long narrow *spit*—like a barrier island but still attached to the coast—from huge quantities of sediment carried to the shore by the large river that later became Chesapeake Bay. By 3000 B.P. a risen sea had drowned much of the former coastline, the northern arc of the Outer Banks diverged southeastward from a more rapidly retreating shoreline, and a southern arc of barrier islands had emerged, fed in places by longshore drift from the south.

I copied these profile diagrams and maps from an impressive report on the Outer Banks produced by U.S. Geological Survey with the support of the National Park Service. Although coauthors Bob Dolan and Harry Lins used many graphics to explain the evolution and hazards of the North Carolina coast, their most dramatic illustration is a map (Figure 4.9) projecting the appearance and position of the coastline for the year 2980.[26] If the sea continues to rise at the current rate over the next millennium, the Outer Banks will have advanced toward the land a distance of roughly twice their width—a valid calculation that justifies the obvious cartographic exaggeration of the islands' width. Because of coastal drowning, the inner shore is receding too, and in places the estuaries will grow wider. The warning to planners and developers is clear, and Dolan and Lins let the map speak for itself.

A thousand years is a long time, though—much longer than the expected life of a second home or beachfront condo with a million-

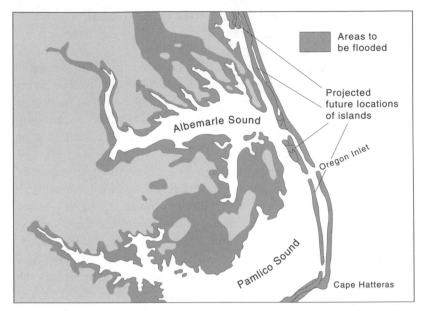

Figure 4.9. Projected North Carolina coastal zone, 2980 A.D. Redrawn from Robert Dolan and Harry Lins, *The Outer Banks of North Carolina*, U.S. Geological Survey Professional Paper no. 1177-B (Washington, D.C., 1985), p. 14, fig. 25.

dollar view, and far too long to discourage real-estate developers and wealthy home buyers eager for profits or scenery. To relate erosion rates to the lifetimes of people and buildings, Dolan and Lins included fifty-four pages of large-scale maps covering the entire length of the Outer Banks at 1:24,000, the basic scale of USGS topographic maps. Despite the capacity for considerable detail, their maps contained few reference features: the left panel of Figure 4.10, a black-and-white replica of the map for a small area on the southeast fringe of Nags Head, shows only a gray shading for land and dashed black lines for the more important roads—far fewer details than the topographic map in the right-hand panel. (More detailed information was available, as the topographic map in the right panel demonstrates, but existing maps were inconsistent in scale and accuracy, and the added production costs would have been enormous.) Against this minimalist background, a thick black line (red in the original) represents the predicted shoreline for the year 2002, and a thick dotted line (blue in the original) shows "the mean storm surge, or overwash penetration distance." Although areas inland from the dotted line might expect "relatively stable vegetation" in the year 2025, land between the solid and dotted lines is prone to flooding and erosion, and locations seaward of the solid line are at much greater risk.

My initial reaction was disappointment. Instead of ominous beach erosion, all the fifty-four pages of maps seem to show is a predicted shoreline slightly west of its position in the early 1980s. And in many places, the red line is farther seaward, indicating deposition and a naturally growing beach! The hazard zone delineated by the blue line depicts a more widespread problem—even where the red line predicts beach accretion, overwash is likely 100 to 1,000 feet beyond the shoreline—but the featureless gray island reveals little about threats to life or property. Moreover, the "mean storm surge" represented by the blue line ignores the devastating temporary inlets that breach dunes during severe storms, thereby suggesting that the Outer Banks are less hazardous than they really are. It is hardly surprising that the U.S. Geological Survey dropped the lengthy appendix from second and subsequent printings of the report.[27] If you buy a copy of *The Outer Banks of*

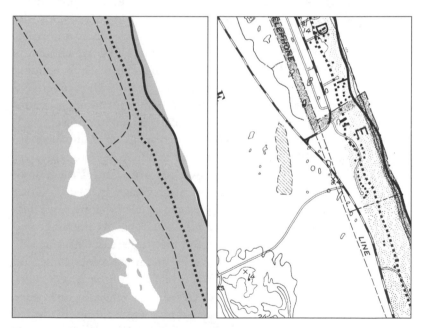

Figure 4.10. Predicted shoreline (heavy solid line) for the year 2002 and mean surge penetration distance (dotted line) for beachfront southeast of Nags Head, N.C. Left panel from Robert Dolan and Harry Lins, *The Outer Banks of North Carolina*, U.S. Geological Survey Professional Paper no. 1177-B (Washington, D.C., 1985), p. 70; map reoriented to place north at the top. Right panel from Oregon Inlet, N.C. 7.5-minute topographic quadrangle map, 1:24,000, photorevised (Reston, Va.: U.S. Geological Survey, 1953, 1983).

North Carolina at National Park Service visitors centers, where it's a consistent best-seller, you'll get a shorter, less expensive booklet that explains the threat of beach erosion to barrier islands but omits local details. Just as well.

D ropping the dubious appendix must have been an easy decision for the U.S. Geological Survey because other federal agencies had already mapped coastal hazards. In the early 1970s, for instance, NOAA prepared a series of Storm Evacuation Maps covering the Atlantic and Gulf coasts at a scale 1:62,500 (one inch represents approximately one mile).[28] At roughly the same time, the Federal Emergency Management Agency (FEMA), which administers the national flood insurance program, launched its series of Flood Insurance Rate Maps, examined in chapter 6.[29] FEMA also produces Flood Hazard Boundary Maps, preliminary hazard-zone maps for communities interested in joining the flood insurance program.[30]

Directed at local emergency management officials, the Storm Evacuation Maps were a cooperative effort of two NOAA divisions: the National Weather Service, which monitors the atmosphere and issues storm warnings, and the National Ocean Service (NOS), which surveys the nation's coasts and produces hydrographic charts. By integrating general information from USGS topographic maps with precise NOS estimates of mean sea level, NOAA hoped to improve community preparedness and increase the effectiveness of hurricane warnings.

Designed for emergencies, NOAA's Storm Evacuation Maps are straightforward and focused. The excerpt and map key in Figure 4.11 reveal efficient symbols and labels as well as a thoughtful selection of features around East Hampton, a coastal village in eastern Long Island. Heavy lines and type identify escape routes by name, large dots with spot elevations mark low points vulnerable to flooding, and hatched area-shading symbols point out built-up areas. Omitting side streets and less important roads makes room for essential details, and highly legible road names correct a shortcoming of the USGS topographic maps used as sources. NOAA's design relies on only three inks: blue for water; black for evacuation routes, elevations, urban areas, and all names and reference areas; and red for three categories of flood-prone areas. Colored a conspicuous dark red, the lowest, most vulnerable areas (0–10 feet above mean sea level) are readily apparent to local officials. But when a large surge height is forecast, police and fire officials can easily recognize higher flood-prone areas shaded more

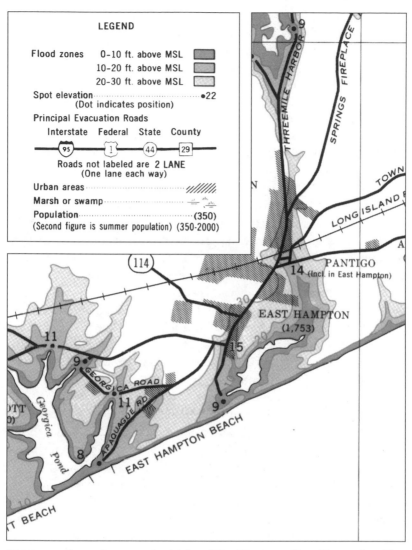

Figure 4.11. Evacuation routes in vicinity of East Hampton village. Excerpt from East Hampton, N.Y., Storm Evacuation Map, T-15030, 1:62,500 (Rockville, Md.: National Ocean Service, 1973).

subtly in medium red and pink. The maps' large size—the East Hampton sheet, covering much of eastern Long Island, measures 29.5 by 35 inches—obviates handling several topographic quadrangle maps. (Five map sheets cover the entire 120-mile length of Long Island as well as New York City.)

Despite an effective design, NOAA's Storm Evacuation Maps relied solely on elevation in forecasting the threat of coastal storms—a naïve view that ignores dominant wind direction and the greater exposure of some locations to storm surge and overwash. Coastlines vary in regularity and orientation, wind and waves often attack from the northeast, and dune ridges protect inland areas from all but the most severe storms. For these reasons and a host of local anomalies, the maps misrepresent their dark red areas as equally hazardous. Although officials familiar with local conditions can use the information effectively, in the hands of a martinet obsessed with evacuating all areas below 10 feet regardless of location, the maps themselves could be dangerous.

NOAA's Storm Evacuation Maps remind me of California's Special Studies Zones, an equally naïve approach to hazard-zone mapping. Like distances of 50 or 500 feet from a surface fault, elevations of 10 feet, 20 feet, and 30 feet above mean sea level are little more than conveniently plausible round numbers: a weak conceptual foundation for an inexpensive, scientifically unsophisticated mapping program designed to do something intuitively useful and do it quickly. And like the earthquake scenarios that have made California's fault-zone maps obsolete, computer-based simulations that consider local and regional conditions as well as a range of hypothetical storms now offer coastal areas a more reliable basis for evacuation planning and hazard mitigation.

For some areas, this improved reliability came not a moment too soon. Substantial increases in the coastal populations aggravated the difficulty of getting people to leave their homes and called for more limited and precise evacuations. One solution was *phasing*, the evacuation of the most vulnerable areas first, then the next most vulnerable areas, and so on.[31] But phasing demands a highly accurate assessment of relative hazardousness. Dependable maps are equally important for *vertical evacuation*, a strategy of moving people from threatened areas to the middle and upper floors of high-rise buildings and other "hardened" structures. Vertical evacuation keeps evacuees within the community and encourages greater compliance with evacuation orders. Reliable maps are necessary, of course, to identify areas appropriate for vertical evacuation and to allocate residents to shelters.

Local officials on Long Island got their new maps in 1993—as an appendix to the New York State Hurricane Evacuation Study supported by FEMA and carried out by the U.S. Army Corps of Engineers with the cooperation of New York's State Emergency Management Agency (SEMO).[32] According to Allan McDuffie, who supervised mapping by the Corps of Engineers, the new evacuation maps rely on the National Hurricane Center's SLOSH model and "the best topographic maps we can get our hands on."[33] By comparison, the earlier NOAA maps were "a mere approximation."

An acronym for Sea, Lake, and Overland Surges from Hurricanes, SLOSH is a set of mathematical equations for estimating floodwater elevations over both land and sea.[34] The equations represent the effects of wind stress and the sharp drop in atmospheric pressure near the center of the storm. Two kinds of information are needed: the position, pressure, and size of the hurricane and the elevations of the land and seafloor at a fan-shaped radial network of grid points. The model for the New York/Long Island coast used a grid of 6,840 fixed elevation points to compute water heights for 533 hypothetical storms with wind speeds between 74 and 155 mph.[35] Although unreliable for operational forecasting—hurricanes can speed up or change direction without warning—SLOSH models are valuable for simulating a broad range of plausible storms based on meteorological records.[36] The National Hurricane Center has developed models for twenty-two "SLOSH basins" along the Atlantic and Gulf coasts, most of which have the new evacuation maps.

Evacuation zones representing expected inundation, not elevation, call for a distinctly different map key. Long Island's new maps show areas likely to be attacked by surge from hurricanes with wind speeds of 74–95 mph, 96–110 mph, 111–130 mph, and 131–155 mph—categories 1 through 4 of the Saffir/Simpson hurricane-intensity scale, used to develop hypothetical storms for the SLOSH simulations.[37] Within each county, analysts delineated evacuation zones with similar inundation potential and bounded by easily recognized features. Useful during an evacuation, these zones also served the study's demographic, transport, and shelter analyses. For Long Island's two eastern counties, the maps combine categories 1 and 2, and assign each evacuation zone to one of three classes:

Category 1–2	Surge Area
Category 3	Additional Area
Category 4	Additional Area

As with NOAA's Storm Evacuation Maps, the darkest, most intense shading marks the first class, which includes the most vulnerable areas, low and close to the sea, while the lightest area symbol indicates zones to be evacuated only for an extraordinarily intense, category 4 storm.

These arcane labels assume two kinds of user: field operations personnel told to evacuate particular zones and emergency management officials familiar with the hurricane study's *Technical Data Report*. While the evacuation-zone maps provide police and fire officials—assumed to do what they're told—with recognizable zone boundaries and graphically logical shading, the technical report advises emergency management officials on when and how to decide whether to evacuate. In addition, the report examines the study's assumptions and limitations. Storm simulation is not precise, the *Report*'s authors point out, and in some areas surge heights might well exceed the calculated maximum by as much as 20 percent. Moreover, inundation estimates do not consider tides. A sentence highlighted in bold type warns: "If astronomical high tide occurs coincidentally with the peak storm surge, the combination could be considerably higher than the SLOSH surge values shown in the inundation maps."[38] However detailed and thorough, evacuation studies are tools, not cookbooks.

Despite more scientifically reliable hazard-zone boundaries, the hurricane study's detailed inundation maps are less visually striking than their predecessors.[39] Although a larger scale—1:24,000 (one inch represents two thousand feet), corresponding to USGS topographic quadrangle maps—affords greater detail, the maps are printed with only two poorly contrasting inks: black for reference features and blue for storm inundation information. But not all SLOSH-based evacuation maps are equally detailed or visually bland. For example, North Carolina's new evacuation maps consist of a blue inundation overlay printed on black-and-white county road maps at 1:125,000 (one inch representing two miles). And some maps for Florida communicate relative risk with shades of red and yellow. According to Allan McDuffie, the maps have not been standardized, and each state chooses its own scale, colors, and format.[40]

Eclectic cartography is merely a symptom of the nation's loosely structured, largely voluntary approach to hurricane preparedness. Emergency planning is primarily the responsibility of local governments—counties, townships, and villages—which rely on federal agencies for storm warnings, technical expertise, and financial assistance.[41] And somewhere in the middle, state governments initiate, support, or help coordinate emergency planning with varying degrees of savvy and commitment. Coordinating the federal efforts is FEMA, the designated "lead agency." But FEMA lacks the expertise of NOAA and the Corps of Engineers. That our hurricane evacuation system works reasonably well in most places is amazing. Stronger federal leadership seems inevitable, though, amid fears of a storm of unprecedented ferocity.[42]

Despite cooperation in emergency preparations, coastal communities often resist federal and state efforts to limit development. This resistance reflects the clout of home owners and developers, who consider themselves the big losers when government restricts land use, abandons seawalls and costly "beach nourishment" schemes, and denies flood insurance to those who rebuild, again and again, at taxpayer expense. Scientists, FEMA administrators, and many state officials support "coastal retreat," a strategy of accepting erosion as inevitable, relocating existing buildings away from the beach, and requiring "setbacks" for new construction. But the Army Corps of Engineers—ever eager to stabilize the existing shoreline with engineering solutions—enjoys the powerful support of construction, real estate, and mortgage interests.

Few critics of "artificial beaches" are as outspoken as Orrin Pilkey, Jr., James B. Duke Professor of Geology and director of the Program for the Study of Developed Shorelines at Duke University. Pilkey considers natural beaches an endangered species and campaigns against "newjerseyization"—a pejorative reference to New Jersey's addiction to "hard stabilization" with groins, jetties, and seawalls.[43] Because the sea has risen a foot in the last century—a rate likely to accelerate—he equates the Corps' strategy with the Maginot Line. An activist who enjoys addressing public hearings, Pilkey strikes a populist chord: should government save the buildings of the rich or preserve scenic beaches for everyone?

Pilkey's most effective platform is the Living with the Shore guidebooks, which he coedits with geologist Bill Neal. Each guide treats a section of coastline, typically within a single state. The books have two goals: to warn of coastal erosion, especially hazards connected with storms, and to be understandable to the layperson.[44] Published by the Duke University Press and sponsored by the National Audubon Society, the series has grown to seventeen titles, with another five planned. Pilkey wrote the first guide, *How to Live with an Island*, in 1975 with support from FEMA, which didn't want the controversial book to be a government publication.[45] Although Pilkey wrote portions of several of the guides, he usually recruits geologists with experience in the state's coastal geomorphology. Some titles reap no royalties, but earnings from the more successful guides, such as *Living with the Coast of Maine*, a best-seller among tourists, are "plowed back in," to support new projects.

In addition to explaining coastal erosion and related issues, each guide provides an illustrated tour from one end of the coast to the other. Discussion focuses on detailed site analyses for short sections of the coast. Because Pilkey never standardized the cartography, site maps vary widely in design from book to book. But all maps are black and white, straightforward in addressing risk, and carefully integrated with their accompanying text. Figure 4.12, an excerpt from *Living with the California Coast*, is typical in its use of numbers to show erosion

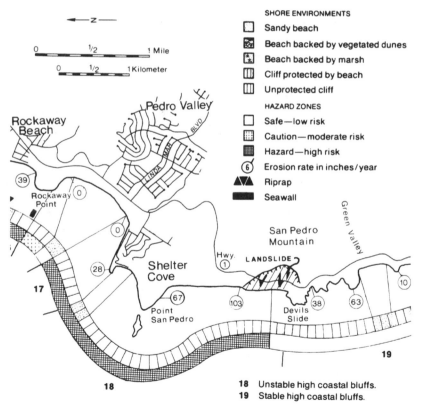

Figure 4.12. Portion of the site-analysis map for the Pacific coast south of San Francisco. Reprinted from Gary Griggs and Laurel Savoy, eds., *Living with the California Coast* (Durham, N.C.: Duke University Press, 1985), p. 149. Reproduced with permission.

rates, intensity shadings to represent relative risk, and point symbols to describe "shoreline engineering" practices. (Susceptible to tsunamis, the Pacific coast is more persistently attacked by waves and storms.) Pictorial symbols describe the shoreline, while zones labeled "hazard" and "caution" point out threats to structures. Diagrams explaining erosion and photographs illustrating damage help readers interpret the costs and aggravation implied by the maps. With this treatment, Pilkey believes, "anyone who can read a map can find out about potential coastal hazards."[46]

Chapter Five

Death Tracks

When this book began to take shape, it had no chapter on tornadoes. There was nothing much to write about, I thought: no
detailed risk maps like those for earthquakes, floods, and other geophysical hazards. I could deal with tornadoes once and for all, cleanly
and creatively, with the pair of distribution and probability maps in
chapter 1. But these most violent and lethal of storms are not so easily dismissed, especially in a book on hazards in the country visited by
three-quarters of the world's tornadoes. And meteorologists, who have
been monitoring tornadoes and posting watches for more than a century, have gained valuable insights by mapping their data.

Calling tornadoes violent and lethal is not hyperbole. In an average
year, these powerful, locally intense storms kill 93 Americans. At least
that's the annual mean for the 3,524 deaths recorded by the National
Weather Service for the period 1953 through 1991.[1] Few years are average, though, and in recent decades the yearly death toll has varied from
less than 20 to several hundred. The atmosphere is notorious for devastating mood swings, which include catastrophic storms like the Tri-
State Tornado of March 18, 1925, which slashed a 219-mile path across

Missouri, Illinois, and Indiana and killed 689 people—3 a minute—in a three-and-a-half-hour rampage.[2] Equally ominous are outbreaks like the 148 tornadoes that on April 3–4, 1974, claimed 315 lives and destroyed more than $600 million in property.[3] Although tornadoes kill fewer people, on average, than either lightning or floods, no other weather-related catastrophe is so thoroughly destructive.

Chapter 1's tornado probability map (Figure 1.3) is a convenient place to begin. Used widely in government publications on tornadoes, this map of average annual incidence points out the "Tornado Alley" stretching from Texas to the Dakotas as well as a dearth of tornadoes in the Rocky Mountains. But annual averages ignore seasonal variations that explain why tornadoes are particularly common in the center of the country. Figure 5.1 shows, for instance, that May is the worst month for tornadoes in Oklahoma and Texas, whereas June and July are the peak twister months in the upper Midwest. This pattern reflects the position of the jet stream, which triggers tornadoes by bringing cool, dry polar air from the Rockies and Far West into contact with warm, moist tropical air from the Gulf of Mexico.[4] In Indiana, Oklahoma, and other high-incidence states, tornadoes are more common in April or May, when the jet stream typically bends south

Months of peak tornado activity

Figure 5.1. Months of peak tornado activity. Redrawn from National Oceanic and Atmospheric Administration, *Tornado Safety: Surviving Nature's Most Violent Storms* (Washington, D.C., 1982), p. 7.

through Oklahoma before swinging northward through Indiana. By contrast, the tornado season occurs a month or two later in the Northeast, after the jet stream has moved farther north.

Regional differences in seasonality suggest a need to map tornado hazards month by month. Although annual averaging is appropriate for casualty underwriters, who sell insurance a year at a time, emergency management officials in the eastern states could benefit from probability maps for individual months like July or August.

Anyone who thinks tornadoes are largely a problem for Tornado Alley states should look closely at Figure 5.2, copied from a NOAA booklet on tornado preparedness. A single sentence explains the map's rationale: "The greatest potential for casualties from tornadoes is not necessarily where the greatest number of tornadoes occurs, but where there is a combination of high tornado incidence and a dense concentration of population."[5] To address these concerns, the map's authors computed a *threat rating* by multiplying tornado incidence (in tornadoes per 10,000 square miles) by population density (in persons per square mile) and then dividing by 10.[6] Surprisingly, the greatest

Threat rating from tornadoes, 1953–1969

Figure 5.2. Threat rating from tornadoes, 1953–1969. Redrawn from National Oceanic and Atmospheric Administration, *Tornado Preparedness Planning* (Washington, D.C., 1978), p. 27.

"threat" occurs in Massachusetts (347), which rates well ahead of Oklahoma (32) and Texas (16), the leading states, respectively, in tornado incidence and tornado deaths. Although threat ratings computed for whole states ignore intrastate differences in population density—thereby underestimating substantial threats near Tulsa, Dallas, and Fort Worth—the index demonstrates a need for tornado preparedness well beyond the Midwest and the Great Plains. To underscore this point, another map in the guide placed Massachusetts tenth in the nation in number of tornado deaths.

Population distribution affects tornado-hazard mapping another way: tornado counts seem suspiciously low in sparsely settled areas, where minor twisters are often not reported. Climatologist Tom Grazulis, who compiled the massive, two-volume catalog *Significant Tornadoes, 1880–1989*, suggests that official tallies miss 2,000 tornadoes a year—many more than the official annual count of roughly 770 tornadoes.[7] Most unreported tornadoes, Grazulis believes, either occur in sparsely settled areas, where they go unnoticed, or are too poorly formed or atypical to be reported as tornadoes—unless they cause significant damage. In both eastern and western Kansas, for instance, half of all reported tornadoes damage buildings, but because sparsely populated western Kansas has fewer structures to damage, the percentage of tornadoes hitting buildings should be smaller—a discrepancy explained only by serious underreporting. Extrapolating across the entire state the incidence rate for Shawnee County, which contains the state capital, suggests that only 41 percent of Kansas tornadoes get into official records. Similar calculations for nine other states yielded reporting rates ranging from only 29 percent for Arkansas to 78 percent for Wisconsin.

Another vexation for tornado cartographers is the substantial variation from year to year in the number, destructiveness, and geographic pattern of tornadoes: How long an averaging period is needed for a stable, geographically reliable map of tornado incidence? Common sense suggests that a representatively stable map requires more than one, two, or even five years of data. But a recent National Weather Service study that compared the 1970s with the 1980s suggests that even ten years is too short.[8] (The 1980s registered 383 fewer tornadoes than the 1970s, but this net change reflects a drop of 665 in the East and an increase of 283 in the West. Naturally, rates of incidence for the two decades produced markedly different maps.) After experimenting with a variety of longer periods, the researchers concluded that to establish a stable geographic pattern requires thirty-five years of data. Well, at least their incidence maps for 1950–1984 and 1955–1989 were highly similar. They had only forty years of data, though, and I wonder whether more data might not yield a longer recommended averaging period. Perhaps a study based on only thirty-five years of data would have found a thirty-year period quite adequate.

The most serious flaw of the tornado-hazard maps in chapter 1 is the implication that all tornadoes are equally powerful—a shortcoming as deficient as equating magnitude 4 and magnitude 7 earthquakes. No more naïve than seismologists, meteorologists need an operational standard for assessing and reporting the magnitude of tornadoes. In 1971, University of Chicago professor T. Theodore Fujita filled the void with the tornado-intensity scale that bears his name.[9] Fujita divided wind speeds between 73 mph (just below hurricane velocity) and the speed of sound into twelve equal intervals, each identified by the letter F (for Fujita) followed by an integer representing the intensity level. In practice, the upper seven levels are unnecessary because even the most powerful tornadoes never exceed level F5, with winds between 261 and 318 mph. To describe portions of a tornado with wind speeds less than 73 mph, Fujita added level F0. In describing tornadoes verbally, some meteorologists collapse levels F0 through F5 into three groups (weak, strong, and violent), whereas others prefer Fujita's six escalating adjectives for the degree of damage:[10]

Weak	Light damage	F0	40–72 mph
	Moderate damage	F1	73–112 mph
Strong	Considerable damage	F2	113–157 mph
	Severe damage	F3	158–206 mph
Violent	Devastating damage	F4	207–260 mph
	Incredible damage	F5	261–318 mph

Because of obvious difficulties in measuring wind speed during a tornado—anemometers are unlikely to survive, much less register, F3-force winds—Fujita related storm intensity to structural damage and other ground marks and developed qualitative descriptions for each intensity category. Useful for coding recent tornadoes, these descriptions are also applied to historical tornadoes, whose intensity can be judged from eyewitness accounts, newspaper stories, and meteorological records. For example, an F2 tornado can be identified by "roofs torn off frame houses; mobile homes demolished; boxcars pushed over; large trees snapped or uprooted; [and] light-object missiles generated," whereas a marginally more destructive F3 tornado leaves behind "roofs and some walls torn off well-constructed houses; trains overturned; most trees in forest uprooted; [and] heavy cars lifted off ground and thrown."[11] To promote consistent coding by observers in different regions, Fujita provided reference photographs describing damage typical for various intensities. Snapshots of the areas visited by an "incredible" F5 tornado reflect his own sober assessment: "Not much is left."[12]

Maps accompanying a NOAA disaster survey team's report on the Tampa Bay tornadoes of October 3, 1992, illustrate the application of Fujita's scale. Within a two-hour period five separate tornadoes ripped through parts of the Tampa–St. Petersburg, Florida, metropolitan area, killing four people and causing $100 million of property damage.[13] As part of its evaluation of National Weather Service forecasts and storm warnings, the disaster team mapped the paths of the two strongest tornadoes by delineating areas of F0, F1, F2, and F3 damage. One of these maps (Figure 5.3) describes the F3 Pinellas Park tornado, which claimed three lives along its 2.6-mile path. The map shows a single patch of F3-level damage, where the tornado completely destroyed several concrete-block and wood-frame houses. Although a small part of the path, this area of "severe" damage is sufficient to characterize the entire tornado as F3. Dots and labels on the map identify the trailer park where two people died and the housing development where a third victim perished in a collapsed garage. All three deaths occurred in the F2 area.

Only the most destructive or unusual tornadoes are studied so intensively, with air photos and ground surveys.[14] Most twisters are coded by the local Weather Service official responsible for compiling reports for the monthly publication *Storm Data*. Compilers typically cover all or part of a state and rely largely on accounts provided by a clipping service.

Caution is needed in interpreting counts, rates, and maps based on Fujita intensity ratings.[15] Although devised to represent wind speed, the F-scale is largely a damage scale, with tenuous links to wind intensity. These links are far from simple for a variety of reasons, including the difficulty in estimating structural vulnerability and the paucity of structures in open country. Distinguishing between an F2 and an F3 tornado, for instance, is highly subjective in the absence of trains, trees, and "well-constructed houses." And what exactly is a "well-constructed house"?—would you ask the owner, builder, or real estate agent,

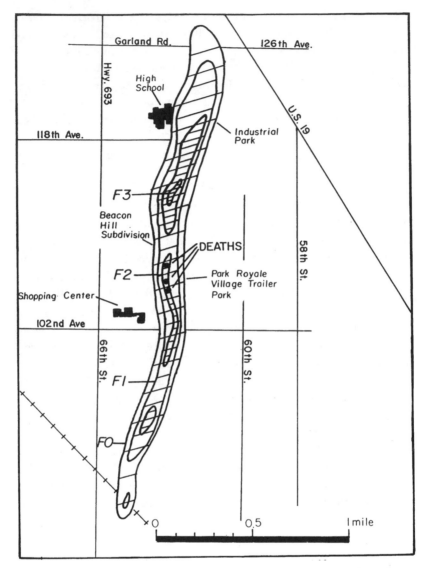

Figure 5.3. Map of the Pinellas Park tornado. Reprinted from *Natural Disaster Survey Report: Tampa Bay Area Tornadoes, October 3, 1992* (Silver Spring, Md.: National Weather Service), p. A-6.

attempt to infer quality from the neighborhood, or send an architectural engineer on a costly search for telltale fragments? If the damage area lacks a broad range of target buildings, visible damage merely establishes a lower bound, as when the tornado that demolishes a mobile home is rated F2 even though its winds might easily have damaged far more substantial structures in its path. Moreover, because estimators not only interpret damage differently but often make educated guesses about intensity, the Fujita scale is more subjective than most tabulations and maps imply. According to Charles Doswell and Donald Burgess, researchers at the National Severe Storms Laboratory, "many tornadoes have inappropriate F-ratings, perhaps by two categories or more."[16]

Intensity ratings are especially troublesome in assessing risk for strong and violent tornadoes. Because the time since Fujita introduced his intensity scale in 1971 is too short to establish stable, meaningful geographic patterns, risk estimates rely, at least partly, on historical data, which are often incomplete or exaggerated. The subjectivity of these retrospective intensity ratings—coded largely by students—was readily apparent to Tom Grazulis, who compared tornado data bases maintained independently by Fujita and the National Severe Storms Forecast Center (NSSFC).[17] Covering the period 1916 to 1985, Fujita's data overlapped by thirty-five years the NSSFC data base, which extended back only to 1950. The period 1950–1970 yielded the most discrepancies: half the tornado ratings differed by more than one category, 2,000 differed by two or more categories, and a few differed by four categories! For the period 1971–1976, the NSSFC evaluators consistently placed many F1 tornadoes in the F2 category, while Fujita's ratings were comparatively conservative. Moreover, after carefully examining original accounts for the period 1916–1949, Grazulis identified 4,028 F2 or greater tornadoes, in contrast to only 3,465 strong or violent tornadoes in Fujita's data base. Although ratings have been more consistent, and presumably more reliable, since 1976, state-to-state inconsistencies are still apparent in *Storm Data* because of under- or overestimates by compilers.

Despite these drawbacks, the Fujita scale allows meteorologists to make maps such as Figure 5.4, which presents a generally reliable geography of strong and violent (F2 or higher) tornadoes. Based on forty years of comparatively recent data, this NSSFC map portrays the number of highly destructive tornadoes in an average year per 10,000 square miles (a square box 100 miles by 100 miles).[18] As with the incidence map in chapter 1, which includes weak (F0 and F1) tornadoes, the rate of occurrence is low in the western half of the nation as well as in the rugged highlands extending from eastern Tennessee and western North Carolina through Maine. Although exceptionally destructive tornadoes are most frequent in central Oklahoma, high rates also occur in northeastern Nebraska (near Omaha), central Arkansas (near Little Rock), southwestern Mississippi (around Jackson), northern Alabama (near Birmingham), northern Indiana (between Indianapolis and Chicago), and a small area in eastern Missouri (near St. Louis). Additional regionally high rates occur along the East Coast in central Florida (between Tampa and Orlando), eastern North Carolina (near Raleigh), and southeastern Pennsylvania (around Philadelphia). On the original map plotted by National Weather Service scientists, *X*s marked other "pockets of enhanced occurrence" in eastern North Dakota (near Fargo), northeastern Colorado (near Denver), west Texas (near Lubbock), and north central Iowa (between Des Moines and Minneapolis). But because most of these locally higher rates occur near cities—with higher densities of damageable structures—the map authors were skeptical of their significance.

Based on slightly older NSSFC data covering only twenty-nine years, Figure 5.5 presents a substantially different geography for vio-

Average annual incidence of strong and violent tornadoes per 10,000 square miles: 1950–1989

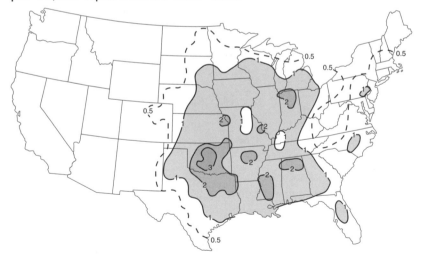

Figure 5.4. Average annual incidence of strong and violent tornadoes per 10,000 square miles, 1950–1989. Redrawn from Joseph T. Schaefer and others, "The Stability of Climatological Tornado Data," in Christopher R. Church and others, eds., *The Tornado: Its Structure, Dynamics, Prediction, and Hazards*, Geophysical Monograph no. 79 (Washington, D.C.: American Geophysical Union, 1993), 459–66; map on p. 465.

Average annual incidence of violent tornadoes
per 10,000 square miles: 1950–1978

Figure 5.5. Average annual incidence of violent tornadoes per 10,000 square miles, 1950–1978. Redrawn from Joseph T. Schaefer and others, "Tornadoes: When, Where, How Often," *Weatherwise* 33 (April 1980): 52–9; map on p. 56.

lent (F4 or F5) tornadoes.[19] Most striking is the map's replacement of the Tornado Alley between the Dakotas and north central Texas with scattered high-incidence regions in northern Iowa–southern Wisconsin, Indiana–north central Kentucky, northeastern Arkansas, and northern Alabama. All four violent-tornado regions are at least moderately populated, with abundant structures vulnerable to tornadic winds—a pattern suggesting a population bias that undermines the map's credibility. Equally troublesome is the comparative rarity of F4 and F5 tornadoes, which account for only 2 percent of all tornadoes. As a recent National Weather Service study noted, "It is quite likely that the relative infrequency of strong and violent tornadoes precludes determining a meaningful occurrence pattern . . . at the present time."[20] Just as well, I suppose.

Insurance firms assess tornado hazards more directly, through their patterns of casualty losses. Large firms have a clear advantage, of course, and with twenty-one million policyholders and more than thirty years of data, State Farm Fire and Casualty Company need not rely on climatological risk maps.[21] According to actuary Dennis McNeese,

State Farm's geography of tornadoes is based principally on *rating territories*, which also reflect fire and accidental injury. Territories vary in number from state to state, depending on state regulations as well as risks and losses. Kansas law, for instance, requires treating the entire state as a single territory, whereas in Colorado the company was free to delineate three regions, based partly on a greater incidence of hail on the eastern slope of the Rockies.

Each territory's rates reflect several risk factors, among which "wind peril," tornadoes, and hail are a major component. Insurance companies typically divide the wind factor into two categories, catastrophic and noncatastrophic. Because noncatastrophic claims are far more numerous, insurers don't rely solely on policyholder data in addressing catastrophic hazards. State Farm, for example, treats hurricanes as a second catastrophic component and uses simulation models and climatological data to refine its rating territories in coastal and tornado-prone locations. Even so, refined territories have yet to reveal meaningful variations in tornado incidence within Oklahoma: in State Farm's experience, according to McNeese, tornadoes "seem to occur pretty much randomly around the state."[22]

Despite uncertainty about intensity ratings and stable patterns, mapping has yielded important discoveries about the behavior of tornadoes. More than two decades ago, for instance, maps revealed that tornadoes avoid heavily built-up central cities, where dense settlement and industrial activity generate a "heat island" of rising warm air, which suppresses tornadoes.[23] Professor Fujita, whose maps detected tornado-free areas in Chicago and Tokyo, even simulated this effect in a laboratory model.[24] Maps have also suggested that the dearth of tornadoes over the Appalachians and relatively rugged parts of the Midwest reflects not only sparse settlement but a disruption of tornadic vortexes by rough terrain.[25]

Detailed maps of individual storms reveal that tornadoes arise from updrafts in thunderstorms, which usually slow down as they form tornadoes. Typically the storm cell and a descending tornado move toward the northeast along a more or less straight course at about 30 mph. But not all tornadoes are so predictable. Snowden Flora's *Tornadoes of the United States* describes an eccentric 1944 twister in Iowa that "remained stationary ten to twenty minutes and then made a U-turn."[26] A thunderstorm that forms, matures, and dissipates over two to three hours might produce a tornado that lasts only twenty minutes.

Occasionally a thunderstorm splits into two or more cells, each forming a separate tornado. Fortunately, tornadic thunderstorms are the exception: the 50,000 to 100,000 thunderstorms that form annually over the United States produce fewer than 3,000 tornadoes.[27]

Particularly fearsome are the powerful, long-lived thunderstorms called *supercells*, which can produce families of strong and violent tornadoes. The Pinellas Park tornado, described in Figure 5.3, formed from one of two supercells that ravaged the Tampa Bay area on a single day. NOAA's disaster survey team confirmed this link by plotting

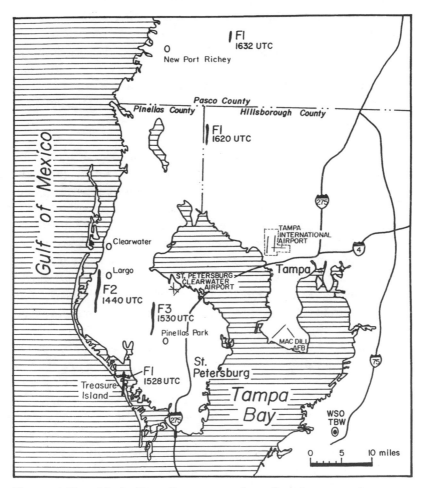

Figure 5.6. Map of the five Tampa Bay–area tornadoes reflects the passage of two supercells. Reprinted from *Natural Disaster Survey Report: Tampa Bay Area Tornadoes, October 3, 1992* (Silver Spring, Md.: National Weather Service), p. A-5.

all five tornadoes on a single map. The first supercell formed the F2 Largo tornado, on the left side of Figure 5.6, which touched down in midmorning. (The Universal Time 1440 UTC corresponds to an Eastern Daylight Time of 10:40 A.M.) Less than an hour later, another supercell moved northward in the same direction nine miles to the east. Traveling at over 50 mph, this second supercell formed four separate tornadoes, with roughly collinear paths.

Local National Weather Service forecasters, who did not expect tornadoes that day, first learned of the Largo tornado when a resident called to report that a twister had just touched down.[28] Equally unaware that another thunderstorm would produce additional tornadoes, the Weather Service merely issued a "Severe Weather Statement" to reinforce a "Severe Weather Outlook" broadcast earlier that morning. Only after the second supercell's first twister struck Treasure Island, on the coast, did the Tampa Bay forecast office issue a tornado warning. By that time, though, an F3 tornado had already sliced through Pinellas Park.

Mitigation efforts that work for earthquakes, volcanoes, and coastal storms are powerless against tornadoes. Fierce tornadic winds and the vast area at relatively high risk make engineering solutions and land-use controls impractical. Resistant structural modifications are prohibitively expensive, after all, and an outright ban on mobile homes, vulnerable as well to hurricanes and other strong winds, would precipitate a torrent of opposition from the manufactured housing industry and populist supporters of low-cost single-family dwellings.[29] Government's only practicable strategy is an integrated system of early detection, efficient warning systems, accessible shelters, and informed citizens.

Technology for more reliable tornado watches and warnings has been around for more than twenty years, but until recently the government has been reluctant to invest in Doppler radar and related improvements. Much of this reluctance reflects ideological priorities: the Evil Empire or the Federal Deficit—take your pick—makes a more portentous villain than the Uncertain Atmosphere. And while it seems cynical to say so, our lack of resolve reflects fatalism about the more violent "acts of God" as well as smug satisfaction with dramatically benevolent disaster relief efforts.

Doppler radar is the heart of NEXRAD, the costly but promising

national network of 160 *Next* Generation *Rad*ar stations scheduled for
full operation in 1997.[30] The concept is named after German physicist
Christian Doppler, who in 1842 noted that sound waves from an
approaching whistle bunch up, increasing in frequency and pitch,
whereas waves from a receding whistle spread apart. Radar exploits
Doppler's discovery by generating and analyzing its own precisely constructed waves.[31] Each WSR-88D station—the acroynm WSR stands
for *w*eather *s*urveillance *r*adar—is a combination transmitter–receiver–image processor that estimates wind velocity and direction by measuring the *Doppler shift* of electromagnetic waves reflected backward,
or *backscattered*, by moisture droplets or flying debris. Although very
little of the transmitted signal is backscattered to the receiver, complex
electronics generate high-resolution color maps of wind velocity, precipitation, and other weather phenomena.

Doppler radar is less a forecasting method than a tool for more
timely detection and recognition of tornadoes and other storms.
Radar-based *nowcasting* is especially useful when the Weather Service
posts a "Severe Thunderstorm/Tornado Watch" for the next six to
eight hours.[32] Within the 20,000-square-mile watch area, weather officials are particularly alert for the 200-square-mile supercells that often
(but not always) form most (but not all) tornadoes. Within a supercell,
radar operators look for vertical structures with rapidly rotating winds.
Although conventional weather radar, unable to measure wind speed,
sometimes shows the distinctive "echoes" of fully mature tornadoes,
NEXRAD promises not only to increase the detection rate and expand
warning times from two minutes to twenty minutes or more but to
yield fewer false alarms.

These improvements are lifesavers. An early NEXRAD system
installed in Norman, Oklahoma, detected 96 percent of all severe
thunderstorms observed from March through June 1991—a marked
advance over the 65-percent success rate of older radar systems.[33] On
June 5, for instance, the system provided a fourteen-minute warning
for a tornado older radars could not have detected. The average warning preceded touchdown by eighteen minutes, sufficient for sirens and
media bulletins to save countless lives. Equally promising are experiments with automated detection algorithms that give the operator
another, more vigilant set of eyes by continuously scanning radar
images for "tornadic vortex signatures" and indicators of non-super-
cell tornadoes.[34]

DEPICTION OF THE TOTAL COVERAGE (AT 10,000 FT ELEVATION) PROVIDED BY THE COMPLETED NATIONAL NEXRAD NETWORK.

DARKENED AREAS OVER THE ROCKY MOUNTAINS ARE GAPS IN COVERAGE AT THE 10,000 FT LEVEL. NEXRAD COVERAGE WILL ALSO BE PROVIDED IN ALASKA.

Figure 5.7. National Weather Service representation of the NEXRAD network. Reprinted from U.S. Congress, House Committee on Science, Space, and Technology, *Tornado Warnings and Weather Service Modernization: Hearing before the Subcommittee on Natural Resources, Agricultural Research, and Environment*, 101st Cong., 1st sess., 1989, p. 77.

A 1989 Department of Commerce announcement of the "modernization and restructuring" of the National Weather Service painted an optimistically impressive picture of NEXRAD.[35] But while an accompanying map (Figure 5.7) promised "total coverage" of the conterminous United States, except for a few gaps in sparsely settled parts of the West, a parenthetical phrase "(at 10,000 ft elevation)" in the map's title glossed over a serious drawback. Tornadoes, it seems, require a denser network of NEXRAD stations than other weather phenomena. Not surprisingly, the National Weather Service Employees Organization, an AFL-CIO affiliate concerned with job losses and transfers, responded with a very different map (Figure 5.8), showing "optimal coverage" of less than a quarter of the U.S., including most of Tornado Alley.[36]

Although neither view is wrong, the Employees Organization map makes Weather Service claims of improved tornado detection appear disingenuous. After all, weather radar scans a circular region with a narrow, slowly rotating beam, which grows wider and less accurate with increased distance. According to Rodger Brown, a tornado expert at the National Severe Storms Laboratory, size (relative to the radar beam) and strength are crucial: "Small, weak tornadoes you can see out to perhaps 20 kilometers (12 miles), whereas large, strong torna-does you might detect out to perhaps 120 kilometers (75 miles), but after that the resolution is not adequate."[37] Another limitation is earth curvature, which obscures the definitive image of a funnel extending below the cloud base beyond 50 miles (80 km).[38] As a result, areas not far beyond the circles in Figure 5.8 must rely upon eyewitness reports of tornadoes invisible to both radar operators and computer algo-rithms. "They really couldn't have [NEXRAD stations] dense enough to cover tornadoes," explains Brown, "but we can see the parent sys-tems where all the really big tornadoes occur."

No one denies that NEXRAD is a notable improvement over older radars. Although the conventional system serving the Tampa Bay fore-cast office gave no warning of the lethal October 3, 1992, tornadoes (Figure 5.6), a NEXRAD system in Melbourne, on the other side of the Florida peninsula, showed the two supercells on the edge of its screen—too briefly, though, to attract the attention of the lone fore-caster on duty. Nonetheless, NOAA's disaster survey team reported enthusiastically that the NEXRAD system scheduled for installation in Ruskin, south of Tampa, "would almost certainly have provided fore-casters with earlier clues to the approaching severe weather."[39] Reas-

NEXRAD SITES WITH OPTIMUM COVERAGE
FOR DETECTION OF TORNADO VORTEX
SIGNATURES (70 km)

Figure 5.8. Alternative view of the NEXRAD network, from the National Weather Ser-vice Employees Organization. Reprinted from U.S. Congress, House Committee on Science, Space, and Technology, *Tornado Warnings and Weather Service Modernization: Hearing before the Subcommittee on Natural Resources, Agricultural Research, and Environ-ment*, 101st Cong., 1st sess., 1989, p. 104.

suring words only if you live in Tampa–St. Petersburg, or near another NEXRAD site.

Simply put, some areas will receive less timely tornado warnings than other areas. Are these differences real? Yes—without timely warnings, areas outside the circles are at greater risk than adjacent zones within the circles. Are these differences important? It's difficult to say: lightning and flash floods, after all, kill many more Americans than tornadoes, and, as Brown points out, NEXRAD is far less geographically constrained in identifying the large storms associated with these more common hazards. Even so, a somewhat denser weather radar network would provide more timely warnings for populous areas with a history of strong and severe tornadoes—areas identified perhaps with a less generalized, more carefully thought-out version of the threat map in Figure 5.2. But at $2.5 million a site, "total coverage" by the little circles on the NWS Employees Organization map seems unlikely outside densely populated tornado-prone regions, where the investment is more readily defended.[40] Whatever solution localities negotiate with the federal government will, no doubt, reflect maps similar to those examined here—risk maps that also serve as propaganda maps.

Chapter Six

Floodplains, by Definition, . . .

Tornadoes and floods have radically different geographies. There's only one Tornado Alley and it stands out on the map (e.g., Figure 1.2) as a broad, elongated zone of concentration that schoolteachers like to call a *belt*. Markedly lower tornado incidence to its west and east reinforces an analogy with the Corn Belt, the Bible Belt, and—my favorite—the Men's Pants Belt (a clothing manufacturing area once active in northern Georgia). By contrast, all parts of the country, even deserts, are subject to flooding, so that instead of a belt, the nation's streams, rivers, and coastlines blanket the country with a dense network of narrow flood hazard zones. Although the news media occasionally report regional flooding, as in the Midwest in spring and summer 1993, only a small part of any region is ever under water. Largely a local phenomenon, flooding is too spatially complex for the small-scale national risk maps used in previous chapters.

Because of its lattice-like geography, flood damage can usually be reduced, if not avoided altogether, by moving back from the water's edge toward higher ground. But persuading people to retreat is difficult; relocating structures is expensive, and flood-prone areas offer aesthetic amenities that other hazards lack—unlike toxic waste sites, for example, it's delightful to have a home and property where "a river runs through it." And there's historical inertia to overcome; riverbanks and coastlines endowed the newly independent United States and its colonial predecessors with invaluable assets: seaports, inland navigation, water power, level cropland, and pathways through mountainous terrain. As the nation expanded westward, gentle gradients along rivers and inland lakes fostered canals and railroads, and floodplains offered rich alluvial soils as well as convenient access to the depot, industrial sites with water power, and inexpensive building lots for factory workers. To enjoy these benefits, farmers, industrialists, merchants, and home buyers bought into a leisurely but potentially catastrophic game of hydrologic roulette. The list of players now includes wealthy retirees and other owners of shorefront property, often second homes. The National Flood Insurance Act of 1968 brought yet another group to the table: through federal disaster relief and subsidized flood insurance, all American taxpayers absorb a substantial share of their fellow citizens' flood losses.

Even so, subsidies encourage participation, reduce flood losses, and thereby serve the national interest. Because widespread flooding is enormously disruptive, a well-subscribed insurance program can lessen shocks to the national economy. Subsidized rates are not just another entitlement, though: to become and remain eligible for federal flood insurance, a community must adopt strict, conscientious floodplain management and ban new construction in highly vulnerable areas. Moreover, subsidies apply only to structures built before flood insurance maps were available. Higher, more actuarially sound premiums for newer buildings discourage future construction in flood-prone areas.

Subsidies are not the only incentive: because flood insurance is required before a bank can grant a federally insured mortgage on property in a flood-hazard zone, nonparticipating communities grow more slowly, if at all. Moreover, disaster relief funds for repair or reconstruction are denied communities that lose eligibility or choose not to participate in the program. Once flood insurance maps are available, a community must either join the program or forfeit a range of grants and disaster assistance.

However coercive an instrument of public policy, flood insurance affords a humane and sensible middle ground between ignoring misery and bailing out willfully negligent victims. It's also both a supplement and an alternative to engineering solutions such as seawalls and artificial levees, which are expensive to build and maintain, and don't always protect against truly catastrophic flooding. Congress addressed these needs by establishing the National Flood Insurance Program, administered by the Federal Insurance Administration, a division of FEMA. Passing the law proved far simpler, though, than setting up a reliable system for estimating and mapping the risk of flooding.

Flood insurance maps are one of the federal government's most detailed cartographic products. Property owners and insurance administrators need to determine quickly and precisely which structures and building sites are in a delineated flood-hazard zone. To relate flood-hazard zones to streets, political boundaries, and other reference features, *Flood Insurance Rate Maps* (FIRMs) have scales ranging from 1:4,800 (1 inch represents 400 feet) to 1:24,000 (1 inch represents 2,000 feet), with the larger, more detailed scales especially common in built-up areas. Keeping these detailed records current is a massive cartographic headache. FEMA must compile, print, warehouse, distribute, and revise nearly one hundred thousand separate map sheets, or *panels*, for the twenty-one thousand communities mapped thus far.[1] The agency prints about six million panels a year, and keeps thirty-five to forty million map sheets in stock. No wonder a recent appraisal of federal floodplain management practices concluded that FIRMs are often out-of-date, erroneous, or difficult to obtain.[2]

A multitude of map sheets makes it difficult to find intriguing examples. Too numerous and narrowly informative for automatic distribution to research libraries in the government's Depository Library Program, flood maps are usually available for inspection only at local land-use planning and emergency management offices in the jurisdictions covered. Although FEMA's Flood Map Distribution Center sells personal copies, the maps' comparatively large scale spreads out interesting symbols, so that a richly informative page-size excerpt is hard to find.[3] Moreover, a minimalist selection of landmarks hides features not easily described by roads or waterways.

Retraced without reduction from panel 10 of the FIRM for the town of Vernon, Vermont, Figure 6.1 illustrates how graphically banal flood insurance maps can be. (Printed in a dark gray, slightly greenish ink, the original artwork is even more drab.) Located five miles south of Brattleboro, Vernon is rural, not thickly settled, and more than adequately portrayed at 1:12,000 (1 inch represents 1,000 feet). Even so,

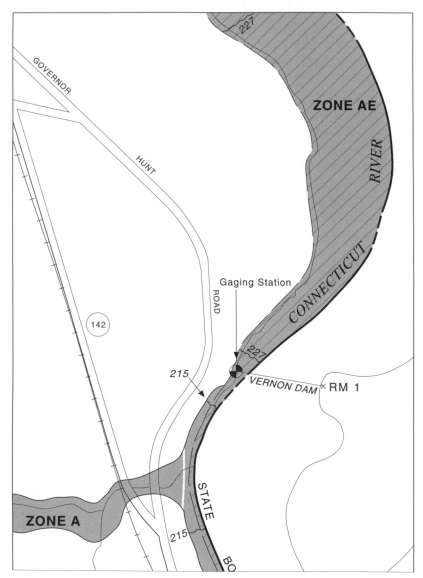

Figure 6.1. Portion of flood insurance rate map for Vernon, Vermont. From Federal Emergency Management Agency, *FIRM, Town of Vernon, Windham County, Vermont*, Community-Panel no. 500137 0010 C, effective 27 September 1991.

the map's double-line road symbols grossly exaggerate the area's two-lane roads—no, they're not 100 feet wide, as the map scale would suggest—and the absence of buildings and driveways ignores a pleasantly picturesque hamlet. That's not all the map doesn't show: between the

"Hunt" in Governor Hunt Road and the "River" in Connecticut River
is the Vermont Yankee Nuclear Power Plant. For a fuller view of what's
at risk here, look at the artist's sketch in Figure 6.2, which shows houses, the power plant, the roads, a dam, and the site on the Connecticut
River floodplain once proposed as a low-level radioactive waste disposal facility—or as opponents called it, an LLRW dump.

Back to the flood map. Although several boundaries cross the area, a
dismal gray shading identifies only two "special flood hazard areas,"
with a 1-percent chance of being inundated in any calendar year. Zone
AE, extending from the top of the map to the bottom, follows the west
bank of the Connecticut River. The suffix *E* means that hydrological
engineers have established *base flood elevations*, used to estimate potential flood damage and calculate insurance premiums for individual
structures. Near the bottom of the excerpt a white boundary separates
zone AE from zone A, on the floodplain of a tributary too small to warrant a flood elevation study. Within zone AE a pattern of parallel diagonal lines identifies the *floodway*, where buildings and other obstructions
would interfere with the river's ability to carry away waters from the
100-year flood.[4] Banning new structures on the floodway safeguards
areas immediately upstream from deeper, more extensive flooding.

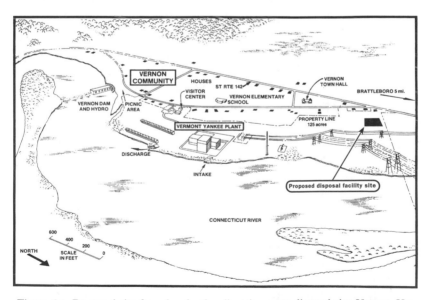

Figure 6.2. Proposed site for a low-level radioactive waste disposal site, Vernon, Vermont. From Vermont Yankee Nuclear Power Corporation public information pamphlet,
ca. 1993.

Printed in a single ink to save time and money, the map wastes no gray shading on floodlands outside Vernon. Identified only by its name and banks, the Connecticut River lies outside the study area below the dam, where its west bank marks the Vermont–New Hampshire border. Vernon Dam, which spans the river, is the sole clue that the thin sinuous line in the lower-right portion of the excerpt represents the east bank. The river is roughly 800 feet across at the dam, but less than a mile downstream (beyond the map excerpt), its width shrinks to less than 300 feet. Meandering slightly to the northeast, the river reveals its east bank again in the extreme lower-right corner. Although the flood insurance map for the New Hampshire community directly opposite Vernon would show a corresponding strip of zone AE along the river's eastern edge, each FIRM treats a specific political unit and offers few details beyond its designated village, town, or county.

While most of the left half of the excerpt is not shaded, areas outside zones A and AE are not flood-free. Labels elsewhere on the panel—the entire sheet measures 37.5 inches by 26 inches—assign these areas to zone X, beyond the 100-year flood line. Flood insurance is available in zone X, but because inundation is less likely, premiums are usually much lower than in the A zones. In the late 1970s, a simpler schedule of rates removed an earlier distinction between zones B and C, found on older FIRMs. (Both zones were deemed safe from the 100-year flood, but zone B was threatened by the 500-year flood.) Although the new rates also erased a complex system of numbered A zones (A1 through A30), the keys of flood insurance maps still include a daunting list of hazard categories (A, AE, AH, Ao, A99, V, VE, D, and X), only a few which appear on most panels.

Flood insurance rates reflect type of construction and elevation above water. Although rates in zone X ignore elevation, premiums in the A zones vary substantially with the elevation difference between a building's lowest floor and the site's base flood level. A difference of a foot can be costly, especially if the lowest floor lies below the flood base. Typically a property surveyor measures the first-floor elevation, and if the building is in zone A, the local floodplain manager determines the base flood elevation. In zone AE, though, information provided on the FIRM helps the owner, builder, or insurance agent determine the flood base. Easy estimation is especially important for new construction, because a well-elevated structure can save the owner higher premiums as well as the grief of water in the living room.

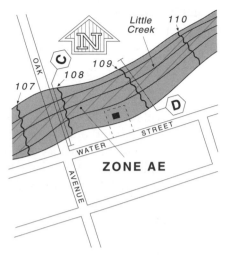

Figure 6.3. Hypothetical FIRM for a narrow floodplain.

Here's how it's done. On Figure 6.3, a portion of the FIRM for a narrow floodplain, squiggly lines perpendicular to the river represent base flood elevations of 107, 108, 109, and 110 feet. Using the map scale and measurements recorded on the property survey, the home owner plots the lot boundary (dashed line) and house (black rectangle) on the north side of Water Street, east of the intersection with Oak Avenue. Although outside the floodway, the building is clearly in zone AE, and not automatically eligible for the lower rates charged in zone X. If the structure were midway between the lines for 108 and 109 feet, its base flood elevation could be interpolated as 108.5 feet. But because it's slightly farther upstream, a more accurate approximation is 108.6 feet.

That tenth of a foot could be important, especially if an accurate property survey indicates the first-floor elevation is higher than 111.6 feet. In the mid-1990s, for instance, if the first floor were 3 or more feet above the site's base flood elevation, a year of flood insurance for a one-floor single-family building without a basement cost $0.14 per $100 of insured value up to the basic limit of $45,000 and $0.06 per $100 thereafter.[5] By comparison, in zone X, regardless of elevation, the rates were $0.25 and $0.07 for basic and additional limits. *Unfair!* you say? Not at all: the first home is 3 feet above the level of the 100-year flood, whereas the zone X residence is nonetheless vulnerable (maybe even more so) to a larger but less frequent flood. In zone AE, the owner of a building within a foot of the base flood level would pay a justifiably higher basic rate of $0.33, and if the elevation of the lowest floor were a foot below the base flood level, the basic rate would jump to $0.86—a big bite out of a weekly paycheck, perhaps, yet quite affordable at $387 annually for $45,000 of coverage.

B ase flood elevations are crucial to understanding flood-hazard maps as well as the fiscal problems of the National Flood Insurance Program. The story gets murky, though, because base floods rely on mathematical notions of frequency and risk sufficiently arcane to alienate many readers, not just those troubled by math phobia or innumeracy. Although I avoided probability theory in previous chapters, flooding demands a more direct approach than earthquakes, volcanoes, coastal flooding, and tornadoes. But don't despair: flood-frequency analysis affords a conveniently graphical, comparatively concrete introduction to abstract concepts important to many kinds of hazard mapping.

A good place to begin is the Vernon flood insurance map (Figure 6.1) and the gauging station at the west end of Vernon Dam. This and other stream gages provide frequent, uniform measurements that illuminate relationships among precipitation, evaporation, infiltration of water into the soil, runoff into streams, size of the drainage basin, and rate of water flowing through a dam's spillway or turbines. An important concern is the likely extremes: the plausible minimum and maximum amounts of water passing a point on a river. For the Vernon hydroelectric plant, hydrologic data helped design engineers specify the size and power of the generators and the height of the dam needed to assure their uninterrupted operation. Extreme flows are a serious concern for electric utilities and municipal waterworks, because too little water can be as troublesome as too much.

A stream gage is a vertical ruler for marking and recording the *stage*, or height, of the water surface above a reference level.[6] Easy to observe, gage height provides a convenient estimate of *discharge*, the rate of water flowing past a point on a river or stream. Recorded in cubic feet per second, discharge reflects water velocity and cross-sectional area. Continuous or frequent recording is necessary because stage and discharge vary from hour to hour, day to day, season to season. Gages are located at dams and other places where the cross section is stable, so that the hydrologist can use gage height and a precise plot of the cross section to calculate the area across which water is flowing at any instant. Multiplying area by current velocity yields dis-

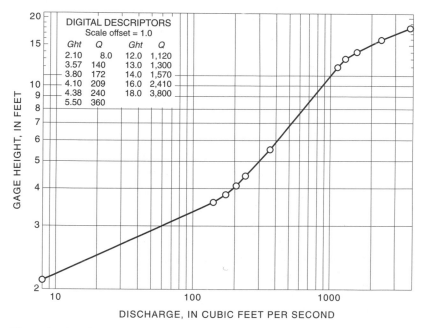

Figure 6.4. A typical rating curve. Redrawn from E. J. Kennedy, "Discharge Ratings at Gaging Stations," *Techniques of Water-Resources Investigations of the United States Geological Survey*, chap. A10 of bk. 3, *Applications of Hydraulics* (Washington, D.C., 1984), p. 11.

charge. Consider, for example, a small river, where the cross-sectional area for a particular stage is 100 square feet. If water velocity is 3 feet per second, discharge is 3 times 100, or 300 cubic feet per second (cfs).

Velocity, unfortunately, is neither easy to measure nor uniform throughout the cross section. Friction retards the flow along the sides and bottom of the channel, and the current is most swift, on average, in the center of the stream at the surface. Fortunately, a reliable relationship between stage and discharge makes continuous measurement of velocity unnecessary.[7] Hydrologists need only measure velocity once, at representative positions along the cross section, to calibrate a *rating curve* graphed with a vertical scale for gage height and a horizontal scale for discharge. With Figure 6.4, a typical rating curve with logarithmic scales, an engineer reading inward from the left axis can quickly convert a gage height of 11 feet to a discharge of 1,000 cubic feet per second. Because it relates stage (the more direct indicator of flood level) to discharge (the more significant hydrologic variable), the rating curve is an essential tool in flood-frequency analysis.[8]

Stream gages typically record stage continuously on a clock-driven cylinder similar to the drum recorders used for temperature, humidity, and seismic energy. Although water-supply analysts archive stage/discharge measurements recorded as frequently as five minutes apart, flood insurance analysts are concerned largely with the maximum stage recorded for a given year. A convenient reporting interval, the year represents a full cycle of snow melt, spring floods, and the drought and cloudbursts of summer; each year's weather (or so we like to think) is independent of the previous and the following year's weather. Moreover, in the insurance industry, where premiums are billed annually, actuaries typically think of catastrophic flooding as a 100-year event—the stage/discharge with only 1 chance in a 100 of being equaled or exceeded in a single year.

The hitch, though, is that hydrologists seldom, if ever, have 100 years of flood data. And even if they did, observations nearly a century old might not reflect present runoff and stage-discharge relationships—contemporary flooding typically is more severe because of recent obstructions on the floodplain, less wetland to retain floodwaters, and an increased accumulation of roofs, roads, and parking lots, all of which accelerate runoff. Moreover, there's no guarantee that the 100-year recording period is typical of current and future weather. So by luck or by choice, the analyst works with peak discharges recorded over 30 to 50 years, and treats these data as a sample drawn at random from a longer set (or *population*) of peak discharges. Some years in this larger, unknown population will have bigger peak discharges than any yet recorded, and some will have smaller. To make an educated guess about yet-to-occur peak discharges, the hydrologist turns to probability theory.

Statistical analysis of flood data begins by reordering the peak discharges Q from lowest to highest and computing the recurrence interval T, a measure of how often, on average, a flood this large appears in the data. The calculation is described by the formula

$$T = (n + 1)/m,$$

where n is the number of years of record and m is the rank order (1 for the highest peak discharge and n for the lowest).[9] In the following table of 36 peak annual discharges, the recurrence interval of the smallest discharge ($Q = 214$ cfs; $m = 36$) is

$$T = (36 + 1)/36 = 37/36 = 1.028 \text{ years (375 days),}$$

whereas the recurrence interval of the largest discharge ($Q = 570$ cfs; $m = 1$) is

$$T = (36 + 1)/1 = 37/1 = 37 \text{ years.}$$

These times estimate that, on average, the larger peak discharge will be equaled or exceeded only once in a 37-year period, whereas the smaller peak discharge will be equaled or exceeded once in 375 days.

Q	T	Q	T	Q	T	Q	T
214	1.028	303	1.370	333	2.056	400	4.111
244	1.057	304	1.423	346	2.176	402	4.625
264	1.088	304	1.480	346	2.312	406	5.286
271	1.121	310	1.542	350	2.467	415	6.167
271	1.156	317	1.609	359	2.643	417	7.400
271	1.194	320	1.682	359	2.846	449	9.250
272	1.233	320	1.762	362	3.083	483	12.333
275	1.276	327	1.850	374	3.364	530	18.500
279	1.321	332	1.947	387	3.700	570	37.000

Recurrence intervals of 1.028 years, 37 years, or 100 years invite misinterpretation by anyone who ignores the "on average." A recurrence interval of 37 years does not forecast a discharge of 570 cfs or larger every 37 years like clockwork. In fact, within a future 37-year period a peak annual discharge this large might occur two, three, or even four times—or perhaps never. After all, a large flood last year does not make a large flood this year any less likely. To avoid possible confusion, many hydrologists prefer the recurrence interval's inverse, the *exceedance probability*. Saying, for example, that a discharge of 570 cfs has a 0.027 probability (1/37) of being equaled or exceeded in any single year is a more precise, less misleading way of stating risk.

(Exceedance probabilities help explain the addition of 1 to the numerator in the recurrence-interval formula. Without the 1, the lowest annual peak discharge in 36 years would have a probability of 1.000—absolute certainty—which ignores the real but remote possibility of an even lower annual peak. Adding the 1 yields a more defensible exceedance of 0.973.)

The next step in flood-frequency analysis is to plot the annual peak discharges and recurrence intervals on the probability graph in Figure 6.5.[10] In contrast to the graph's uniform, arithmetically scaled vertical axis, the uneven horizontal scale pulls the highest and lowest recurrence intervals toward the center so that the points fall more or less in line. Because the spacing of vertical grid lines is based on the so-called

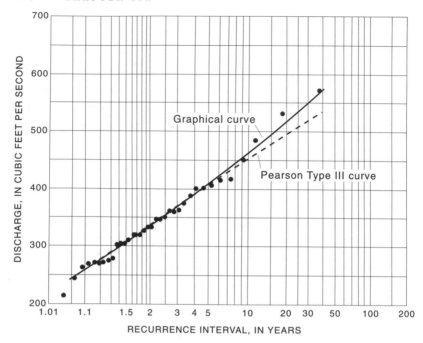

Figure 6.5. Normal-probability representation of the thirty-six annual peak discharges. Redrawn from H. C. Riggs, "Frequency Curves," *Techniques of Water-Resources Investigations of the United States Geological Survey*, chap. A2 of bk. 4, *Hydrologic Analysis and Interpretation* (Washington, D.C., 1968), p. 9.

normal curve, this type of graph is called a normal-probability plot. (There's nothing at all "normal," though, about this symmetric, infamous bell-shaped curve; statisticians more properly call it the Gaussian curve, after its inventor, the brilliant mathematician-physicist Karl Friedrich Gauss.)

To examine the general relationship between annual peak discharge and recurrence interval, the hydrologist draws a straight line or gentle curve through the pattern of points. Curve fitting is inherently inexact, though, so hydrologists use a precise numerical technique to fit a straight-line or other "model" to the data. Figure 6.5 compares the dominant flood-frequency model, based on the Pearson Type III probability distribution, with a smooth graphical curve plotted by eye. Although the graphical curve better represents the larger, less frequent peak discharges, the Pearson Type III model affords more statistically reliable estimates of discharge for the 100-year and 500-year floods.[11]

Extending both curves in Figure 6.5 to the right as far as the 100-year recurrence interval illustrates the uncertainty of extrapolating beyond the data. Projecting the graphic curve upward and to the right yields a discharge of approximately 645 cfs, whereas a Pearson Type III extrapolation forecasts a 100-year flood discharge of roughly 590 cfs. If this lower, probability-based estimate seems too close to the 570 cfs yearly peak observed during 36 years of record keeping, remember that the mathematical model treats the 36 data values as a random sample—a sample that might well include the 50-year, the 100-year, or even the 500-year flood!

One more equation before moving on. The basic Pearson Type III formula,

$$\log Q = M + (KS),$$

estimates the flood-level discharge Q (expressed as a logarithm) from the mean M and standard deviation S of the logarithms of the recorded peak discharges. The equation includes a frequency factor K, based

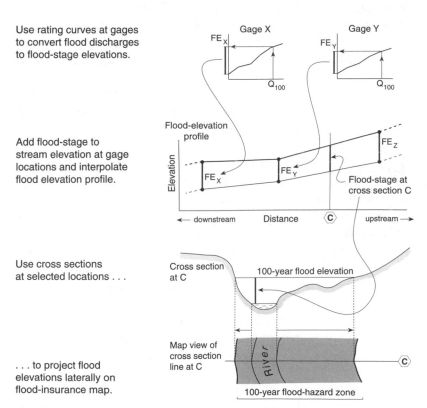

Figure 6.6. Using estimated flood-level discharge to delineate the 100-year flood boundary.

on both a specified recurrence interval (usually 10, 25, 50, 100, or 500 years) and a regional "skew coefficient," which hydrologists can look up on a map.[12] It's a complex process, but the goal is straightforward: linking a short but presumably representative set of data with a regionally calibrated probability model to predict the magnitude of rare occurrences. Other, equally sophisticated techniques estimate flood discharge-frequency relationships for ungaged streams as well as for ungaged points along a gaged stream.[13] Incorporating regional trends into flood-stage estimates helps educate the guess.

Delineating the 100-year flood boundary requires three more steps, described in Figure 6.6. First, for each gaging station, engineers use a rating curve (Figure 6.4) to convert predicted flood-level discharge to a flood elevation. Next, they link together these flood elevations in a *flood-elevation profile*, which runs lengthwise along the stream or river and establishes the base flood elevations shown on flood insurance maps. And finally, they delineate the area of likely inundation by using topographic maps to project flood elevations laterally outward from the river onto the floodplain at representative cross sections.[14] (On some FIRMs, letters similar to *C* and *D* in Figure 6.3 mark the locations of these cross sections.) Although detailed specifications and a precise methodology guide the engineering firms hired for FEMA's flood insurance studies, the resulting FIRMs are no better than available stream-gage and topographic data allow.[15]

For FEMA cartographers, the flood hazard means far more than riverine flooding from rain and snowmelt. As defined in the fine print of flood insurance policies, a flood is "[a] general and temporary condition of partial or complete inundation of normally dry land areas from overflow of inland or tidal waters or from the unusual and rapid accumulation or runoff of surface waters from any source."[16] Delineating comparable 100-year flood zones throughout the country requires a flexible repertoire of customized techniques. In northern states, for instance, where thick ice accumulates on rivers during winter months, FEMA derives stage-frequency relationships that also consider ice-jam flooding.[17] Special hazard analyses are also used for inland lakes, alluvial fans at the base of mountains, land threatened by the failure of dams and levees, arid and semiarid areas subject to mudflows and mud floods, and shorelines. Along the Atlantic and Gulf coasts, subject to persistent and costly damage from hurricanes and northeasters, risk mapping is based on the SLOSH (Sea, Lake, and Overland Surges from

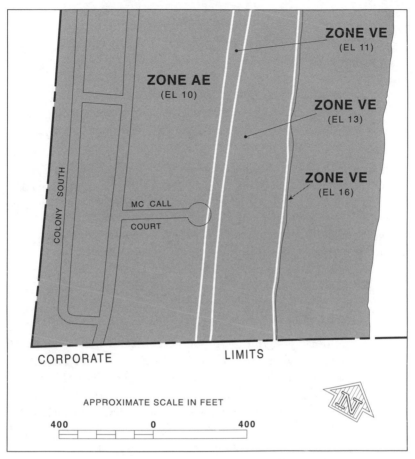

Figure 6.7. Portion of flood insurance rate map for Nags Head, North Carolina. Shaded area includes an AE zone, three VE zones (one quite narrow), and a broad, uniformly wide belt extending seaward from the shoreline at mean sea level. From Federal Emergency Management Agency, *FIRM, Town of Nags Head, Dare County, North Carolina,* Community-Panel no. 375356 0016 C, map revised 2 April 1993.

Hurricanes) and other simulation models described in chapter 4—another strategy for carefully controlled, educated guesswork.

Similar in appearance to their riverine counterparts, flood insurance maps for coastal areas are distinguished by the addition of V or VE zones. *V,* for "velocity hazard," reflects the likelihood of waves of 3 feet or more, whipped up by high-velocity storm winds, and *VE* identifies a V zone with a base flood elevation, used to determine insurance premiums. A uniform base elevation throughout the zone simplifies premium calculation, but accuracy and fairness demand multiple zones. The excerpt in Figure 6.7, from the southern tip of the town of Nags Head, North Carolina, shows three VE flood elevations (16, 13, and 11 feet). (To provide an informative page-size illustration, I lightened the dismally dark shading for the 100-year flood hazard area—on the original the type is barely legible—and moved several zone labels downward into the excerpt.) To qualify for lower insurance rates, houses here are built on stilts. Just inland, a somewhat wider AE zone, with its own base elevation (10 feet), extends westward as far as, and probably beyond, the town boundary. A 100-year flood would inundate both A and V zones, but wave action would damage the V zones more severely. V-zone rates are more than double A-zone rates, and because homeowners tend to underinsure, an even higher rate is assessed any structure insured for less than half its replacement cost.[18]

Take a moment to compare the excerpt with the smaller-scale topographic and storm-surge maps in Figure 4.10. The center of Figure 6.7 is roughly three-quarters of an inch down from the top of Figure 4.10's right half, and slightly to the right. McCall Court, a cul-de-sac from the main road toward the beach, is readily apparent on both maps, as are Nags Head's southeast and southwest boundaries. Orientations differ because true north is up on the topographic map, whereas the flood insurance panel is aligned with the coast. Because an inch represents 2,000 feet on the topographic map but only 400 feet on the FIRM, Figure 6.7 covers only a small fraction of the area shown in Figure 4.10. (I fudged further by moving the scale bar and north arrow just below the excerpt.) Although the dotted storm-surge line on the topographic map and the stark white zone boundaries on the flood insurance map warn of danger and inescapable loss, their slightly different positions underscore the inherent uncertainty of flood-hazard mapping.

B ecause uncertainty encourages landowners and community officials to challenge the cartographer's delineations, flood insurance maps are far from static. The Nags Head map, for example, lists four official revisions in its legend:[19]

Flood Insurance Rate Map Effective:
November 10, 1972

Flood Insurance Rate Map Revisions:
Map revised July 1, 1974, to change zone designations.
Map revised October 17, 1975, to reflect curvilinear flood boundaries.
Map revised February 19, 1986, to change special flood hazard areas.

Map revised April 2, 1993, to change corporate limits and special flood hazard areas, base flood elevations and zone designations to reflect wave setup, wave runup, and storm induced erosion.

In addition to revisions based on updated flood insurance studies, FEMA provides two other kinds of adjustment.[20] The first type, a Letter of Map Amendment (LOMA), removes a specific property from the 100-year flood zone. Property owners can appeal a flood boundary by submitting scientific or technical data, and if a change is warranted, FEMA issues a LOMA. (A LOMA not only confers a lower insurance rate but might allow a new owner to obtain a mortgage without buying flood insurance.) The second type, a Letter of Map Revision (LOMR), orders an official revision of flood boundaries, elevations, or other map features. Requested through the community's mayor or chief executive, these changes typically reflect local public improvements, new development, annexation, or technical corrections to flood elevations and zone boundaries. After issuing a LOMR, FEMA revises and reprints the affected panels. Because of a backlog of work, republication often takes a year or longer.

So how accurate are zone boundaries on FEMA's flood insurance maps? Few studies address this question, and none offers a definitive or reassuring answer. In a comprehensive report on the disastrous Midwest floods of 1993, the Interagency Floodplain Management Review Committee, chaired by Gerry Galloway (a brigadier general in the U.S. Army with a Ph.D. in the geography of water resources), accepted the conclusion of a 1978 U.S. Geological Survey study that estimated an average flood-elevation error of 23 percent.[21] For the 100-year flood on an average river, this represents a depth of three feet. On a broad, gently sloping floodplain, though, three feet vertically can make an enormous difference horizontally. As the Galloway report noted: "In flat areas, structures located within several hundred feet (horizontally) of the 100-year floodplain also may be at risk."[22] However troubling to home owners and local officials, such disparities don't torment FEMA officials.[23] Maps can overestimate as well as underestimate risk, after all, and the premium paid for a relatively safe structure in an A zone at least partly offsets the loss incurred by a comparatively unsafe building in an X zone.

Estimation errors need not average out, though. Because the credibility and solvency of the flood insurance program depends on its maps, FEMA monitors reliability by overlaying flood-zone boundaries on aerial photography and satellite imagery showing the area inundated by major floods. Although imprecise, these "verification reports" suggest that FIRMs are doing what they intended to do. According to Dan Cotter, FEMA's expert on geographic information systems, the financial solvency of the flood insurance program further attests to the reliability of its maps.[24]

But is the numerical correctness of flood insurance maps even a valid question? Gene Stakhiv, a hydrologist with the U.S. Army Corps of Engineers, doesn't think so.[25] "You can't use the words *accuracy* and *precision* in this area because of a whole range of uncertainties," Stakhiv argues. Every big flood alters the statistical relationships, and continuing changes in land cover, channel characteristics, and other factors, including climatic change, undermine projections based on 30 or more years of data. "We don't have a good idea of what the 100-year flood is—it's a moving target."

A moving target perhaps, but a necessary one. Frank Thomas, the FEMA geographer who oversaw flood-zone mapping during the 1980s, considers the 100-year flood primarily a "policy issue"—an acceptable compromise because Flood Insurance Rate Maps must be based on a "national standard."[26] Emmett Laursen, professor of civil engineering emeritus at the University of Arizona, was equally incisive if not more cynical:

> A procedure for obtaining numbers for the magnitude of these floods was needed that everybody would agree on. The numbers would not need to be the real magnitudes of those rare floods; they would only need to be numbers interpreted and extrapolated from a given set of inadequate data in an agreed-upon procedure, thereby discouraging any questioning since everyone would get the same numbers.[27]

However much Laursen's truth might hurt, the localized pains of a nation without flood insurance would be less bearable.

S elling flood insurance is easy—perhaps too easy. Under FEMA's "Write Your Own" program, a local agent of an approved insurance company can issue a policy on the spot.[28] The insurance company, which neither makes a profit nor incurs a loss, keeps a third of the premium to cover its costs.[29] Unfortunately, insurance agents sometimes misread their maps and assign incorrect rates. In 1982 the U.S. General Accounting Office (GAO) concluded that "many" premiums were based on "erroneously designated" zones.[30] An examination of flood insurance policies for five communities revealed 12 incorrect flood-zone ratings in a sample of 91 properties. But careful inspection of the flood insurance maps could validate only 30 of the remaining ratings. For a variety of reasons, none reassuring, government auditors could not confirm the other 49 ratings: 21 addresses were inadequate, 14 policies mentioned streets not shown on the map, and 14 properties were on a street that crossed more than one zone or were too close to a zone boundary. The GAO recommended strongly that FEMA "establish appropriate management controls to detect and correct flood zone misratings."[31] Flood insurance administrators seem to have taken the advice; although the Galloway report on floodplain management condemned out-of-date and inaccurate flood maps and a 1994 GAO audit found flood insurance rates far too low, neither critique mentioned misrating by insurance agents.[32]

Misrating by banks is equally troublesome. A home buyer seeking a federally insured FHA or VA mortgage usually pays twenty dollars for a local or regional company to look up the property on a Flood Insurance Rate Map and certify its flood-hazard zone.[33] (For a structure built since publication of the community's first FIRM, the owner must also obtain an elevation certificate from a licensed surveyor, professional engineer, or local official.) But for nonfederal loans, some banks are too tolerant. In 1990 Thomas Della Torre, president of Transamerica Flood Hazard Certification, complained to the House Committee on Banking, Finance, and Urban Affairs about mortgage officers who misidentify locations, use flood insurance maps ten or twelve years out of date, or consult the wrong kind of map altogether.[34]

> There is more to making an accurate zone determination than opening a flood map—assuming that you have a current map.
>
> Flood maps do not identify property boundaries, show relatively few streets, fewer street names and they are drawn to comparatively large scales.
>
> To be consistently accurate, lenders must obtain a complete set of flood maps, they must maintain that set of maps and those maps are constantly changed, they need some other set of maps which assist them in actually locating properties such as tax maps or parcel maps, and they need to commit space and they need to train personnel to do this task.

Many banks, Della Torre argued, turn to appraisers, who lack the incentive and tools to make reliable determinations. Hired to review several lenders' mortgage portfolios, Transamerica found that between 5 and 30 percent of properties in flood-hazard areas lacked flood insurance.

In the late 1970s, FEMA turned to computer technology for accurate, prompt flood-zone determinations. The idea seemed feasible and efficient: link flood-zone boundary data with the Census Bureau's computerized street map so that insurance agents and banks could call a toll-free 800 number and receive by mail or teletype a computer-generated "certificate of determination" indicating whether a property was in a special flood-hazard zone.[35] If so, a letter from the newly established Flood Insurance Map Information Facility stated the property's zone identification and base flood elevation, a concise interpretation of the flood hazard, the panel number and date of the latest flood map, and whether the community was participating in the insurance program. Intended to promote the sale of flood insurance, the service was free. A $17 million contract called for phasing in the entire country, state by state, over three years, starting in October 1979. Although expensive to set up and operate, the system would recover some of its cost by eliminating map distribution to banks and insurance agents.

A nice idea but decades ahead of its time. However reliable the hardware and software, inaccuracies in existing flood insurance maps and the Census Bureau's geographic data files proved frustrating. Moreover, because not all communities had electronic Census maps, slow and costly manual determinations were more frequent than anticipated. In September 1980, FEMA lowered the expected "hit rate" for automated determinations from an optimistic 98 percent to an unimpressive 67.5 percent.[36] In February 1981, with 144 communities in seventeen states already in the system, the newly inaugurated Reagan administration pulled the plug. Although the project demonstrated the feasibility and eventual efficiency of automated certification, FEMA planners had underestimated the political clout of commercial flood-zone certifiers, who loudly denounced the project as a threat to free enterprise.[37]

In the mid-1990s, FEMA gave electronic flood insurance mapping another try. This time, in addition to converting FIRMs to an electronic format, FEMA is integrating flood-boundary data with detailed topographic information from the U.S. Geological Survey and the Census Bureau, and offering the information on CD-ROMs to flood-zone certification companies.[38] The entire mapping process will eventually be computerized, with contractors submitting flood insurance studies in digital format and on-demand laser printing obviating the massive inventory of paper maps. Although the CD format cannot confer increased reliability, the vast improvement in storage and retrieval will make these electronic flood-hazard maps more beneficial than their paper counterparts.

Data conversion is expensive, though, and at the present rate, conversion to electronic format could take forty years—far too long in the opinion of Gen. Galloway and his panel. Currently, a twenty-five-dollar surcharge on flood insurance policies covers most of the cost of compiling, distributing, and maintaining the printed FIRMs; a small fee (sixty cents on average) recovers the remaining cost. Because an increased surcharge might discourage sales of flood insurance, the Galloway report recommended a direct supplementary appropriation to increase the reliability, timeliness, and effectiveness of "NFIP products," including a national inventory of flood-prone structures.[39]

The National Flood Insurance Program faces many challenges—political, technical, and actuarial. Among its more pressing problems, areas behind artificial levees need more actuarially exact risk analyses, and a generous five-day waiting period allows floodplain dwellers along *mainstem* rivers, which rise slowly, to safely delay buying insurance until flooding is imminent.[40] But digital flood mapping, if it works well, can address two key, persistent obstacles: obsolete, cumbersome paper maps and too few customers. Currently, the participation rate (the proportion of flood-prone properties with flood insurance) hovers around 20 percent—embarrassingly low market penetration for a program three decades old.[41] Through more reliable hazard ratings and more rigorous supervision of local floodplain management, the new maps are certain to increase participation and reduce flood losses. Moreover, the proposed inventory of flood-prone structures, if carefully compiled and maintained, can eliminate map-reading errors, provide better data for estimating risk and setting fair and reliable rates, and support an efficient, effectively targeted direct-mail marketing campaign. If potential customers won't come to the map, the map might come to them.

As a fixed, detailed image of an elusive abstraction (risk of flooding), the Flood Insurance Rate Map is like a historical novel: an easily grasped fictionalized representation that helps readers understand (or think they understand) a complex phenomenon. However politically and societally useful, flood-zone boundaries are not real—as moving targets, they can't be. But by describing risk as plausibly greater here than there, these cartographic caricatures provide a geographic foundation essential for the actuarial framework that makes flood insurance possible. And as part of a larger process (loss reduction), flood maps initiate a dialogue for limiting damage in flood-prone areas. Without risk maps of some sort, even poor ones awaiting better terrain data or an updated hydrologic study, flood insurance would be impossible.

Chapter Seven

Subterranean Poisons

D o bears defecate in the woods? Most certainly, and so do numerous other critters, including humans. Fecal bacteria carried by surface runoff into streams and rivers readily contaminate surface water, which must be routinely treated and rigorously tested. Most cities draw their water from reservoirs or by direct intake from rivers and lakes, and public health laws demand careful monitoring, chemical purification, upstream sewage treatment, and land-use restrictions around reservoirs. By contrast, groundwater has long been a reliable source of potable drinking water. When fallen rain percolates slowly downward into a water-bearing geologic formation called an *aquifer*, soil bacteria break down organic waste and plants recycle nutrients. Because the soil is a widespread, highly efficient filter, well water is almost always safer than surface water. At least until recently.

There are two threats to groundwater: depletion and contamination. Depletion results from climatic change or, more typically, from overuse. Temporary depletion, common in many rural areas during dry spells, is usually remedied by drilling a deeper well—which often reduces the water available to neighbors, who react by deepening their wells. In the High Plains region stretching from South Dakota to Texas, ever deeper wells that "mine" groundwater for irrigated farming have dangerously depleted the Ogallala aquifer, especially in north Texas, where the water table dropped nearly two hundred feet between 1930 and 1980.[1] Depletion also threatens America's outer suburbs, beyond the reach of municipal water systems, as more and more home owners and businesses drill wells and lower the water table. Concerned about contamination as well as depletion, state and local officials place rigorous restrictions on wells and rural septic systems and control population growth by insisting on large building lots.

Groundwater contamination, the subject of this chapter, is more insidious than depletion. Unless betrayed by a foul odor or peculiar taste, biological and chemical pollutants can migrate undetected into nearby wells, including public wells supplying thousands of homes. Pollution sources vary widely, though, and uncertainty about contamination makes any region dependent on groundwater a hazardous geographic environment. That includes most of the country, particularly rural areas. But because subterranean poisons rarely travel great distances, a national hazard-zone map is even less appropriate for groundwater contamination than for flooding.[2] As a result, surveillance is largely the responsibility of state and local officials, rather than the federal government. Even so, citizens are wary of contamination from nearby industries, and water quality is a contentious issue around proposed regional landfills and newly discovered hazardous waste sites.

Compared to earthquakes and tornadoes, groundwater contamination is a hidden yet often manageable threat, requiring monitoring and remediation rather than warning systems and insurance. Without competent monitoring, waterborne poisons might evade detection for years, even after local youngsters start dying of leukemia. Surveillance is especially difficult because of complex three-dimensional relationships among geologic formations, pollution sources such as waste dumps and gasoline storage tanks, and the location, depth, and pumping rate of drinking-water wells. Difficult but not impossible, though, because health and environmental protection officials can identify likely contaminants, monitor water quality, and base shrewd predictions on meager data and powerful simulation models. Before examining the uses and limitations of modeling, this chapter offers a quick course in groundwater hydrology.

Groundwater is part of the hydrologic cycle, an amazingly efficient solar-powered engine that recycles water, sustains life, and sculptures landscapes. Elementary texts in physical geography explain the hydrologic cycle with intriguing graphic flow models: networks of arrows and numbers linking the system's principal processes (evaporation, transpiration, cloud formation, atmospheric circulation, precipitation, runoff, and groundwater movement) and reservoirs (oceans, ice caps, clouds, rivers, lakes, soil, plants, and groundwater). For this brief treatment, though, I jump directly to Figure 7.1, which zooms in on groundwater's role in the larger hydrologic system.

Permeable bedrock and surface deposits provide both a reservoir and a flow medium for groundwater. Unconsolidated materials such as sand and gravel, which are highly porous and allow water to move freely, make an ideal aquifer. Consolidated bedrock that has been fractured or partly dissolved also provides the necessary connected openings. Water percolating downward from the soil fills the aquifer's pores, fractures, and other voids and moves laterally under the influence of gravity. The *water table* marks the boundary between an upper zone of *aeration*, where movement is downward, and a lower zone of *saturation*, where movement is largely horizontal.

Equally significant are *aquicludes*, comparatively impermeable rock formations, such as clay or shale, that retard movement. Figure 7.1 describes how juxtaposition of permeable and impermeable layers affects the delivery of groundwater to springs, rivers, oceans, and other discharge areas. Springs occur where the upper boundary of an aquiclude intersects the land surface. On the left side of the large hill, for instance, a spring discharges water moving to the left in the upper aquifer. On the opposite side, a small, elevated aquiclude forms a

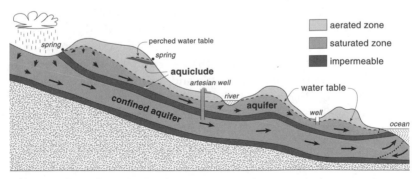

Figure 7.1. Groundwater's role in the hydrologic cycle.

Moderate pumping produces
small cones of depression.

Vigorous pumping at middle well
lowers water table below other wells.

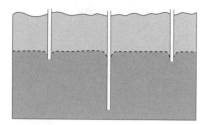
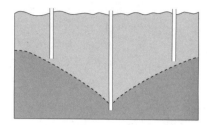

Figure 7.2. Vigorous pumping of deep well (right) can divert water from otherwise bountiful neighboring wells (left).

perched water table, with a local zone of saturation discharging at another hillside spring. Groundwater also feeds the river near the center of the diagram. For the most part, though, the long upper aquiclude, sloping downward to the right, diverts most of the region's groundwater toward underground springs along the coast. Beneath this aquiclude, a *confined aquifer* carries water from a recharge area on the far side of the hill to an even deeper interface with the ocean. Water travels slowly through confined aquifers, which can take several centuries to discharge water from a recharge area more than a hundred miles away.

Aquifers also discharge water through wells dug or drilled downward to the saturation zone of an unconfined aquifer. Especially valuable is an *artesian well* (named for Artois, in northern France), at which water rises naturally to the surface without need of bucket or pump. A confined aquifer fed from a distant, substantially higher recharge area has sufficient hydrostatic pressure to force water upward in the well, all or part of the way to the surface. Not all wells tapping confined aquifers are artesian, but even partial hydraulic assistance is helpful.

As Figure 7.1 illustrates, groundwater moves toward rivers, valley bottoms, and the sea so that the water table, as a three-dimensional surface, generally reflects surface terrain. Pumping distorts this relationship by lowering the water table and forming a *cone of depression* around the well. Figure 7.2 describes the effect of especially vigorous pumping, which creates a broad, deep depression that can cause neighboring wells to go dry. In the High Plains of north Texas, deep wells pumping huge quantities of water from the Ogallala aquifer have disturbed wells as far as forty miles away.[3] Because groundwater flows toward low points on the water-table surface, pumping not only disrupts well yields and groundwater movement but might even divert a plume of contamination from its normal path. Coastal areas, where water moves from sea to aquifer under enormous hydrostatic pressure, are especially vulnerable to salt-water intrusion if vigorous pumping inland reduces the supply of fresh water needed to hold back the salt-water interface.

Figure 7.3 illustrates the migration of contaminants percolating from a landfill atop an aquifer supplying local drinking water. Polluted groundwater ordinarily follows a path of least effort downward from higher to lower elevations on a water table that vaguely mimics surface topography. Unlike surface water, in which turbulence thoroughly disperses pollutants within a short distance, groundwater follows streamlines, which might converge or diverge slightly but generally carry the plume along intact, with little mixing. In the left-hand diagram, a plume originating at a landfill moves downhill to the right, away from the water-supply well at A, upslope from the landfill. But a substantially deeper well pumping large amounts of water for cooling overheated machinery, say, can reverse the plume's direction. The right-hand diagram illustrates this effect by adding a new well at B, upslope from A; the deep cone of depression changes the hydraulic slope and diverts the contamination plume toward the drinking-water well. Because pumping can reverse the direction of flow, environmental health officials often place monitoring wells both upslope and downslope from hazardous sites.

Rural residents and public health authorities are especially wary of septic systems and cesspools located near drinking-water wells. Because highly permeable soils expedite migration of semitreated

Water table following slope of terrain directs contamination from landfill away from water-supply well A.

Vigorous pumping of new well B draws contamination plume toward well A.

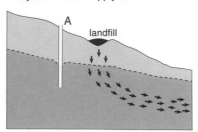 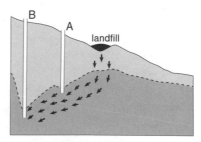

Figure 7.3. Vigorous pumping (right) can divert plume of contamination from its normal, gravity-influenced path (left).

sewage into groundwater, some states prohibit new homes in areas without municipal sewers unless the soil passes a "perc test." The lot owner who wants a septic-system permit chooses a site and hires a professional engineer, who digs or drills a test hole, perhaps six feet deep, to check for a high water table—which would automatically rule out a permit. If visual inspection suggests adequate drainage, the engineer then makes two or more comparatively shallow holes, widely separated on the proposed site, fills them with water several times, and watches them drain. Local or state regulations specify the depth, diameter, and number of test holes, the number of fillings and drainings, and a procedure for estimating percolation rate with a watch and tape measure. If the lot owner is lucky, testing takes half to three-quarters of a day: any shorter would mean a high percolation rate and vulnerability to groundwater contamination, and much longer might reflect slow percolation and the threat of overflowing wastewater. For the most part, though, septic treatment works nicely, with more than twenty-two million residential and commercial septic systems serving a third of the national population.[4] But substantial increases in rural nonfarm households, requiring a half million new septic systems each year, demands careful review of test results, proposed locations, lot sizes, and local conditions.[5]

Bacterial and viral contamination is particularly troublesome in *karst* areas, characterized by sinkholes, collapsed caverns, disappearing streams, and relatively little surface drainage. Derived from *kres*, a Slavic word for stone, karst refers to landscapes like the eastern Adriatic coast, where water percolating through limestone and other easily dissolved bedrock has carved out caves, caverns, and other conduits for the ready movement of groundwater.[6] In karst terrain, rainfall infiltrates rapidly, with little or no percolation near sinkholes. Runoff carrying fecal matter or particles of decaying animal carcasses escapes filtering by the soil, and pesticides that enter the groundwater reservoir with little resistance can emerge an hour or so later at a spring or shallow well. Nearby areas contribute further impurities, carried in by surface streams that leak—and often disappear altogether—into the aquifer below.

In central Pennsylvania's Nittany Valley, where I attended graduate school, I observed firsthand another karst hazard, the discharge at limestone springs of chemical waste leaking from industrial holding ponds. The professor leading our geology field trip paused at a chemical plant outside town and asked students to sniff the air near a lagoon full of foul stuff with a distinctive odor. We then drove a mile or so to an undeveloped area where water flowing from a hillside carried the same stench. Apparently natural conduits led all too efficiently from the waste lagoon to the rural spring. The chemical company was lucky, the prof observed: if the aroma had entered a farmer's drinking-water well, there would have been a far bigger stink. Water pollution, it turned out, was not the company's only hazard; one night several months later the plant blew up and seriously damaged a barn a quarter mile away.

Municipal and industrial landfills are perhaps the most obvious and closely monitored sources of groundwater contamination. Intended for household and yard waste, the municipal landfill receives whatever toxins residents and businesses throw in the trash: spent batteries containing mercury, rusty cans of lead paint, highly poisonous cleaning solvents, the rat killer "we really ought to get rid of," and other chemical discards from basement or garage. Campaigns to establish separate waste streams for hazardous substances, as most localities now require, are never completely successful, and in many cities the damage was done years ago, when the municipal landfill routinely accepted industrial waste. Manufacturers often used private landfills, many of which are now Superfund sites, or "stored" waste materials on site in bulldozed lagoons, excavated pits, and other allegedly sound "surface impoundments." Storage ponds built with a plastic liner or floor of clay to prevent leakage are notorious *point sources* (environmental-science jargon for specific, individual sites) of groundwater pollution, as are underground tanks for storing gasoline, fuel oil, and toxic chemicals. Among the more outrageous industrial polluters, mining and mineral processing firms that dump wastes in tailings ponds or huge mounds have poisoned thousands of wells and streams. Old wells are another significant point source of contaminants: unless carefully sealed, abandoned oil, gas, and water wells provide comparatively direct downward pathways for animal waste and chemical spills.

Perhaps the most ethically objectionable disposal technique is the discharge or injection of industrial hazardous wastes through wells into deep-lying confined aquifers.[7] Isolated "indefinitely" from aquifers supplying drinking or irrigation water, injected waste can be dealt with, if at all, thousands of years from now by our intelligent offspring and their far superior technology. Maybe, maybe not. Confined aquifers are not thoroughly impermeable, after all, and earthquakes and volcanism could reexpose this toxic legacy more quickly than intended.

"Nonpoint" sources of contamination, occurring across broad areas, are less easily traced or controlled than point sources such as landfills and lagoons. Old, badly corroded sewer pipes can leak profusely, for instance, and locating the worst leaks requires extensive testing and digging. Similarly, a stream carrying raw or partly treated sewage from a neighboring county might leak into an underlying aquifer at various places along its channel, wherever and whenever vigorous or widespread pumping, or even a drought, has lowered the water table below the stream bed. Obvious and widespread nonpoint groundwater pollution is often too politically sensitive for timely action. In northern states, for instance, roadsalt dissolved in percolating rainwater has poisoned thousands of wells, and in prime farming areas, pesticides and fertilizers pose a serious threat to health.[8] But drivers depend on salt spreading for ice removal, and farmers are reluctant to sacrifice crop yields and income.

Decades ago, when aquifer protection mostly meant keeping the well away from the cesspool, groundwater maps did little more than describe the locations of wells and springs. For example, the Maryland Geological Survey's 1956 report on water resources in Baltimore and Harford Counties was largely a do-it-yourself kit for property owners and well drillers.[9] Instead of mapping aquifers, the study includes two intermediate-scale topographic (1:62,500) maps, one for each county, showing the locations and inventory numbers of 1,174 wells and springs. Sixty-three pages of tables list such attributes as completion date, altitude (surface elevation), depth, yield, and the name of the water-bearing geologic formation. Even so, the report seems incomplete, listing few wells dug or drilled before 1944 and providing a chemical analysis for only 61 wells. The book also includes a generalized geologic map of the area and a large sheet of well-log diagrams describing the thickness, depth, and various materials encountered in drilling 68 wells. In Figure 7.4, the logs for two Baltimore County wells illustrate the imprecise driller descriptions ("red dirt" and "brown rock") that still permeate much of the groundwater hydrologist's basic data.

Although modern groundwater maps describe the areal extent and depth of individual aquifers, there is neither a typical groundwater

Figure 7.4. Column diagrams for two well logs. Reprinted from R. J. Dingman, H. F. Ferguson, and Robert O. R. Martin, *The Water Resources of Baltimore and Harford Counties*, Bulletin no. 17 (Baltimore: Maryland Department of Geology, Mines and Water Resources, 1956), plate 4.

map nor a systematic nationwide effort to map subterranean water resources. The broadest attempt at a national cartographic inventory is the series of more than seven hundred Hydrologic Investigations Atlases published by USGS's Water Resources Division.[10] Not a bound book of maps like a world atlas or road atlas, the typical Hydrologic Atlas consists of a 9-by-11 ½-inch map jacket containing one or more large folded map sheets. The format, content, and number of maps reflect each project area's size and geologic complexity. Mostly printed in color, the maps range in scale from 1:24,000, the typical scale of USGS topographic and geologic mapping, to 1:250,000 or smaller for comparatively generalized maps covering large study areas.[11] The series addresses both surface water and groundwater, but coverage is far from complete because state and local governments initiate most projects by agreeing to pay half the cost of collecting data and compiling maps. Even so, because of graphic complexity, lack of standardization, and a focus on raw data, the maps are not readily accessible to local officials and citizens.[12]

While Hydrologic Atlases largely provide detailed maps for scientists and engineers, another Geological Survey series, the Water-Resources Investigations Reports, often includes maps that interpret analyses and describe recommendations. Seldom longer than seventy letter-size pages, a typical report uses maps, diagrams, and tables to assess a problem and propose a solution. Symbols and labels are usually straightforward, as in Figure 7.5, which shows the extent of hydrogen sulfide contamination in three counties in northern Ohio. Rivers, county and township boundaries, and a few tick marks representing meridians and parallels provide a geographic frame of reference, large dots locate wells where water was sampled, and a gray shading focuses the reader's attention on areas where groundwater has been found to smell of rotten eggs. This generalized pattern of high-concentration areas reflects not only well locations but also the region's principal aquifers and general directions of groundwater movement, described on maps near the beginning of the report. The accompanying text points out that no safety standard exists for hydrogen sulfide, which corrodes pipe and plumbing fixtures as well as assaults one's olfactory sense.[13] For readers interested in more detail, the pocket at the back of the report contains four hydrogen sulfide maps covering parts of the region at a much larger scale.

For multiple groundwater hazards, which can be numerous and varied, both general and scientifically trained readers appreciate a composite map of relative risk. This need must have been abundantly clear to the USGS hydrologist who mapped aquifers and examined known and potential sources of contamination for a water-resources investiga-

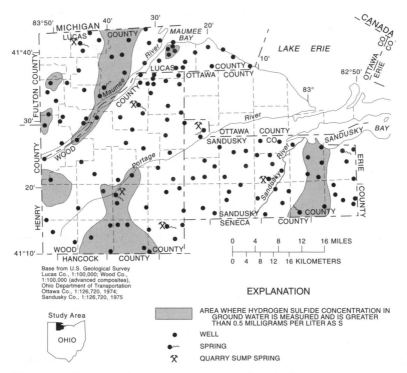

Figure 7.5. Risk map relates areas with potentially objectionable concentrations of hydrogen sulfide to network of monitoring wells. From Kevin J. Breen and Denise H. Dumouchelle, *Geohydrology and Quality of Water in Aquifers in Lucas, Sandusky, and Wood Counties, Northwestern Ohio,* U.S. Geological Survey Water-Resources Investigations Report 91-4024 (Columbus, Ohio: U.S. Geological Survey, 1991), p. 99.

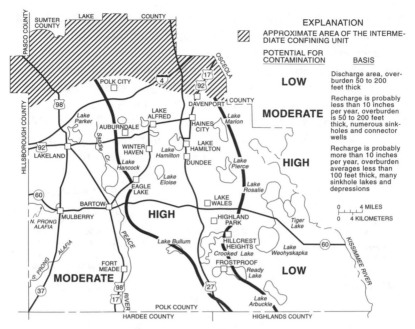

Figure 7.6. Estimated potential for contamination of the intermediate aquifer system in Polk County, Florida. Reprinted from G. L. Barr, *Ground-Water Contamination Potential and Quality in Polk County, Florida*, U.S. Geological Survey Water-Resources Investigations Report no. 92-4086 (Tallahassee, Fla.: U.S. Geological Survey, 1992), p. 34.

tion report for Polk County, Florida—his report had much to summarize: twenty-seven maps, eighteen tables, and seven diagrams describing geology, aquifers, population growth, well locations, water quality, hazardous waste sites, phosphate mines, chemical plants, and citrus farms, where pesticides, solvents, fuels, and other chemicals were a serious threat.[14] In this case, though, local geology and water use demanded a pair of composite risk maps: one for the confined "intermediate aquifer system," lying directly below a more vulnerable unconfined aquifer, near the surface, and the other for the even deeper Upper Floridian aquifer. Although the Upper Floridian aquifer supplies most local drinking water, both the intermediate and Upper Floridian aquifers are accessible to most of the county's wells. The surficial aquifer, highly vulnerable if not already contaminated, supplies relatively little drinking water and did not warrant a cartographic summary.

Figure 7.6, the risk map for the intermediate aquifer, is blunt, visually effective, and readily understood. Thick boundaries and the large, bold labels "high," "moderate," and "low" highlight the geography of

danger—a parsimonious if not perfect solution for a black-and-white
risk map with only three categories, simple boundaries, and numerous
reference features that might conflict graphically with color or graytone shadings. At the upper right, a crowded map key describes for all
three risk categories the potential for recharge from sinkholes and the
overlying and widely contaminated surficial aquifer; like many "confined" aquifers, Polk County's intermediate aquifer is vulnerable in
places to leakage from above. The map key also links the diagonal, parallel-line pattern in the northern part of the county to the cryptic

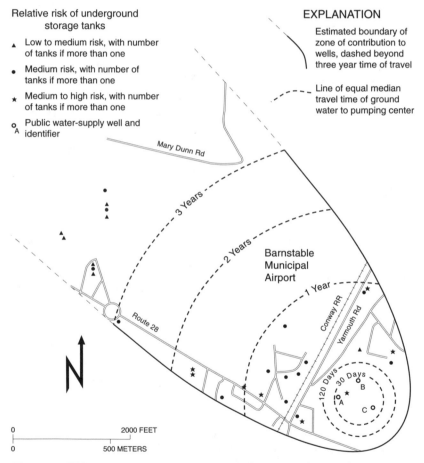

Figure 7.7. Risk to groundwater from underground storage tanks near the Barnstable,
Massachusetts, public-supply well field. Reprinted from Julio C. Olimpio and others,
*Use of a Geographic Information System to Assess Risk to Ground-Water Quality at Public-
Supply Wells, Cape Cod, Massachusetts*, U.S. Geological Survey Water-Resources Investi-
gations Report no. 90-4140 (Boston, Mass.: U.S. Geological Survey, 1991), p. 25.

"intermediate confining unit," revealed earlier to readers of the report as a portion of the intermediate deposit too impermeable for use as an aquifer.

Maps of relative risk can be much more specific. In the late 1970s, for instance, the Massachusetts Division of Water Pollution Control joined with the U.S. Geological Survey, the Environmental Protection Agency (EPA), and local governments to assess threats to public water-supply wells on Cape Cod.[15] Among the numerous, highly detailed maps produced by the long-term study, Figure 7.7 depicts the threat from underground storage tanks near three wells in the town of Barnstable. The map's geographic scope is the well field's *zone of contribution*, that portion of the local aquifer supplying water under normal pumping. Elongated toward the northwest, the zone's shape and orientation reflect the southeastward movement of groundwater in the vicinity. Point symbols representing degree of risk show the locations of buried storage tanks, and dashed lines labeled in years or days portray the approximate time required for leaking liquids to reach the wells.

To assess relative danger for each tank, hydrologists constructed a composite risk ranking based on six factors: distance from the closest well; land use; the age, size, and contents of the tank; and the construction material.[16] Concrete tanks, judged most likely to leak, were assigned a rank of 3, with lower ranks for tanks built of steel (2), double-wall steel (1), or fiberglass (0). The land-use factor addressed the structural loading (weight) from buildings and roadways as a potential cause of tank failure by ranking highway businesses (6) well above residential areas (0). The other four factors conferred their highest ranks on tanks older than thirty years, within 638 feet of a water-supply well, and containing 20,000 gallons or more of waste oil, used oil, or gear oil, rich in small metal fragments or toxic additives. By contrast, a tank with the lowest potential risk would be more than 4,400 feet away from a well, younger than five years, and hold no more than 1,000 gallons of heating oil—or be empty. Summing each tank's rankings for all six factors yielded a composite risk score, which the hydrologists divided into the three categories represented in Figure 7.7 by small triangles (low risk), circles (medium risk), and stars (medium-to-high risk). Although this or any other composite index is merely an educated guess, a single map assessing the combined effects of various risk factors offers planners and environmental health officials an overview of dangers to local drinking water. Composite ratings can be equally useful in broader, small-scale assessments, demonstrated in the Cape Cod study by a countywide map that related zones of contribution for all public drinking-water wells to the locations of landfills rated as moderate, moderate-to-high, or high risk.[17]

Addressing potential sources of pollution is the easy part: inspectors can examine buried storage tanks, after all, but the landscape reveals little about the shape and areal extent of the zone of contribution and even less about the time required for leaks to contaminate local water wells. To confront these challenges, hydrologists have devised mathematical models based on the theory of fluid dynamics.[18] Relying on information about surface hydrology, subsurface geology, and data on pumping rates and water-table elevations, computer models simulate water movement through soil, surface sediments, and permeable rock from areas of recharge to points of withdrawal.[19] Like simulations of earthquakes, hurricanes, and other natural processes, numerical models of aquifer systems promote preparedness by helping communities assess vulnerability.[20]

Figure 7.8 relates elements of groundwater flow to the problem of estimating a well's zone of contribution. Smooth, gently curving flow lines describing the movement of groundwater intersect at right angles *equipotential lines* representing gravity and artesian forces—the hydraulic head, or pressure, that propels the water forward. When pumping produces a pronounced cone of depression (Figure 7.2), the lowered water table diverts several flow lines backwards toward the well. For given rates of withdrawal and recharge, the flow network will stabilize, with some flow lines converging toward the well to define the zone of contribution and others moving ahead. Flow lines converging from opposite directions at the *point of stagnation*, where in principle nothing moves, mark the boundary of the zone of contribution. Darcy's Law, derived experimentally by French engineer Henry Darcy in 1856, describes the flow network mathematically by relating direction and velocity of flow to differences in hydraulic pressure and the "transmissivity" or "hydraulic conductivity" of a porous medium.

A second network, with rectangular cells arranged in rows and columns, helps groundwater models manage data and orchestrate calculations.[21] Each cell is an accounting unit that receives water from some neighbors and loses water to others, or perhaps to a well or spring. In a model for an unconfined aquifer, cells are recharged from above, and designated *stream cells* receive additional water from an

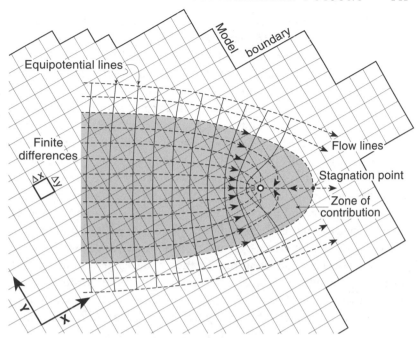

Figure 7.8. Finite-difference model simulates groundwater flow by solving flow-line equations and accounting for exchanges of groundwater among adjoining cells in a rectangular network.

overlying watercourse. For a confined aquifer, the model might include cells that gain or lose water through a locally leaky aquiclude. Accounts must balance, with the water leaving a cell exactly matching the water received. Of course, when recharge or pumping conditions change, a cell may increase or decrease its storage—its hydrologic "savings account"—by raising or lowering the water table.

A simulation model's most important task is estimating the overall effect of proposed land development and other threats. If, for instance, pumping were increased, the model must assess the well's effects throughout the region, not just at the bottom of the cone of depression. Adjustment begins at the cell containing the well, where a hypothetical lowering of the water table increases the flow from neighboring cells, in which water elevation also drops. Because surrounding cells must, in turn, offset their own contributions to the well by collecting more water, the model propagates this decline in the water table outward, adjusting the balance between immediate neighbors and their neighbors, and then between neighbors once removed and their neighbors, and so on. In addition to expanding the cone of depression, the model requires further calculations and waves of adjustment to address secondary effects at other cells with streams or wells. The hundreds or thousands of cells needed to model groundwater for a large or complex area with numerous streams and wells demand a powerful computer.

Simulation is especially useful in evaluating development plans, as illustrated by a water-resources investigation for the Red Clay Creek drainage basin, astride the Delaware-Pennsylvania border about thirty miles southwest of Philadelphia.[22] On the fringes of both Philadelphia and Wilmington, the area has grown steadily since the 1950s. To cope with increasing demand for groundwater, county planners wanted to compare several pairs of alternative strategies: large (one-acre) or small (half-acre) minimum building lots, on-lot wells or public-supply wells, and on-lot septic systems or community sewage treatment. And the sewer option required further choices: pumping treated wastewater outside the area or returning it to the basin through either spray irrigation or discharge into local streams. To explore these approaches, Geological Survey hydrologists designed a simulation model with 1,454 square cells, each 1,000 feet on a side.

Figure 7.9 describes a portion of the model, around development area A, at the northern end of the basin. The upper-left excerpt illustrates the strings of stream cells needed to model the recycling of treated wastewater, and the other excerpts describe three diverse development strategies. Contour lines portray the simulated lowering of the water table, in feet, with tiny hachures pointing toward areas of even greater decline. In the upper-right excerpt, three 0.1-foot contours predict minimal effects for one-acre lots with on-site septic systems, estimated to return 90 percent of wastewater to the basin. By contrast, in the lower-left excerpt, deep cones of depression around hypothetical public-supply wells (black dots) reflect the demands of a population twice as large living on half-acre lots. Because treated wastewater is pumped out of the basin, this scenario portrays a broad decline in the water table. At the lower right, negative contours show a predictably higher water table around a hypothetical spray-irrigation project returning 40 percent of wastewater to the aquifer.

Reliable simulation requires careful calibration, based on detailed records of pumping rates as well as water-table measurements taken periodically over a representative period of time for a suitably dense network of monitoring wells.[23] What's representative and suitable depends, of course, on the range and type of change the hydrologist

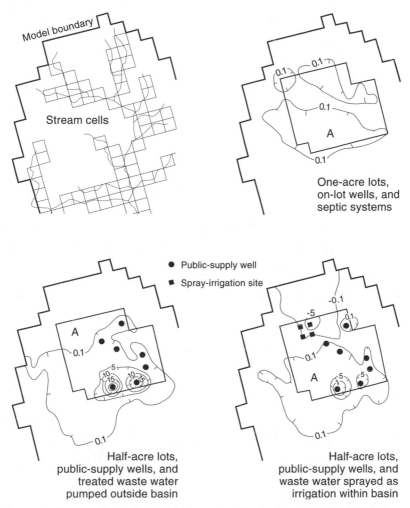

Figure 7.9. Excerpts illustrating change in water table simulated with finite-difference model (upper left) for three different development plans. Contour lines describe lowering of the water table, in feet. Development area A is at north end of the basin. From Karen L. Vogel and Andrew G. Reif, *Geohydrology and Simulation of Ground-Water Flow in the Red Clay Creek Basin, Chester County, Pennsylvania, and New Castle County, Delaware*, U.S. Geological Survey Water-Resources Investigations Report no. 93-4055 (Lemoyne, Pa.: U.S. Geological Survey, 1993), pp. 31, 52, 58, and 59.

intends to simulate: a model concerned largely with exploring the effects of pumping on flow direction and the water table is less demanding, for example, than a model for tracking a toxic plume. Moreover, the cautious hydrologist "field-checks" the model's performance by experimenting with pumping rates and comparing observations from monitoring wells with results from the model.[24] In calibrating and validating the Red Clay Creek model in Figure 7.9, hydrologists collected water-level measurements at 375 wells and conducted pumping tests (to estimate transmissivity) at 77 wells and more extensive "aquifer tests" at 6 wells.[25] Experimental trials can be useful not only in evaluating and refining the model but in identifying the need for additional monitoring wells.[26]

Useful in assessing vulnerability, modeling is equally valuable in managing remediation. Getting rid of a contaminant means not only cleaning up the source at the surface but also extracting or treating the poisoned water.[27] Remediation often requires drilling a new, high-capacity well, suitably situated to capture the plume—akin to a leech sucking stale blood from a bruise. It's often helpful, if not essential, to divert the plume from operating supply wells. Cleanup relies on

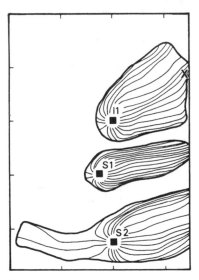
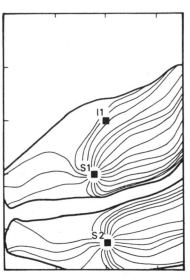

During remedial action   Normal pumping resumed

Figure 7.10. Simulated flowpaths and one-year capture zones during remediation of contamination from source at X (left) and after resumption of normal pumping (right) at water-supply wells S1 and S2. Interceptor well I1 was drilled to capture contaminated groundwater. Tick marks 1,000 feet apart indicate that the interceptor well is less than a half mile from the source. Redrawn from E. Scott Bair, Abraham E. Springer, and George S. Roadcap, "Delineation of Traveltime-Related Capture Areas of Wells Using Analytical Flow Models and Particle-Tracking Analysis," *Ground Water* 29 (1991): 387–97; illustrations on p. 393.

modeling in several ways: simulating the spreading plume's forward movement, finding efficient locations for an interceptor well and additional monitoring wells, and steering the plume toward the interceptor well through reduced or increased pumping at existing supply wells.

Simulation's most demanding task in groundwater remediation is estimating capture areas around interceptor and supply wells. The concept is similar to the zone of contribution used to study groundwater vulnerability: a well pumping at a given rate for a specific period of time apprehends all contaminants within its zone of capture. Thus, if the offending landfill, lagoon, tank, or spill site is within a well's one-year zone of capture, pumping throughout the year following surface cleanup should remove all remaining contaminated groundwater. Interactive modeling allows officials to experiment with drilling location, rate and duration of pumping, and appropriate adjustments at nearby water-supply wells. In the left panel of Figure 7.10, for instance, flow lines and one-year capture zones illustrate the planned removal of contaminants that threatened drinking water in Wooster, Ohio.[28] By the time remediation began, monitoring wells revealed that a plume of DCE (dichloroethylene) and TCE (trichloroethylene) originating at source X had spread westward nearly half a mile. Diversion was an important part of the cleanup: reduced pumping at supply wells S1 and S2 allowed interceptor well I1 to extract all contaminants within a capture area designed to include most of the plume. The right panel represents groundwater flow following remediation, when well I1 stopped pumping and the supply wells resumed their normal pumping rates, deepened their cones of depression, and enlarged their zones of capture.

As the converging flow paths in Figure 7.10 illustrate, estimating the zone of capture requires tracking hypothetical particles of water backward in time, in this case through a full year. Remediation sometimes requires forward tracking as well, to simulate a zone of contamination around a known source.[29] If officials know when and where a release occurred, they can model the spreading plume and assess the advantages of rapid remediation. Modeling is useful even when the source is unknown, to suggest efficient locations for monitoring wells, reconstruct the plume's history, and suggest plausible sources.

However convincing a map of simulated flow lines and capture zones, particle tracking yields an approximation no more reliable than the model's geologic assumptions, calibration data, and validation

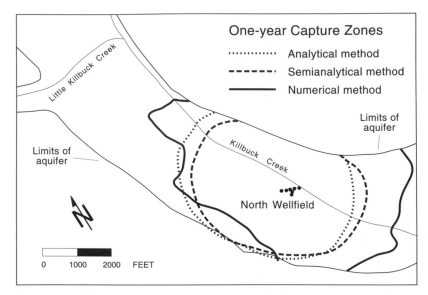

Figure 7.11. Comparison of one-year capture zones for three comparatively different flow models. Redrawn from Abraham E. Springer and E. Scott Bair, "Comparison of Methods Used to Delineate Capture Zones of Wells: 2. Stratified-Drift Buried-Valley Aquifer," *Ground Water* 30 (1992): 908–17; illustration on p. 915.

tests. To assess sensitivity to measurement error, hydrologists often experiment with slightly different transmissivity coefficients, water-table elevations, and pumping rates. Sensitivity experiments that yield generally similar, stable patterns reduce uncertainty. Because reliability also depends on the model's theoretical, mathematical, and computational foundations, employing an inappropriate model can be as risky as trusting a single set of assumptions and data. Ohio State University hydrogeologists Abraham Springer and Scott Bair demonstrated the importance of choosing a suitable model by superimposing capture zones computed for an elongated unconfined aquifer with three conceptually different models.[30] As Figure 7.11 illustrates, their analytical and semianalytical groundwater models yielded markedly smoother and less precise capture zones than a grid-based numerical flow model, which accounted for local differences in aquifer thickness and hydraulic conductivity. Although EPA guidelines for wellhead protection recommend all three predictive models, only the numerical method, relying on detailed data rather than mathematical sophistication, avoided oversimplification.

Looking months ahead and seeing the otherwise unseen is but one of mapping's contributions to the fight against groundwater contamination. In addition to timely assessments of the present and revealing reconstructions of the past, mapping and modeling afford insightful portrayals of groundwater flow that can help scientists, public officials, and citizens appreciate how an aquifer behaves, why it's vulnerable, and how intervention can remove contaminants and safeguard drinking water.

Numerical modeling is a recent development. Although rooted in mid-nineteenth-century discoveries, computer simulation of groundwater flow was first attempted in the early 1960s—shortly before the nation awakened abruptly to the less benevolent spin-offs of motor vehicle fuels, cleaning solvents, and thousands of other well-intentioned chemical inventions with potentially disastrous impacts on the hydrologic cycle.[31] Since then, simulation has become a standard tactic for tracking and containing unseen, unseemly seepage from landfills, industrial sites, and neighborhood gasoline stations.

But simulation is no substitute for surveillance. Communities largely dependent on wells for drinking water need two complementary strategies: a computer model to simulate contaminant migration within the local aquifer and a network of strategically placed monitoring wells, for routine sampling and testing. Those with a sampling network have a fighting chance; those without play a dangerous game of pollution roulette. Do you know where your local monitoring wells are?

Chapter Eight

Ill Winds

G rowing up in a suburb with tall trees, little traffic, and no smokestacks, I never saw air pollution as a threat. Red dust regularly rained down on backyard clotheslines across town, near Bethlehem Steel, but Larchmont's residents rarely (if ever) complained of bad air. We innocently burned leaves in our streets and pumped leaded gas into our cars. Tobacco smoke was not yet a recognized hazard, a neighbor's trash incinerator was at most an eyesore, and the black exhaust from Baltimore Transit's new buses irritated streetcar buffs more than anyone. Years later I instinctively think of air pollution as something you see. Every winter morning on the drive to work, I am awed by the huge gray plume above the university's steam plant—a frightening sight until I recall that its furnaces burn natural gas (the safest fossil fuel around), thereby creating water vapor, which condenses instantly in the frigid Syracuse air. Less visibly ominous is a pharmaceutical complex less than three miles from my home— despite an innocuous name ending in "Laboratories," this factory is the second leading source of toxic air emissions in New York State.

Visible or not, air pollution is a serious health hazard, especially for infants, the elderly, and anyone with chronic heart or lung disease. Much of the problem is "particulate matter," a catchall for tiny pieces of soot, dust, sulfates, nitrates, and other easily inhaled substances known to incite lung cancer, emphysema, and asthma.[1] Larger particles and droplets are trapped by the cilia and moist mucous membranes lining the nose—unless asthma or a head cold makes you a mouth-breather—but particles smaller than 10 microns (a micron, or micrometer, is a thousandth of a millimeter or about 0.000039 inch) easily reach the bronchial tubes, and "fine" particles, with diameters less than 2.5 microns, penetrate deep into the lungs. So much for judging air quality by seeing how much disgusting residue a tissue or clean handkerchief can scrape from the vestibule of the nose. Although the handkerchief test detects gook in the air, microscopic particles that sneak past our nasal sentry are more insidious.

Weather can increase the danger. Winds stir up dust from freshly plowed fields and construction sites as well as transport smoke from power plants, wood stoves, and forest fires. But because atmospheric turbulence mixes clean and dirty air, thereby diluting the concentration of inhalable particles, asthma victims suffer more when the air is calm. In many regions stagnant air reflects a recurrent *temperature inversion*, in which air above several hundred feet or so is warmer than the air immediately below. Normally, air temperature declines with altitude, so that warmer, lighter, more buoyant air near the ground creates a rising current, which carries smoke and auto exhaust upward. High-altitude winds then spread the crud across a wide area and partially corroborate the claims of incinerator advocates that the solution to pollution is solution. Not so, though, when a layer of warmer air aloft traps cooler air below. Temperature inversions are especially troublesome in the Los Angeles basin, surrounded on three sides by mountains.[2] Inversions are common in the morning, when ocean breezes reinforce cool, heavier air that drained down overnight from higher elevations. Air enriched by the auto emissions of several million highly mobile consumers rises no higher than the inversion layer to create an effect dubbed *smog*—part smoke, part fog. Tourists from the dreary, humid East easily misinterpret L.A.'s hazy morning sky as a sign of imminent rain. I spent a week there in the mid-1980s, lecturing at UCLA, a short walk from my hotel. The first few mornings I walked outside, looked up, hurried back to my room for an umbrella, and wondered why these crazy Angelenos didn't mind getting wet.

Air pollution is only partly particulate. Invisible gases that enter the bloodstream through the lungs can cause cancer, heart disease, and a variety of respiratory ailments. Under the Clean Air Act, the Environmental Protection Agency established standards for particulate matter and five other "criteria pollutants": carbon monoxide (CO), lead (Pb), nitrogen dioxide (NO_2), ozone (O_3), and sulfur dioxide (SO_2). The EPA and its counterparts in state government also monitor a variety of other known or suspected airborne toxins, including asbestos and mercury.[3] This chapter examines efforts to monitor, map, model, and control air pollution. More so than for hazards treated earlier, air pollution has a cartography shaped by regulation and negotiation, economics and politics. Controversies likely to affect the political geography of clean air include indoor air pollution, the battle over smokers' rights, racial justice, and emissions credits, whereby government can trade cleaner air in one region for continued pollution elsewhere.[4]

Ambient air requires greater vigilance than groundwater, which moves more slowly and predictably. Although water supplied to consumers might be tested several times a day, monitoring wells are often sampled once a month or less. By contrast, air quality must be measured daily, hourly, or by the minute, in accord with standards reflecting each pollutant's source, sensitivity to changing traffic or weather conditions, and impact on humans.[5] Carbon monoxide, a result of incomplete combustion of gasoline and other carbon fuels, has two standards: 1-hour and 8-hour thresholds that may not be exceeded more than once a year. One "exceedance" is acceptable, but two or more bring restrictions and penalties. Particulate matter and sulfur dioxide have dual standards based on 24-hour intervals and annual averages, and ozone concentration may not exceed a maximum hourly level.[6] By contrast, lead levels are governed by average concentrations estimated for each calendar quarter. Because lead's health effects are cumulative rather than immediate, its maximum quarterly average may not exceed 1.5 micrograms per cubic meter, a tiny but deadly concentration if inhaled regularly by infants and children. And nitrogen dioxide, a yellowish gas especially troublesome to respiratory sufferers in large cities, may not surpass an annual average level estimated from hourly measurements.

Although individual monitoring stations often measure more than one toxin, each pollutant has its own carefully tailored monitoring network, designed to reflect unique origins, impacts, and official limits. For example, ozone, which reaches dangerous levels in the warmer months, when intense sunlight triggers chemical changes in volatile organic compounds and nitrogen oxides, is not monitored during the winter in northern states. And carbon monoxide, closely related to vehicular traffic, is most efficiently monitored at "hot spots" with slow, heavy traffic and restricted air movement.

Air sampling is a complex, highly automated process, in which a fan draws air through a filter or electronic sensor.[7] Manual monitors, used widely for checking particulate matter, operate for a fixed interval. Typically a timer turns the fan on at midnight and turns it off 24 hours later. An operator stops by in a day or so, retrieves the filter for laboratory analysis, and resets the timer. Sampling the air every six days cuts the cost of lab testing by six-sevenths yet covers all days of the week within a six-week period. Although manual monitors are suitable for particulate matter, carbon monoxide and other pollutants with more temporally exacting standards require continuous monitors that measure concentration every minute of every day. Usually based on a photometer sensitive to a particular gas, continuous monitors can sound an alert when conditions approach a dangerous level or store their results for periodic transmission by telephone to a central data bank.[8]

Within Syracuse, a city of roughly 150,000 people, a single continuous monitor measures carbon monoxide and two additional manual monitors, located on rooftops, measure particulates.[9] In an adjacent suburb, another continuous monitor checks for sulfur dioxide, ozone, and "soiling," a form of particulate matter, and records atmospheric conditions as well. Air sampling is expensive, and the New York Department of Environmental Conservation (DEC), with EPA approval, monitors only locally troublesome pollutants. Hence the nearest monitor for nitrogen dioxide is 120 miles to the west, in Amherst, a suburb of Buffalo. In contrast, New York City, with several million inhabitants and heavily congested traffic, has numerous continuous and manual monitors. According to Gordon Clickman, DEC's environment engineer responsible for monitoring the Syracuse area, the Big Apple is "the real problem in the state."

Locating a monitoring station is a scientifically critical, if not politically sensitive, decision. Environmentalists insist on a site that reliably reflects maximum exposure of the region's population, ethnic leaders argue for special attention, and the chamber of commerce, the manufacturers' association, and the convention and tourist bureau resist any location that might affect their members' pocketbooks or the area's reputation. But the EPA's siting guidelines, published in the *Code of*

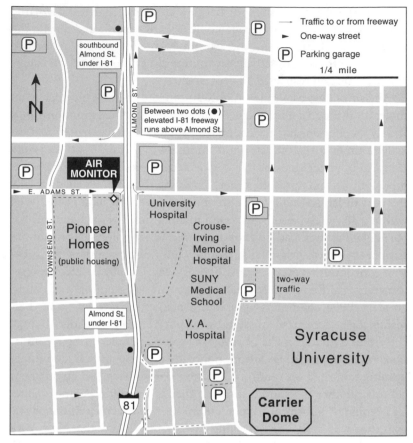

Figure 8.1. Syracuse's carbon monoxide monitor is located at a CO hot spot near a busy intersection.

Federal Regulations, are explicit and detailed.[10] Localities have little leeway to fudge results or overrule alleged misinterpretations of legislative intent.

Monitoring becomes controversial when a slight exceedance at a single station affects a wide area. This happened in Syracuse in 1992, when state and EPA officials required oxygenated gasoline throughout Onondaga County as a result of two out-of-bounds carbon monoxide readings during the winter of 1989.[11] Because petroleum suppliers cannot customize their products for small markets, the mandate affected several adjoining counties as well and sparked the intense resentment of local officials, filling station operators, and countless motorists, for whom the price of gas jumped several cents a gallon. In the eyes of many, the culprit was a small white trailer parked ten feet away from five lanes of idling traffic. The trailer housed a continuous air monitor that measured CO levels every minute and transmitted its readings to Albany sixteen times a day.[12] State officials installed the monitor in 1982, after a traffic-flow model identified the site as a carbon monoxide hot spot.

As the 1989 exceedances demonstrated, this location is troublesome on cold winter evenings when early arrivals for university basketball games or concerts at the Carrier Dome add to rush-hour traffic.[13] Figure 8.1 describes the monitor's situation, near an elevated section of Interstate 81, which skirts the downtown business district, and East Adams Street, a broad one-way thoroughfare headed east toward a large medical complex and the university. Directly beneath the elevated freeway, Almond Street carries southbound traffic that just exited I-81 a block north of the Adams-Almond intersection, where most cars turn left for the university or the hospitals. Also entering the intersection are two streams of northbound traffic, one from I-81's other exit ramp and the other on Almond Street below the freeway (Figure 8.2, left-hand panel). A steep hill immediately east of the

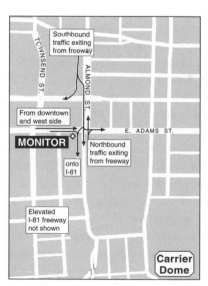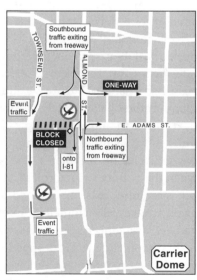

Normal traffic past air monitor   Rerouting plan for Dome events

Figure 8.2. To avoid further exceedances in the vicinity of the air monitor (left), local police reroute traffic (right) during basketball games and other winter events at the Carrier Dome.

freeway as well as moderately tall buildings throughout the neighborhood retard atmospheric circulation. "Cold starts" are another factor: because a car's catalytic converter doesn't operate efficiently for several minutes, especially in winter, several large parking garages in the area further aggravate the problem. But the real trigger is weather: when stable air accompanying a high-pressure cell discourages mixing, the Adams-Almond neighborhood becomes a mini–Los Angeles basin in which the CO level can rise quickly.

Did the DEC and EPA overreact by requiring oxygenated gasoline throughout the county? Yes, says David Coburn, director of Onondaga County's Office of the Environment, who compared the regulators' decision to "a farmer who reacts to a small patch of crabgrass by dousing a 100-acre field with Round-Up."[14] No, say environmentalists committed to strict enforcement of air-quality regulations. Besides, they contend, a few steps south of the trailer is a large public housing project—people live there and need protection. Nonsense, argues Coburn: health department tests indicate that indoor air in the neighborhood is safe; if anyone is at risk, it's the cop directing Carrier Dome traffic below the freeway.

To persuade the EPA to take the county off the "nonattainment" list, the Syracuse Metropolitan Transportation Council devised a strategy for reducing traffic near the monitor.[15] In addition to adjusting the timing of traffic signals and arranging for a police presence during rush hours, planners rerouted Carrier Dome traffic around the intersection on Townsend Street, a long block to the west. As the right-hand panel of Figure 8.2 illustrates, the scheme further reduces idling traffic by completely closing the entire block of Adams Street next to the monitor. Winning public acceptance was not difficult: a quicker but slightly longer route was better than the increased price of oxy-gas. During the plan's first year, officials rerouted traffic only when the county's meteorological consultant warned of stagnant air, but after motorists complained about the confusing off-again, on-again strategy, the council adopted the plan for all winter basketball games, rock concerts, and other Dome events. Not surprisingly, peak carbon monoxide levels dropped and the EPA agreed to lift the oxygenated-gas requirement after a single season so long as local officials controlled traffic congestion. Although the county's strategy smacks a bit of trickery, charges of cheating seem pointless because folks living near the monitoring station can now breathe cleaner outdoor air during high-traffic periods.[16]

How widespread are air-quality violations, and where are they concentrated? Answers to these questions are found in the *National Air Quality and Emissions Trends Report* the EPA issues each fall for the preceding calendar year. For the six pollutants covered by the National Ambient Air Quality Standards, tables list maximum observed levels, graphs describe recent trends, and maps point out nonattainment areas. Especially interesting and provocative are estimates, by state, of the population living in counties not meeting the standards. In 1992, for instance, over 21 percent of the national population lived in a nonattainment county.[17] These estimates can be misleading, though, because tabulations ignore relative severity as well as overestimate exposure in counties with comparatively small nonattainment areas. Nonattainment is, in practice, a regulatory matter, and once placed on the EPA's watch list, an area redeems itself only through atonement and attainment—a convincing pollution-reduction plan as well as several years of satisfactory performance.[18] Although minor violators like Onondaga County usually remove themselves from the list quickly, some metropolitan areas have been on the nonattainment list for decades.

Although the *Trends Report* contains numerous time-series graphs describing steady, fitful, or uncertain progress for the nation as a whole, its maps are snapshots of a single calendar-year. When I asked whether EPA had a composite multiyear map of persistent or severe offenders, the official then in charge of the annual publication replied, "No—not that I am aware of." So I compiled my own highly generalized display, in the spirit of the six very-small-scale natural hazard maps (Figure 1.1) in chapter 1. EPA cartographers didn't make the task easy: the *Trends Report* varies widely from year to year, with little standardization in map design aside from an obsessive use (and abuse) of color.[19] Even so, persistent nonattainment or dangerously high levels are not hard to recognize, and despite its greatly simplified boundaries, the crude six-part compilation in Figure 8.3 is a useful caricature of the worst air in America in recent years.

For serious students of environmental geography, my thumbnail cartographies of danger hold few surprises. Although large-scale urbanization is the dominant influence on the six pollutants, noteworthy exceptions abound. For instance, particulate matter is troublesome in many parts of Colorado, northern California, and the Pacific Northwest because of widespread heating with wood. Regional industrial patterns are equally prominent in the maps for lead and sulfur

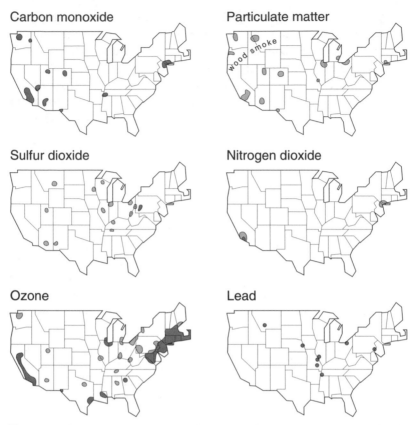

Figure 8.3. Highly generalized patterns of comparatively poor conditions for the six pollutants covered by the National Ambient Air Quality Standards. Compiled on an appropriately vague base map from the *National Air Quality and Emissions Trends Report* for recent years.

dioxide, associated respectively with metal processing plants and power plants burning high-sulfur coal. Ozone, which forms massive plumes that migrate rapidly and dissipate slowly, is a markedly more widespread problem than, say, nitrogen dioxide. Although few areas appear on more than two maps, the topographically impaired Los Angeles basin registers seriously high levels of carbon monoxide, nitrogen dioxide, ozone, and particulate matter. Despite measurable progress since 1980, L.A. has the foulest air of any large metropolitan area in the United States.

Botched color and problematic map design aside, the *Trends Report* includes a few visually effective exemplars like Figure 8.4, a dramatic pseudo-three-dimensional black-and-white picture of hot spots for atmospheric lead. Thin vertical prisms represent maximum quarterly mean concentrations of lead recorded during 1992 in the vicinity of thirty-one point sources, mostly smelters. Along the back of the diagram, each prism appears again, against a vertical scale similar to the horizontal rulings in a police lineup. Across this unorthodox map key a dashed line represents the air-quality standard, 1.5 micrograms per cubic meter. In 1992 most areas met the standard, but four sites had violations (diagrammatically at least) off the scale. Lead concentration near one smelter, Master Metals in northern Ohio, was so large (37.4), its prism had to be truncated to save space.

I like this chart for several reasons. Clearly the best map in the 1992 report, it demonstrates that black-and-white cartography can be less confusing than color maps as well as less expensive to reproduce. Its larger, more detailed view reveals noteworthy patterns and differences too detailed for my six minimaps, which exclude many marginal yet potentially serious areas and in their omissions paint rosy pictures that ignore numerous regional and local struggles, many far from victorious, to reduce emissions. And in focusing on industrial sources, the

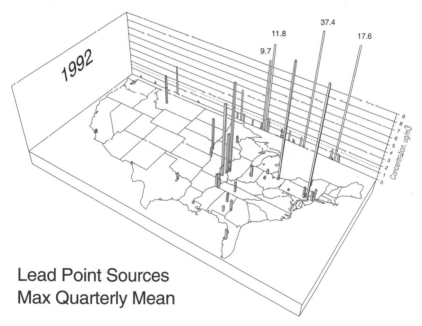

**Lead Point Sources
Max Quarterly Mean**

Figure 8.4. Prism map of maximum quarterly mean lead concentration near lead point sources. Reprinted from U.S. Environmental Protection Agency, Office of Air Quality Planning and Standards Research, *National Air Quality and Emissions Trends Report, 1992*, EPA publication no. 454/R-93-031 (Research Triangle Park, N.C., 1993), p. 3-12.

map highlights a remarkable triumph of air-quality management: a 97-percent reduction in atmospheric lead between 1970 and 1990, largely through phasing out unleaded gasoline.[20] Although getting the lead out seems easy in retrospect, the same pragmatic resolve might eventually reduce health risks imposed by other airborne toxins, including deadly substances omitted from the EPA's list of criteria pollutants.

Reining in anything as wispy as air pollution is impossible without estimating its magnitude and identifying major contributors. Although monitoring helps regulators judge the effectiveness of air-quality standards and assess penalties for nonattainment, mandatory emissions reports from individual factories, power plants, and incinerators provide the refined information base essential for pollution control programs as well as emergency management and epidemiological research. Thanks to the Emergency Planning and Community Right-to-Know Act of 1986, local officials and private citizens have ready access to the Toxics Release Inventory (TRI), a massive computer data bank, now covering more than three hundred chemicals.[21] All manu-

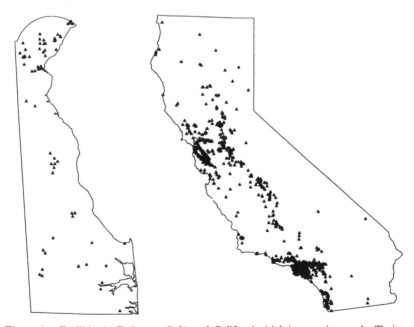

Figure 8.5. Facilities in Delaware (left) and California (right) reporting to the Toxics Release Inventory for 1992. Reprinted from U.S. Environmental Protection Agency, *State Fact Sheets: 1992 Toxics Release Inventory: Public Data Release*, EPA publication no. 745-F-94-001 (Washington, D.C., 1994), n.p.

facturing sites with ten or more full-time employees processing over 25,000 pounds annually of a listed substance must report air, land, surface-water, and underground-injection releases as well as transfers to recycling or waste-treatment plants. Because air emissions account for nearly 60 percent of the more than 3 billion pounds of releases reported each year,[22] the TRI data are an important tool for environmental planners and health officials as well as a valuable supplement to the separate reporting networks for waste incinerators and power plants.[23]

Annual summaries of the TRI data offer intriguing insights on the regional geography of air pollution. Prepared by the EPA's Office of Pollution Prevention and Toxics to assist citizens interested in local air quality, *State Fact Sheets* includes two-page summaries for the fifty states, Puerto Rico, and the Virgin Islands.[24] In addition to listing telephone numbers of contact persons and reporting various statewide totals, each state summary contains four lists of superlatives: Top Five Chemicals for Total Releases, Top Five Chemicals for Air/Water/Land Releases, Top Ten Facilities for Total Releases, and Top Ten Facilities for Air/Water/Land Releases. Each list provides a separate breakdown for the air, water, and land categories, and the two facilities lists report firm names and locations. For each state a small map represents with a tiny triangle every facility filing a TRI report. The maps are roughly the same size and, as Figure 8.5 demonstrates, provide more detail for small states like Delaware than for large states like California. Even though triangles clustered around cities and along important transport routes offer few surprises, the maps occasionally point out unexpected pollution sources in thinly settled areas. Largely rhetorical, these cartographic vignettes encourage readers to request more detailed data for specific areas. Be wary, though, of occasional positional errors—the coordinates of the triangles, like other TRI data, are self-reported by individual polluters.

A thicker TRI report, which EPA staffers call the "national summary," includes several country-wide maps among its numerous statistical tables.[25] Some maps present rates of change between the current and the previous year, others offer single-year snapshots of total or specific releases. Figure 8.6, the most intriguing map in the 1992 report, describes state-to-state differences in air emissions of known or suspect carcinogens. A darker-is-more cartographic metaphor exploits black's traditional association with death and doom by pointing out (in the map's top two categories) states releasing over 5 million pounds of cancer-causing emissions. Hazardous emissions are most pronounced in two longitudinal bands: the old northeastern manufacturing belt running from Illinois through New York and a newer industrial region stretching from Texas and Louisiana to the Carolinas. Although hazardous emissions in the Rust Belt is no surprise, the bigger threat is in the Southeast, to which oil and gas wells, decent ports, lower taxes and labor costs, and comparatively lenient environmental regulations have attracted many petrochemical and pesticide producers. For anyone interested in a single highly generalized map of hazardous air emissions, the EPA's airborne carcinogens map is as good as any.

Although Figure 8.6 reports noteworthy regional variations, it's geographically vague as well as cartographically flawed. Markedly worse air does not suddenly envelop eastbound travelers crossing from New Mexico into Texas, for instance, and all of Texas, Indiana, and North Carolina are not equally and uniformly polluted. More egregious in the eyes of academic cartographers is the EPA's portrayal of

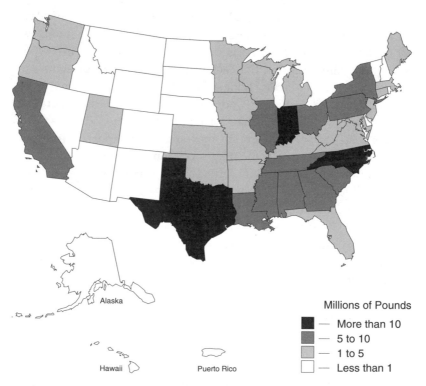

Figure 8.6. TRI releases of known or suspect carcinogens to air, by state, 1992. Reprinted from U.S. Environmental Protection Agency, *1992 Toxics Release Inventory: Public Data Release*, EPA publication no. 745-R-94-001 (Washington, D.C., 1994), p. 80.

count data with a *choropleth map*—the name for maps that fill states, counties, and similar areal units with light-to-dark shading symbols. Because their symbols vary in darkness or intensity, choropleth maps are ideal for showing variation in intensity, rate, or density. By contrast, air emissions measured in millions of pounds are sums or magnitudes, more logically portrayed by graphic symbols (such as circles) varying in size. Choropleth maps of count data are potentially misleading because large areas, as a consequence of their size, are likely to include more of whatever the map is about, and thus contain darker symbols than smaller areas with equal or greater densities. The result is a big area with a dark symbol that looks more ominous on an air-emissions map than a cluster of several small areas separated by historical accident. Oregon and Washington, for example, have separate totals less than that of California, but merging them could create a new state with over 5 million pounds of cancer-causing emissions, putting the entire Pacific coast in a league with the Northeast and Southeast. Conversely, splitting California in two could make the entire American West seem uniformly bland and healthy. Even so, the pattern in Figure 8.6 accords with other data in suggesting that the nation's two manufacturing belts warrant careful epidemiological surveillance.[26]

More relevant to human health are patterns of dispersion and deposition showing what happens to air emissions after winds and air masses spread them around. Depending on the pollutant, the impact of dispersion is either highly local or broadly regional. Local patterns are important to state and county health departments concerned with a specific source, whether proposed or currently operating. In addition to estimating the facility's stack emissions, regulators need to determine where the pollutants are going, in what amounts, and with what effects on humans. Officials know that heavy metals and particulates are more concentrated within a few miles of the source and use mathematical models to account for the influence of wind speed and direction, terrain, and smokestack height. Transport models also address sulfur dioxide emissions and other pollutants with broader, perhaps international, impacts.

Dispersion models, designed to assess local impacts, typically are more complex and computationally demanding than deposition models, applied nationally or internationally. There are numerous dispersions models—the EPA uses at least forty-eight different ones—each tailored to specific geographic conditions, data, time interval, and reg-

162   CHAPTER EIGHT

ulatory requirements. Some models address only urban areas, for instance, whereas others are largely rural. Short-term models estimate concentrations for periods of one, eight, or twenty-four hours, whereas long-term models estimate seasonal or annual averages. Long-range models assess impacts outside a five- or ten-mile radius of the source, and mobile-source models treat highway emissions. Modeling often is a two-stage process, in which a less demanding screening model provides an initial, appropriately conservative estimate to determine whether more refined modeling is needed.

Dispersion modeling's greatest challenge is complex terrain, including the cliffs and canyons of urban high-rise business districts as well as the hills and valleys of open country. Rough topography creates chaotic turbulence and complicates estimating the plume's center-line trajectory, which can shift and bifurcate unpredictably as streamlines dodge obstructions.[27] Not surprisingly, a U.S. General Accounting Office study of the reliability of EPA dispersion models found that seven of nine complex-terrain models overestimated concentration on average between 140 and 1720 percent.[28] To describe these models as conservative is to mix skepticism with gross understatement.

One way of grappling with uncertainty is to use more than one model.[29] Several years ago, when Onondaga County was preparing to build a "waste-to-energy facility" (the current euphemism for an incinerator that generates electricity), environmental impact consultants based the air-quality analysis on two dispersion models. For lower elevations, they used the Industrial Source Complex model's short-term version (ISCST), one of the EPA's more generally reliable "preferred" models. Suitable for regulatory use, the ISC model accommodates flat or gently rolling terrain in both urban and rural settings.[30] To better reflect conditions in higher elevations south of the city, the consultants adopted the Complex I model, an EPA screening model deemed more dependable for areas higher than the stack and its typical plume.[31] Guided by a year of meteorological data (wind direction and speed, temperature, and precipitation) and a generalized elevation map, computer models simulated emissions from a plant operating at 115 percent of capacity. (The extra 15 percent was added as a safety factor.) The environmental impact statement described these simulated emissions with maps showing annual average concentration as well as the second highest concentrations for 1, 3, 8, and 24 hours.

Centered on the site of the proposed incinerator, all five maps use principal highways and the city boundary as a geographic frame of reference. As Figures 8.7 and 8.8 illustrate, the maps represent concentration with isolines similar to the contour lines on a topographic map. These lines connect points with equal estimated concentrations, identified intermittently by labels. Although the pattern of isolines and labels indicates a general decline in concentration with increasing distance from the incinerator site, numerous irregularities illustrate the influence of terrain. In Figure 8.7, for instance, peak values occur about two miles directly south of the site, as identified by a closed contour labeled "8" inside another closed contour labeled "6"—nowhere else on the map are concentrations this high. Within the "8" contour are two closed contours too tiny for their own labels: local peaks masking concentrations of 10, which I estimate by extrapolating upward (6-8-10) from lower contours with a vertical interval of 2.[32] Two factors account for a maximum concentration in this area: north-to-south winds and high elevations close to the site. Although elevated areas to the east are downwind a bit more frequently, the hilly area two miles due south is closer and more vulnerable.

A second impact map illustrates the need for long and short time intervals. When compared to Figure 8.7, which portrays the second highest 1-hour concentration, Figure 8.8 reveals a notably different pattern for average annual concentration. Note the nested pair of closed contours four miles east-southeast of the site, just below the Route 173 label. These isolines represent the annual map's highest concentrations, ranging from 0.06 to over 0.08—a potential threat to residents of some of the region's most expensive homes as well as inmates of the county prison. By comparison, two miles south of the incinerator the estimated annual concentration doesn't get much above the 0.04 value represented by the unlabeled dumbbell-shaped isoline.

Why do peak concentrations occur in markedly different places? The answer lies in the difference between short-term and long-term conditions. Figure 8.7's second highest 1-hour peak falls closer to the site because at least twice a year winds blowing directly south favored the nearby hill with comparatively rich concentrations. In contrast, Figure 8.8's annual average peak occurs farther east because winds from the west-northwest are more common than winds from the north. Understandably, the peak concentration directly south of the site diminishes in prominence on the 3-hour, 8-hour, and 24-hour maps as its rival to the east-southeast grows progressively more prominent.

These maps are not simple snapshots of a single simulated bad-air day. For the first map the computer model estimated pollutant con-

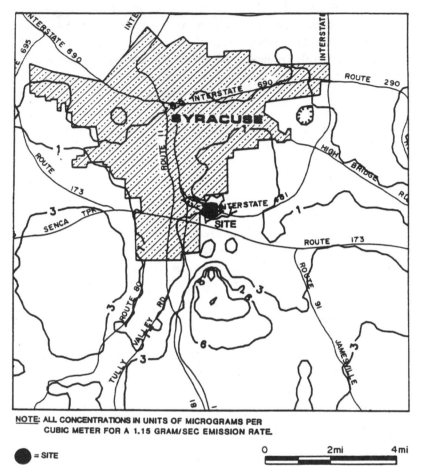

NOTE: ALL CONCENTRATIONS IN UNITS OF MICROGRAMS PER
CUBIC METER FOR A 1.15 GRAM/SEC EMISSION RATE.

● = SITE

0   2mi   4mi

Figure 8.7. Predicted second highest one-hour unitized concentrations of emissions from proposed incinerator. Reprinted from *Onondaga County Solid Waste Management Program. Appendix J: Air Quality Impact Analysis* (1988), fig. S-5.

centration at various locations throughout the area for every hour for an entire year. At each location the computer saved the second highest value (as the EPA does in determining carbon monoxide exceedances) and used these results to draw the contours in Figure 8.7. For the second map the computer based the isolines on annual averages by summing all hourly estimates at each location and dividing by 8,760, the number of hours in the simulated year.

Because air flow disperses all substances more or less similarly, one map conveniently accommodates pollutants with widely different con-

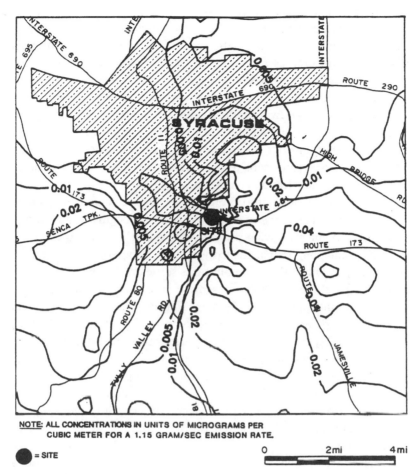

NOTE: ALL CONCENTRATIONS IN UNITS OF MICROGRAMS PER
    CUBIC METER FOR A 1.15 GRAM/SEC EMISSION RATE.

● = SITE

0   2 mi   4 mi

Figure 8.8. Predicted annual average unitized concentrations of emissions from proposed incinerator. Reprinted from *Onondaga County Solid Waste Management Program.
Appendix J: Air Quality Impact Analysis* (1988), fig. S-1.

centrations. For computational simplicity, the isolines portray "unitized concentrations," which environmental engineers can readily
apply to specific pollutants. Each map assumes a hypothetical toxin
leaving the stack with a *unit concentration* of 1 gram per cubic meter
(1.15 g/m³ after applying the 15-percent safety factor to the unit concentration of 1 g/m³) and shows the dispersed concentration in micrograms per cubic meter (μg/m³). (Air flow does indeed dilute pollution,
and with a million micrograms to the gram, using micrograms avoids
cluttering the map's isoline labels with extra zeros.) Applying the uni-

tized map to specific substances is straightforward. For example, if the estimated concentration of a pollutant leaving the stack is 2 g/m³— twice the unit concentration—the peak value 10 on the 1-hour map (Figure 8.7) thus represents 20 μg/m³. But for a less dense toxin with an estimated stack concentration of only 0.1 grams per meter—a tenth the unit concentration—the same mapped peak represents only 1 μg/m³. For these two hypothetical stack concentrations the mapped peak of 0.08 on the annual average map (Figure 8.8) translates into mean concentrations of 0.16 and 0.008 μg/m³, respectively. By applying unitized dispersion maps to estimated stack emissions, engineers can project a concentration level for any pollutant.

Health officials don't stop there. Projecting peak concentration is just one step in a *quantitative risk assessment* that begins by identifying hazardous chemicals leaving the stack as well as estimating exposure through inhalation, dermal absorption, or ingestion of local beef, fish, vegetables, and milk. Because pollutants settle on skin, plants, and soil as well as in the lungs, risk assessment must consider all plausible "pathways," including breast-feeding. For each hazardous chemical, risk assessors determine a worst-case exposure by relating likely stack emissions to dispersion models and considering momentary peak concentrations as well as lifetime exposure to average conditions and the special vulnerability of fetuses and infants. Applying dose-response models based on animal or epidemiological studies, they estimate amounts harmful to humans and derive numerical estimates of risk, stated as a probability. Not a single number, though, but separate estimates for all hazards.

Unless these probabilities are flagrantly ominous, the risk assessment usually concludes with a reassuring interpretation rich in waffle words that unavoidably feed fears its authors hope to allay. For example, Onondaga County's air-quality study was quite explicit in its final paragraph: "Estimates of carcinogenic risk have been prepared using the best available estimates of potency for various carcinogens. Estimates for similar plants show a worst-case estimate of carcinogenic risk in the range of one to ten cases of cancer in a million persons."[33] To skeptics, though, "best available" and "similar plants" raise questions about the reliability of dose-response models as well as the soundness of the proposed plant's design and operation.[34] And because risk assessors themselves cannot agree on the severity of a worst-case scenario, public hearings on environmental impact are often lengthy and contentious.[35] That's what happened in Syracuse, where opponents delayed the incinerator for several years. Although legislative leaders determined to build the plant withstood intense local opposition, officials grudgingly agreed to a comprehensive program of emissions testing and air monitoring.

If computer simulations of air flow are the high-tech missiles in the environmental arsenal, interpolation models are the more traditional heavy artillery. Regulators depend on interpolation models to construct isoline maps from concentrations measured at widely separated monitoring points. Although contours sketched by hand can incorporate intuition and expert knowledge, atmospheric researchers prefer rigorously consistent, comparatively unbiased computer algorithms that quickly convert scattered measurements into a convincing assemblage of isolines.

Interpolation is especially useful in studies of acid rain, which has slowly been destroying fish and forests throughout much of eastern North America.[36] By comparing isoline maps of precipitation acidity, deposition of sulfates and nitrates, and the sensitivity of soil and surface water to acid deposition, scientists have identified both victims and culprits. They now know, for instance, that acid precipitation is especially troublesome in the Middle Atlantic and New England states, which not only receive large amounts of sulfates and nitrates but typically lack limestone and other bedrock providing a natural buffer against increased acidity.[37]

To illustrate how interpolation works, I have enlarged in Figure 8.9 the northeastern quarter of the nitrate wet-deposition map in the National Acid Precipitation Assessment Program's *1992 Report to Congress*.[38] Dots on the map represent sampling points, and their adjoining numbers report the amount of nitrates that rain, snow, clouds, and fog deposited in one year on vegetation, soil, and structures. Look closely at the isoline labeled "18," which the computer has threaded between lower and higher values to define a boot-shaped band extending from eastern Ontario through central New York into southern New England. Within this ridge of high deposition a value of 31 east of Lake Ontario marks the Tug Hill plateau, parts of which receive over 225 inches of acid snowfall a year.[39] Although the pattern varies somewhat from year to year, this band across the southern Adirondacks is a salient feature worrisome to foresters and wildlife biologists.[40] Note, though, the measurements 10 and 22, in southern New York and Pennsylvania, respectively, between the lines labeled "14" and "18"—on this

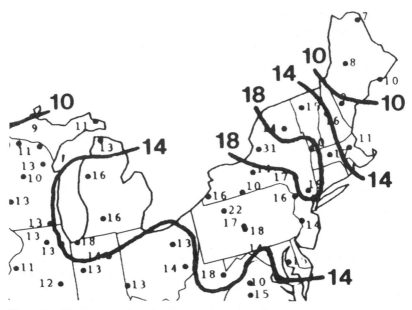

Figure 8.9. Northeast portion of 1991 map of wet nitrate ion deposition, in kilograms per hectare. From National Acid Precipitation Assessment Program, *1992 Report to Congress* (Washington, D.C., 1993), p. 41; excerpt enlarged from fig. 10.

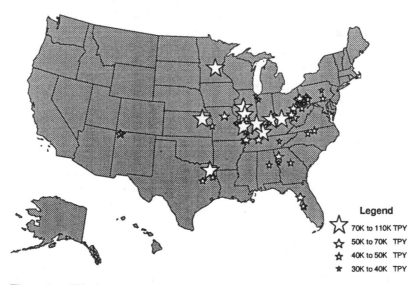

Figure 8.10. Fifty largest sources of nitrogen oxides, 1992. Emissions in thousands of short tons per year. Reprinted from U.S. Environmental Protection Agency, Office of Air Quality Planning and Standards, *National Air Pollutant Emission Estimates, 1900–1991*, EPA publication no. 454/R-92-013 (Research Triangle Park, N.C., 1992), p. 29.

rather generalized, small-scale map, the cartographer (appropriately, I think) omitted contours around these local anomalies.

Because airborne nitrates come mostly from cars, trucks, and coal-fired power plants, maps of wet and dry deposition show generally similar geographic patterns, with high values in the Northeast and comparatively minor amounts in the western half of the nation.[41] Not completely similar patterns, though, because wet deposition also reflects above-average precipitation in places like the Tug Hill, hundreds of miles downwind from major sources. Figure 8.10, a cartographically correct EPA map on which bigger symbols reliably represent bigger contributors, identifies the fifty largest point sources, most of which generate electricity. Air masses generally move across North America from west to east, and storm tracks frequently connect the Ohio Valley with New England. In contrast, dry deposition occurs closer to the source and is thus somewhat less villainous than "acid rain."[42] Sulfur dioxide, an even bigger contributor to acid precipitation than oxides of nitrogen, exhibits a strong correlation between emissions from coal-fired power plants largely in the Midwest and a broad region of high deposition extending from the lower Ohio Valley through Virginia and eastern Massachusetts. Although motor vehicles surpass power plants in producing nitrates, electrical utilities account for more than two-thirds of the nation's sulfate emissions.

Maps of emissions and deposition document a clear and politically contentious pattern of net exporters and net importers. Midwestern power plants prefer low-cost high-sulfur coal from nearby mines, for instance, and even though motor vehicles throughout the country emit objectionable amounts of nitrates and sulfates, the eastern states receive more pollutants than they pass along to areas farther east. Coastal New England, in particular, is very much a net importer, whereas Minnesota, with comparatively little power generation or concentrated settlement to the west, exports emissions from metropolitan Minneapolis-St. Paul to Wisconsin, Illinois, and beyond. And since air masses ignore international borders, emissions from the Midwest penetrate Ontario, while emissions from Ontario invade New England.

In the 1980s Congress decided to fight air pollution with a market-based strategy in which incentives complement restrictions.[43] In addition to pollution allowances with penalties for excess emissions, the new approach established emissions credits that can be saved, transferred to another unit in the same corporation, or bought and sold on the Chicago Board of Trade.[44] Each plant receives an allotment based on past fuel consumption and emissions. At the end of the year a delinquent firm pays a stiff per-ton penalty for its excess emissions, whereas a company in compliance receives a per-ton credit for any surplus improvement. Because credits can be traded as a commodity, an excess polluter can avoid the penalty by buying credits and thus rewarding the company that invested in new technology or switched to cleaner fuel.

Although supporters call emissions trading a flexible solution, critics fear bad air downwind from utilities that buy credits, pass along increased costs to customers, and continue to pollute.[45] Former senator Eugene McCarthy, who likened emissions credits to the "medieval practice of selling indulgences," noted that obnoxious industries can even increase emissions.[46] Regulations prohibiting trades that increase toxic emissions or adversely affect local air quality only partially address these concerns.[47] Environmentalists also questioned initial awards of large allowances to major polluters, which were better able to earn and sell emissions credits for making changes long overdue.[48] Equally troubling were additional allowances created by the 1990 Clean Air Act—a loophole closed only after New York's attorney general, fearing increased acid rain caused by Ohio Valley utilities, sued to prevent the EPA from selling up to 800,000 tons of extra excess emissions.[49]

More controversial perhaps are schemes for cutting motor-vehicle emissions and transferring the credits from one area to another. Typically one or two localities shoulder the burden of foul air, as when Unocal, a California petroleum refiner, avoided the cost of reducing industrial emissions by buying and destroying pre-1971 automobiles—gas-guzzling, smoke-belching clunkers whose demise lowered total statewide emissions but did little for neighbors of refineries allowed to pollute at previous levels.[50] Equally contentious are clever schemes to promote economic development in one part of a state by imposing restrictions on motorists in other regions. In 1994, for instance, an emissions-testing program in seven southern Maine counties helped Louisiana-Pacific Corporation create seventy new jobs two hundred miles north in Aroostook County.[51] Ironically, the irate self-professed victims in the southern counties, forced to undergo testing, buy cleaner gas, or repair their jalopies, reap the benefits of cleaner air, while the alleged beneficiaries farther north shoulder the risk of reduced longevity.

However much motorists resent efforts to control auto emissions, few groups are as aggrieved as cigarette smokers forced into the cold (quite literally) by legislation promoting clean indoor air. Figure 8.11, a map showing incursions on what the tobacco industry touts as smokers' rights, is my final exhibit to support the argument that air pollution, more so than the hazards addressed earlier, has a largely political geography reflecting attitudes toward individual and collective responsibilities as well as struggles among competing interest groups.[52] Although the Surgeon General and the EPA have led the campaign against secondhand smoke in offices, restaurants, and other public places, it's the states and localities that control where smokers may legally light up.[53] And how state legislatures exercise this power says a lot about their constituents' position on the liberal-conservative spectrum. For example, the four states most reluctant to restrain smoking are in conservative parts of the Southeast, whereas the states most eager to make smokers butt out include such liberal bastions as Minnesota and New York. But there are noteworthy exceptions, such as fiercely libertarian New Hampshire—its license plates assert "Live free or die"—where legislators with far grassier roots than their New York counterparts favor the rights and lungs of nonsmokers. Moreover, many western states that instinctively vote Republican and reject federal initiatives have imposed restrictions EPA's cartographers endorse as "moderate." The map will no doubt evolve further as nonsmokers learn to value clean air as much as smokers value smoking.

State support for "smokers rights"

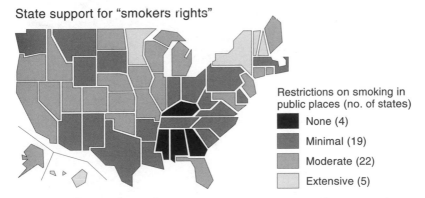

Restrictions on smoking in
public places (no. of states)

None (4)

Minimal (19)

Moderate (22)

Extensive (5)

Figure 8.11. EPA evaluation of states' restrictions on smoking in public places. Redrawn from map in Fran Du Melle, "Laws Protecting Nonsmokers," *EPA Journal* 19 (October–December 1993): 21–2.

Despite pervasive political overtones, air-quality mapping is a triumph of scientific ingenuity in getting by with limited data. By making bad air visible on paper or video screen, maps help scientists and regulators identify cause-effect relationships, assess impacts, and suggest solutions. And as geographic snapshots, air-quality maps encourage citizens and elected officials to compare places, understand threats, and demand action. What's so remarkable about air-quality mapping is that the process works at all: in bridging gaps in our sparse but costly monitoring networks, maps based on air-flow or interpolation models extract meaningful patterns that guide policy and garner public support. In effect, the atmosphere has become an interactive map on which decision makers can detect threats and fine-tune responses—a slow video game in which the prize is survival.

Chapter Nine

Short-Lived Daughters and ELF Fields

President Truman's decision to end World War II abruptly by dropping atomic bombs on two Japanese cities laid the foundation for Americans' dread of radiation. In 1946 journalist John Hersey's best-seller *Hiroshima*, published originally in *The New Yorker*, described in minute detail the unprecedented destruction witnessed by six residents of this medium-size industrial city, where more than sixty-eight thousand people perished and many more suffered radiation poisoning and severe burns.[1] The discovery in 1949 that our ex-ally, the Soviet Union, had its own nuclear weapons led to air-raid drills, fallout shelters, and a spate of atomic monster films. Typical of this genre was the 1954 thriller *Them!*, in which giant ants, the genetic progeny of atomic testing in the New Mexico desert, fly west, colonize the Los Angeles sewer system, wreak havoc on a few nocturnal forays, and break a Hollywood rule by killing off the film's hero.[2] In 1979 an accident at the Three Mile Island nuclear plant ended widespread optimism about safe, clean, and cheap electricity generated with the "peaceful atom." And in 1986 a far more serious release of radiation at Chernobyl reinforced fears of nuclear power, nuclear waste, and electromagnetic radiation in general—anxieties reflected in public apprehension (and occasional hysteria) about indoor radon and magnetic fields around high-voltage power lines.

Treating radon and magnetic fields in the same chapter affords a convenient comparison of two new and highly controversial hazards, both involving electromagnetic energy. Radon, the more serious threat, is a colorless, odorless gas that seeps into buildings from the soil. A radioactive element with a short half-life—half a given amount of the more common isotope (atomic variant) ^{222}Rn disintegrates in only 3.82 days—radon is the second leading cause of lung cancer, albeit well behind smoking in number of deaths. Trouble arises when a radon atom disintegrates into a polonium atom—its "daughter"—and releases an alpha particle, a burst of matter that can damage genes when radioactive decay occurs inside the lungs. Even though the polonium atom, with a half-life measured in minutes or seconds (depending on the isotope), quickly releases another deadly alpha particle, its more easily measured parent, radon, takes the blame. Less well understood is the health hazard posed by extremely low frequency (ELF) magnetic fields around high-voltage transmission lines and transformers. Biophysicists are uncertain how these fields affect cell tissue and genetic material, but limited data suggest links with childhood leukemia and brain tumors. This chapter examines the role of maps in understanding these hazards and warning the public.

INDOOR RADON

Although aware of radon as a source of normal, ever-present background radiation, epidemiologists largely ignored the gas until 1985, when radiation technicians tested the home of Stanley Watras, an engineer at the Limekiln nuclear power plant in eastern Pennsylvania. Facilities handling substantial amounts of radioactive materials monitor workers for contamination with walk-through scanners larger and more complex than the metal detectors at airports. Whenever Watras left the plant, the portal monitor registered extraordinarily high levels of radioactivity, much higher than for fellow workers. Unable to find a radiation leak near his work area or elsewhere within the plant, health physicists tested Watras when he reported for work and found even higher levels than at the end of his shift. In searching for the source, state environmental officials checked his home and discovered radon concentrations greater than 2,000 picocuries per liter (pCi/L)—the highest level ever recorded in a residence and the equivalent of smoking 135 packs of cigarettes a day or having 260,000 chest X rays a year.[3] Advised to evacuate, Watras and his family left their home on January 6th, leaving behind the children's contaminated Christmas presents.[4]

Physicists measure amounts of ^{222}Rn and other radionuclides in curies, named to honor Marie and Pierre Curie, who discovered radium.[5] Because 1 curie (Ci) equals 37 billion disintegrations per second—a prodigious rate of decay associated with a gram of highly radioactive radium—public health officials report the much lower concentrations of radionuclides found in air and water in picocuries (trillions of curies) per liter. One pCi/L thus represents the decay in a liter of air of only two radon atoms per minute. In 1986 the U.S. Environmental Protection Agency adopted 4 pCi/L as the "action level" at which building owners should undertake mitigation. Much of the controversy over radon revolves around whether 4 pCi/L is too low or too high.

The link between radon gas and lung cancer reflects a chain of nuclear disintegrations following the breakup of a radium atom.[6] Each isotope has its own decay series. For example, ^{226}Ra, a relatively stable form of radium with a half-life of 1,600 years, is the parent of ^{222}Rn. Yet birth of a ^{222}Rn atom is merely the first of six disintegrations

$$^{226}\text{Ra} \xrightarrow{\alpha} {}^{222}\text{Rn} \xrightarrow{\alpha} {}^{218}\text{Po} \xrightarrow{\alpha} {}^{214}\text{Pb} \xrightarrow{\beta} {}^{214}\text{Bi} \xrightarrow{\beta} {}^{214}\text{Po} \xrightarrow{\alpha} {}^{210}\text{Pb}$$

1600 yrs   3.82 dys   3.11 min   26.8 min   19.8 min   0.0002 sec

in a chain that produces four short-lived daughters and ends with ^{210}Pb, a lead isotope with a half-life of 22 years. Each disintegration yields an isotope of radon, polonium, lead, or bismuth as well as an alpha(α) or a beta (β) particle. Once a radon atom breaks up, the remaining disintegrations occur relatively rapidly. Two other naturally occurring isotopes of radon, ^{220}Rn (thoron) and ^{219}Rn (actinon), have markedly shorter half-lives than ^{222}Rn and more rapid decay series—part of the reason why they're less common.

Several factors make radon a health hazard. As a decay product of radium, radon is a granddaughter of uranium, which occurs widely in bedrock and soil, albeit in the minute quantities of a "trace element." As a gas, radon moves more readily than its parents through soil and foundation material, and its four-day half-life is long enough to survive the slow migration into a building yet short enough to produce lethal radioactivity. Once inside, radon escapes the winds and convection currents that dilute outdoor radon, which is far less hazardous. When decomposition occurs, the ionized, molecularly attractive radon daughters readily attach to tiny airborne dust particles (aerosols) or water molecules, enter the bronchial passages and lungs, and settle into the outer mucous layer.[7] Although alpha particles have a shorter range than beta particles—0.005 cm in contrast to 0.15 cm, in soft tissue—and cannot penetrate the skin, once inside the lungs, the comparatively massive alpha particles, which can strip electrons from atoms and create free radicals, are far more likely to kill cells and damage genes.[8] Radon contributes less lethal radiation than its daughters, which not only invade the lungs by hitching rides with aerosols but emit two of the three alpha particles in their decay chain.

Health physics explains why radon is dangerous, but decay chains and particle ranges tell us nothing about the lung-cancer mortality rate, which the EPA estimates as approximately 14,000 deaths nationwide each year.[9] To derive this figure, epidemiologists turned to data on the radiation exposure and mortality of uranium miners on the Colorado Plateau.[10] (Radium is a daughter of uranium, and radon levels have been extraordinarily high in uranium mines.) In extrapolating data on miners to ordinary home-owners, statisticians assumed a cumulative effect and a strict proportionality between dose of radiation and risk of death. For example, if a long-term dose of 100 rad produces 2,000 cancers in a population of 10,000 workers, a cumulative dose of 1 rad should yield only 20 cancers. Because a strictly proportional relationship plots as a straight line on the dose-risk graph in Figure 9.1, this assumption is called the *linear hypothesis*.[11] Anchored at the graph's origin, a straight line through the few data points—perhaps only one data point—is a straightforward device for assessing the health risks of low-exposure populations.

However convenient, the linearity assumption is not without critics.[12] Some physicists and radiologists have suggested as alternatives the linear-quadratic and linear-threshold relationships described in Figure 9.1. A quadratic relationship, with risk increasing as the square of exposure—a relationship that seems to fit data from animal experiments—forecasts a comparatively reduced danger at low exposures, whereas a threshold assumes nonzero exposures posing no added risk. Both options project a lower risk than the linear hypothesis, but the curvature and inflection point are difficult, if not impossible, to determine, at least for human populations. As a further complication, a few

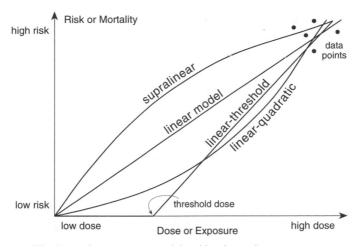

Figure 9.1. The linear dose-response model and its alternatives.

scientists have suggested the possibility of a supralinear relationship forecasting somewhat higher risks for low exposures.[13] No wonder, then, that most epidemiologists, for lack of better data, grudgingly accept the linear model. While the possibility of a nonlinear response is occasionally invoked to question, if not discredit, risk assessments deemed too high—or too low—most health physicists consider the linearity assumption a safe approach to the uncertain dangers of small doses.

Several sources of uncertainty obviate a precise estimate. Although radiation levels in uranium mines were dangerously high, especially before federal standards improved conditions in the late 1960s, ventilation, radon level, and length of the work week varied significantly.[14] Because miners were often exposed more than forty hours a week to concentrations well above the accepted "working level" of 100 pCi/L, the dose received was not necessarily proportional to the number of years a miner worked. And because many miners had not yet died—data for Hiroshima and Nagasaki show new cancers at higher than average rates decades after exposure—cumulative, long-term effects were not fully known. In addition, radon was not the only cause of lung cancer: air in the mines was rich in silica dust, known to cause lung cancer in animals. And because an estimated 72 percent of uranium miners smoked, even nonsmoking miners could not escape the dangers of passive, secondhand smoke.[15] As Figure 9.2 illustrates, by underestimating the exposure of miners, the EPA pushed the upper end

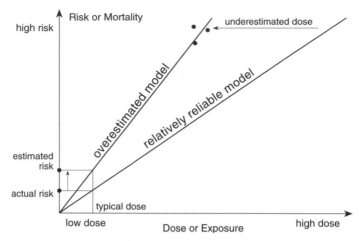

Figure 9.2. An underestimated dose associated with high mortality increases the slope of the dose-response model and exaggerates the mortality risk for a given low-dose exposure.

of the dose-response relationship closer to the graph's origin, increased the line's slope, and exaggerated the risk to the general population.

The lower end of the graph was equally troublesome. High mobility and wide variation in the amount of time spent at home thwarted efforts to estimate the lifetime dose received in a home with a radiation level of, say, 10 pCi/L. In blaming indoor radon for 14,000 of the nation's 130,000 annual lung-cancer deaths, the EPA wisely noted that the toll could be as high as 30,000 or as low as 7,000. Yet in advocating remediation for concentrations as low as 4 pCi/L, the agency was encouraging perhaps as many as 10 percent of home owners to spend several hundred to over two thousand dollars for improvements of questionable value.[16] By comparison, far fewer American homes, perhaps only 0.1 percent, have concentrations greater than or equal to Canada's action level of 20 pCi/L.[17]

Some scientists were openly skeptical. Among the EPA's most vigorous critics were Philip Abelson, a former editor of *Science*, and Anthony Nero, a prominent radiation physicist at Lawrence Berkeley National Laboratory. In a paper entitled "Radon Today: The Role of Flim-Flam in Public Policy," Abelson ridiculed an earlier EPA estimate of 43,000 radon-induced deaths, accused the agency of deliberately frightening the public, and noted that some studies "seem to demonstrate that, if anything, moderate levels of radon are beneficial to the public health."[18] In a column in *Science*, he charged that the "EPA has no solid evidence that exposures to 4 pCi/liter of radon causes lung cancer in either smokers or nonsmokers. Indeed, there is abundant evidence to the contrary in the fact that in States with high levels of radon, inhabitants have less lung cancer than those in States with low levels."[19] Yet fellow physicist Nero was equally adamant that the danger, though overstated, was real. If the federal government wants an efficient solution, he argued, it should help states identify high-radon areas requiring revised building codes and careful testing of all homes.[20]

If 20 pCi/L might be a more appropriate action level, why is an alarmist EPA calling for remediation at 4 pCi/L? Political scientist Leonard Cole, who sees radon as "a unique opportunity" for politicians and regulators, suggests a motive:

> Unlike most environmental hazards, radon is largely a natural phenomenon with no industry to blame for its presence. Moreover, present policy urges that homeowners pay for testing and fixing out of their own pockets. The financial cost to government is negligible, and "protecting" homeowners against the gas has been more politically convenient than for other alleged hazards. The absence of an industrial "culprit" means the absence of an interested party who might be expected to underscore the uncertainties about the issue. This has made it easier for political and regulatory leaders to push aggressive policies of questionable warrant. For public officials, radon is environmentalism on the cheap.[21]

Despite his cynical interpretation of EPA policy, Cole agrees with Nero on the need to focus on high-radon areas.

Maps hold a small but sometimes conspicuous role in the EPA's public information campaign. The most prominent cartographic portrait to date is the multicolor EPA county-unit map of radon zones released in 1993.[22] Printed on a 28-by-24-inch sheet, the map apportions the nation's 3,141 counties among three hazard zones portrayed in bright red, orange, and yellow. Figure 9.3, a smaller black-and-white rendering, shows the same zones without county boundaries. A single zone covers each county, even the huge counties found in the West. Zone assignments reflect "predicted average screening level," that is, the average radiation level if all homes in the county were tested.[23] Zone 1, colored red to represent an estimated radon level above the 4 pCi/L action level, forms an elongated, irregularly shaped high-risk belt stretching across the northern states from New Jersey to Washington, with noteworthy outliers in Maine and Nevada as well as tentacles extending southward into Alabama and New Mexico. Viewers might reasonably infer that this high-radon zone contains an above-average share of homes with radon levels above 20 pCi/L. By contrast, much of the Southeast, particularly along the Atlantic and Gulf coastal plains, falls in zone 3, with average levels below 2 pCi/L. Despite lower average levels in this lower-risk zone, many homes here will test above 4 pCi/L, with some well above 20 pCi/L. According to a fact sheet distributed with the map, "All homes should be tested for radon, regardless of geographic location or zone designation." Appropriately, zone 3 is cloaked in yellow, suggesting caution, rather than colored green, which might imply no risk. Instead of describing areas of high, medium, or low risk—too easily interpreted as high, so-so, or no risk—the map's authors clearly want viewers to think of radon zones as either worst, bad, or serious.

Covering the entire country, the three-zone map replaced an earlier map showing deposits of uranium and other geologic formations considered "areas with potentially high radon levels."[24] Issued in 1987, the year after the EPA set up a radon program based on the 4 pCi/L action level, this "interim" map (Figure 9.4) required an extensive dis-

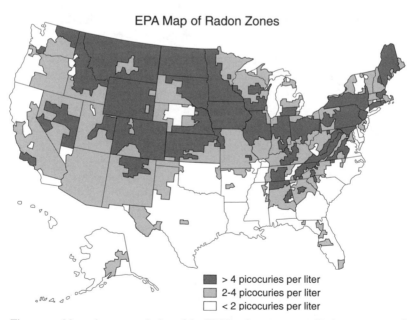

EPA Map of Radon Zones

> 4 picocuries per liter
2-4 picocuries per liter
< 2 picocuries per liter

Figure 9.3. Monochrome rendering of the EPA's radon zones map. Redrawn, at a much reduced scale, from U.S. Environmental Protection Agency, "EPA Map of Radon Zones," map accompanying EPA document 402-R-93-071 (1993).

Areas with Potentially High Radon Levels

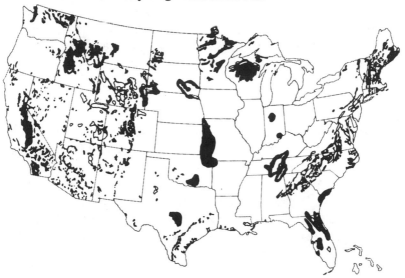

Figure 9.4. Interim radon-hazard map used by the EPA in 1987. Reprinted from Michael Lafavore, *Radon, the Invisible Threat: What It Is, Where It Is, How to Keep Your House Safe* (Emmaus, Pa.: Rodale Press, 1987), p. 30.

claimer. Agency experts in risk communication—the personnel have changed, so I can only guess—apparently felt the media needed a map of some sort, even an incomplete map that understated the hazard's spatial extent. Despite accompanying admonitions that dangerous levels can also occur outside the high-potential zone, the map provided a convenient excuse for home owners eager to ignore the problem. By contrast, the new map warns of a wider, more troublesome hazard.

Were indoor radon not a geographically erratic phenomenon, the radon-zones map (Figure 9.3) could be dismissed as alarmist propaganda. The fact is, though, that two identical dwellings built on adjacent lots by the same contractor and sharing the same bedrock and soil often have very different levels of radon. For instance, although the Watras house was by far the hottest dwelling in Boyertown, Pennsylvania, the house next door tested safely under the 4 pCi/L threshold.[25] But even though the most dangerously elevated levels in the neighborhood were atop a small, local uranium-enriched formation, geology alone could not predict radon concentration.[26] This variability was by no means unique: data from around the country soon revealed a wide variety of factors affecting indoor radon concentration.[27] In particular, new, well-insulated buildings tend to trap whatever radon their comparatively porous cinder-block foundations let into the basement. Soil permeability and geologic faults are important too, and small pressure differences caused by wind, heating, fans, and fireplaces can draw radon into a house.[28] Test data also revealed higher concentrations in the basement than on the first floor as well as higher levels in winter, when people keep windows closed and open doors less frequently.[29]

Despite radon's spatial and temporal variability, Congress wanted a map of radon potential. Directed by the Indoor Radon Abatement Act of 1988, U.S. Geological Survey scientists worked with state geologists and EPA officials to compile a risk map based on five factors: geologic formations rich in uranium and radium, radioactivity measurements collected in the 1970s and early 1980s by planes flying low-altitude traverses with a spectrometer, soil survey data on permeability, indoor radon measurements based on charcoal canister detectors and tabulated by various state and private surveys, and geographic estimates of the proportion of houses with basements.[30] To integrate these diverse measurements, USGS researchers partitioned the country into large but generally homogeneous geologic provinces and computed composite scores taking into account each measure's estimated local reliability. The geologists then used levels of 2 and 4 pCi/L to delineate three hazard zones, and the EPA converted the map to the county boundaries shown in Figure 9.3.

However rhetorically useful in persuading home owners to test for radon, the small-scale national radon-zone map is of limited value to state and local officials, who require more details on relative risk. To address these needs, the Geological Survey prepared reports on radon potential for each of the EPA's ten administrative regions. The 194-page report for region 5, covering Illinois, Indiana, Michigan, Minnesota, Ohio, and Wisconsin, is representative.[31] The first section describes how geologists evaluated available data, delineated zone boundaries, and estimated both a Radon Index gauging relative risk and a Confidence Index reflecting data quality.[32] After a short chapter summarizing broad variations in radon potential throughout the region, chapters addressing individual states examine source materials and geographic patterns of radon potential.

Intended as booklets for local officials, the state sections share a common outline and content, including caveats warning that each

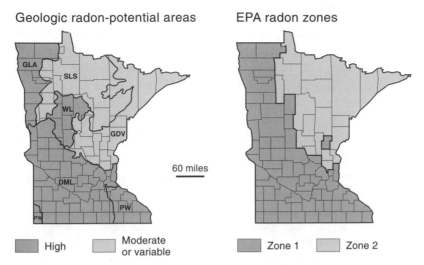

Geologic radon-potential areas   EPA radon zones

60 miles

High   Moderate or variable   Zone 1   Zone 2

Figure 9.5. Geologic radon potential in Minnesota (left) and EPA county-unit radon zones from the national map in Figure 9.3 (right). Left-hand map reprinted from R. Randall Schumann and Kevin M. Schmidt, "Preliminary Geological Radon Potential Assessment of Minnesota," in R. Randall Schumann, ed., *Geological Radon Potential of EPA Region 5*, U.S. Geological Survey Open File Report no. 93-292-E (Denver, Colo., 1994), pp. 122–47, fig. 12.

assessment is "generalized," its scale is "inappropriate for . . . neighborhoods, individual buildings, and housing tracts," and its maps are "no substitute for testing individual homes."[33] Although the state chapters vary according to geology, the Minnesota section is typical in its presentation of twelve maps, the last of which apportions six geologically similar areas among two hazard zones. As illustrated in the left half of Figure 9.5, dark shading indicates "high radon potential," light shading indicates "moderate or variable radon potential," and acronyms identify geologic areas described in the text. The right half of Figure 9.5 demonstrates that the EPA cartographers, in transferring geologic boundaries to their national county-unit map, further exaggerated radon's risk.

An attractive Geological Survey general information booklet entitled *The Geology of Radon* confirms the feasibility of even more detailed radon maps.[34] Among the booklet's numerous illustrations is a map showing zones of high, moderate, and low radon potential for three counties surrounding Washington, D.C.[35] As reproduced in Figure 9.6, the zone boundaries reflect substantial within-county variation—considerably more than the state maps for Maryland and Virginia. Even so, this map is a much generalized compilation from 1:62,500 (inch-to-a-mile scale) maps based partly on measurements of soil gas radon, soil uranium, and indoor radon, and portraying radon potential in far greater detail for each of the three counties.

As often happens in mapping, more detail means more uncertainty. Typical of other large-scale USGS radon-potential maps, the 1:62,500 Montgomery County map carries a string of caveats warning that[36]

> 1) . . . no area of the county is free from the potential for indoor radon levels greater than . . . 4 pCi/L. . . .

> 2) All boundaries . . . should be considered approximate due to imprecisions [sic] in mapping and because of the often gradational nature of geologic contacts. . . .

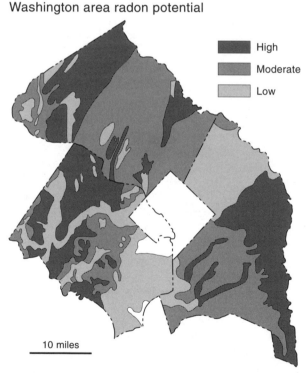

Figure 9.6. Radon potential map for three counties surrounding Washington, D.C. Redrawn from James K. Otton, *The Geology of Radon*, a U.S. Geological Survey general interest publication (Washington, D.C.: U.S. Government Printing Office, 1992), p. 25.

3) . . . [because of] house architecture construction techniques, lifestyle, and user patterns by the residents . . . small localized areas of higher or lower radon potential are likely to occur. . . .

4) The indoor radon measurements shown on this map are winter screening tests [that might not reflect] other seasons of the year. . . . [Moreover,] this data was volunteered and is not a true random sampling.

In short, the map offers no assurance that any home in the county is safe.

Despite mandatory homage to the 4 pCi/L threshold, the map authors broke with the EPA in describing categories that acknowledge the 20 pCi/L level. Notably, their high-risk zone, although based on the EPA criterion, is also defined by the portion of homes exceeding the higher threshold: "A high rating corresponds to a greater than 60 percent chance of a home having elevated levels [above 4 pCi/L] of radon. Indoor radon will fall in the 1 to 20 pCi/L range but as many as 30 percent of houses measured in this area can be greater than 20 pCi/L."[37] However subtle, this dual definition also reflects scientific opinion that the EPA threshold is inefficient and too geographically diffuse. While remediation at 4 pCi/L will surely reduce the number of radon-induced lung cancers, according to physicist Anthony Nero a focus on 20 pCi/L would "help the folks with the most urgent need."[38]

The story doesn't end here. An ambitious research program promises more accurate subcounty radon predictions and more useful radon-potential maps. A joint venture of EPA, USGS, and the Department of Energy, the "Hottest Homes Project" is developing statistical models its supporters hope can focus remediation efforts on localities with substantial proportions of homes above either 4 or 20 pCi/L—or any other threshold, for that matter.[39] Drawing on both physical (geology, soils, meteorology, architecture) data and carefully controlled indoor-monitoring measurements, these analytical models will also address the inherent uncertainty of short-term (two- to seven-day) readings based on radon test kits available from the local hardware store, the National Safety Council, and other sources. Despite the EPA's advice that all homes should be tested, short-term measurements can underestimate the danger. Although home owners with readings above the action level willingly test again before undertaking expensive remediation, those with readings just below 4 pCi/L seem content with a single measurement. This complacency is all too common, according to a General Accounting Office study reporting that 85 percent of home owners with an initial radon reading between 2 and 4 pCi/L ignore recommendations for a second, follow-up test.[40] Analytical statistical models that predict local variability as well as average radon level should encourage repeat testing in areas with high uncertainty.

Anthony Nero, who directs work on the study at Lawrence Berkeley National Laboratory, believes that statistical models will soon be able to refine the 1993 radon zones map, perhaps down to the census tract level. A better map—"a sensible guide for action"—will be expensive, he warns, with a full-scale national program requiring a substantial share of the EPA radon budget.[41] Even so, the results should be worth it, especially for the estimated one-tenth of one percent of households with dangerously high radon levels lurking undetected in the red, orange—and even yellow—zones of the colorful but largely rhetorical EPA radon map.

EXTREMELY LOW FREQUENCY MAGNETIC FIELDS

Lumping radon gas and magnetic fields into one chapter might puzzle readers who know their physics. The juxtaposition is useful, though, because of revealing similarities and differences. Electromagnetic radiation, the arcane and varied phenomenon that underlies both hazards, is something folks who don't know their physics typically dread. In the late 1980s, the media avidly and shamelessly fed these fears with articles entitled "Lung Cancer's Gassy Ally"[42] and "Power Lines and Cancer: the Evidence Grows."[43] With radon, though, some fear is justified: although scientists quibble over the numbers, thousands of Americans die each year because of exposure to indoor radon. By contrast, magnetic fields around transmission lines remain only a "possible" hazard, dangerous (if at all) largely to electric company workers. And equally revealing is the dearth of maps warning of cancer risk near power lines.

Maps had an important role, though, in a widely cited study that instigated much of the controversy. In the mid-1970s, a Denver epidemiologist, Nancy Wertheimer, started visiting the homes of children who died of cancer. "Power lines weren't really on my mind," she told a writer for *Discover*.[44] "I was taking a gunshot approach, looking for pollution, chemicals, formaldehyde from insulation, things like that."

But after repeat visits she noticed that many victims lived near electrical transformers mounted on utility poles. Her curiosity aroused, she collaborated with physicist Ed Leeper in matching each of 344 "cases" (victims) with a "control" (nonvictim) born about the same time in the Denver area.[45] Because they considered residential history important, Wertheimer and Leeper noted the parents' address when the child was born as well as a "death" address—the victim's residence at the time of diagnosis and the matched control's home on the same date. At each address, the researchers sketched the configuration of nearby power lines and transformers, noted the voltage of the distribution wires feeding the transformer, measured the distance from the pole to the house, and coded the residence as "high-current configuration" (HCC) or "low-current configuration" (LCC).[46]

Published in 1979 in the *American Journal of Epidemiology*, their analysis revealed a moderate but distinct association between electric current and cancer, with cases more common where the wiring configuration predicted higher-than-average currents. Moreover, victims were more likely than nonvictims to reside near substations (large ground-based transformers), expected to have the highest currents, and less likely to live near the local circuit's end pole (farthest from the transformer), assumed to have the lowest currents. Different tabulations of the data demonstrated a risk of childhood cancer higher by a factor of two or three in HCC homes, but the researchers could only speculate that magnetic fields were somehow involved, directly or indirectly.[47]

However ill-defined, the possibility of a causal link between electromagnetic fields and cancer triggered a barrage of studies, ranging from laboratory experiments with mice to statistical analyses of mortality among power-company and telephone workers. In addition to transmission lines, researchers looked at subways and high-speed electric railways as well as electric blankets, electric shavers, hair dryers, and other small appliances used close to the body. Although the titles of their reports suggest a broad concern with electromagnetic fields, which have separate electrical and magnetic components, most investigators focused on magnetic fields (related to current) rather than electrical fields (related to voltage). Power lines, electric motors, and Mother Nature produce both types of fields, but only magnetic fields penetrate the body.[48]

Driving the research was the notion that electric currents acting like a magnetic seesaw might interfere with genetic material or make cells more vulnerable to real carcinogens. Although the earth's magnetic field is generally stronger, it is also comparatively constant, whereas the alternating current in power lines and home appliances reverses its magnetic field sixty times a second—an extremely low frequency (ELF) field in comparison to most electromagnetic phenomena. But hundreds of studies and dozens of authoritative reviews have yet to demonstrate a consistent, definitive link between ELF fields and cancer. And even though medical and technical experts judge the results "inconsistent and inconclusive," they typically call for "prudent avoidance" and better data.[49]

Always in search of a good story, journalists are easily intrigued by tenuous research findings with ominous implications. Paul Brodeur, a novelist and correspondent for the *New Yorker*, saw the ELF controversy as not only a public hazard but a damage-control effort by electrical utilities and government scientists.[50] His book *The Great Power-Line Cover-Up: How the Utilities and the Government Are Trying to Hide the Cancer Hazard Posed by Electromagnetic Fields* (1993) followed his earlier exposé *Currents of Death: Power Lines, Computer Terminals, and the Attempt to Cover Up Their Threat to Your Health* (1989), which had also implicated magnetic fields around cathode ray tubes. Brodeur is no stranger to uncovering dastardly plots involving electromagnetic radiation: a decade earlier, in *The Zapping of America: Microwaves, Their Deadly Risk, and the Cover-Up* (1977), he warned of a government-industry conspiracy to conceal the hazards of microwave ovens and telecommunications systems.

Although *The Great Power-Line Cover-Up* contains no illustrations, maps figure prominently in Brodeur's narrative. In particular, his account of an acrimonious public meeting in Guilford, Connecticut, focuses on a map presented by a state epidemiologist who had plotted the homes of twenty-nine town residents with eye, brain, or nervous system tumors diagnosed between 1968 and 1988.[51] Introduced to show that cancers were not clustered within the town, the map was viewed differently by local residents concerned about tumors along a high-current power line leading to a local substation. Although state officials would display the same map two days later at a Rotary Club gathering, they refused a local news reporter's request for a copy. By mistake, though, a health department employee gave a copy to another newspaper, which superimposed the tumor locations on a power-company map and found a dramatic geographic association between

cancer and distribution lines. Although public health and utility officials vigorously denied a meaningful link, the news map suggested
both a causal connection and a cover-up.

In a chapter entitled "Some Red Dots on a Map," Brodeur presents
another cartographic correlation, discovered when Doris Buffo, an
elementary school teacher in Fresno, California, plotted the work locations of colleagues who had died of cancer or been treated for tumors
over the previous ten years.[52] Using an overhead projector, Buffo
enlarged a floor plan of the school—a sprawling one-story structure
with four "pods"—onto a sheet of poster board.

> At the end of the day, she took the poster board home and affixed red
> circular stickers to it to mark the workplaces of cancer victims. When she
> finished, there were seven red circles at various locations in Pods A and
> B, and three circles in the administrative offices just behind Pod A. The
> next morning, she took the diagram to school and showed it to her
> teacher's aide, who reminded her that one of her fifth-grade pupils had
> developed a brain tumor and had died of it in 1986. At recess, she
> brought the diagram into the faculty room and placed it on a table. "All
> the teachers who saw it were shocked by what it showed," she remem-
> bers. "Almost everybody said that we should do something about it."[53]

What shocked Buffo's coworkers was a striking similarity between
the pattern of red dots and a recent report by power-company personnel who had measured higher-than-average magnetic fields in the
part of the school closest to a high-voltage line. A week later, the teachers presented the map at a school board meeting, to which the utility
and the county health department were also invited. A power-company representative maintained that ELF fields within the school were
weaker than those next to most appliances, but he had no explanation
for the clusters of dots. Health officials, who promised a detailed study,
were openly skeptical—case clusters need not be statistically significant or medically meaningful, they noted. For many connected with
the school, though, Buffo's map was more than sufficient evidence:
fourteen teachers requested transfers, and parents demanded that the
school be closed.

Brodeur's tale of a health department reluctant to recognize a problem and a utility willing to make only inexpensive alterations in the
immediate vicinity of the school reads more like bureaucratic foot-
dragging than conspiratorial cover-up. But by weaving selected epidemiological findings into the saga of a power company compelled by
public opinion to modify its distribution system, Brodeur also builds a

case that science and industry really do know that power lines cause cancer.

Not so, argue critics who call his evidence one-sided and accuse him of manufacturing fear. In an *Atlantic Monthly* article on public anxiety generated by the ELF-field controversy, journalist Gary Taubes summarizes the puzzlement, if not outrage, of many scientists and engineers.

> For whatever reason—perhaps the influence of Hollywood—Americans have a marked willingness to distrust any scientist who professes to know better than the lay experts. The best place to go for scientific expertise, however, is often the scientific experts. What makes Brodeur's conspiratorial interpretation all the harder to understand is the level of alarm raised compared with the maximum possible level of the threat. Scientists on both sides of the issue say that they are dealing at the very most with rare diseases and an increased risk that is almost infinitesimal—especially compared with all the other risks of everyday life, from driving and smoking to choice of diet.[54]

In "Currents of Misinformation," a witty attempt to debunk Brodeur's exposés, conservative science writer Michael Fumento blames the media for "[t]errifying the hell out of readers or viewers with talk of children with 'unexplained' brain cancer when in fact *all* brain cancer is unexplained."[55] And even if a link is eventually demonstrated, he argues, the risk to the general population is simply too small to justify the enormous cost of "erring on the side of caution."

Fear and uncertainty about ELF fields indeed have a price—over \$1 billion a year, according to Resources for the Future analyst Keith Florig.[56] In a "Policy Forum" piece in *Science*, Florig identified six economic impacts of the ELF controversy: (1) increased costs to customers unable to import cheap power because of delays in constructing new transmission lines, (2) devalued property along power-line routes, (3) out-of-court settlement of tenuous lawsuits, (4) costly retrofitting and redesigning of facilities to avoid future litigation, (5) modifications to reduce magnetic fields at "existing ELF hot spots" (like the Fresno school), and (6) more expensive "low field" consumer appliances of dubious value.

Another cost is research. Even though people living along power lines have lower exposures than utility workers, electric companies fear not only scientific validation of a causal link between ELF fields and cancer but also judicial confirmation that property owners' fears, however ungrounded, merit compensation.[57] No wonder, then, that the Electric Power Research Institute (EPRI) and other industry consortia, motivated by an unavoidable mix of economic, engineering, ethical, health, and legal concerns, sponsor most research on the possible health effects of magnetic fields.[58] Despite muckrakers charging cover-up, the industry needs as much as anyone to understand the nature and areal extent of whatever risk power lines impose on their neighbors.

Regulatory agencies also want more information. In 1993, for instance, New York's regulators ordered the state's power companies to find out how many people live sufficiently near primary transmission lines to be exposed to above-average magnetic fields. Triggering this request was a new study from Sweden, where researchers had found childhood leukemia rates several times higher than average for children who had lived within 300 meters (660 feet) of a 220 or 400 kV (kilovolt) power line.[59] In response, the Empire State Electric Energy Research Corporation (ESEERCO), a consortium of utilities, engaged a team of geographers and epidemiologists at Rutgers University to plot the routes of 345 kV and higher transmission lines on a detailed electronic map of block-level census data.[60] By spreading buffers of various width outward from the power-line routes, as in Figure 9.7, their geographic information system created long, thin hazard zones reflecting different levels of exposure. By calculating the proportions of each block within the various buffers and relating these calculations to magnetic field strengths measured as representative locations, the Rutgers researchers hoped to make educated guesses about the number of people exposed—or likely to perceive themselves exposed—to ELF fields.

Like the EPA's three-color map of radon zones, intensity bands plotted along power lines are merely a graphic construction—an artificial representation of risk created to promote action or show awareness. But maps of dubious dangers like ELF fields can have real consequences, for instance, making transmission-line corridors even less desirable in the eyes of potential buyers and sparking renewed demands for compensation. Moreover, a cartographic focus on high-voltage lines can be a dangerous distraction. Because the maps portray proximity to transmission lines as the principal threat, viewers can easily ignore symbols representing differences in current flow, height, and wire configuration—exposure factors of no small importance. And like California's earthquake-fault maps, these narrow, linear views of the

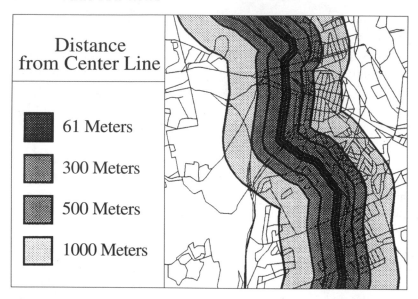

Figure 9.7. Linear buffer zones representing degrees of likely magnetic-field exposure along a power-line route. Courtesy of Daniel Wartenberg, Environmental and Occupational Health Sciences Institute, Piscataway, N.J.

ELF-field hazard too readily imply that if anyone is at risk, it is people living near high-voltage lines. But voltage alone, we know, has little to do with magnetic fields. If New York's regulators are serious about estimating the magnitude of the problem, their research won't end with primary transmission lines.

Chapter Ten

Nuclear Nightmares

Once promoted as a clean alternative to fossil fuel, nuclear power has proved an albatross for investors, ratepayers, and regulators. Skyrocketing construction costs and unforeseen operating expenses discredited claims of efficiency, while embarrassing shutdowns and bureaucratic handwringing over waste disposal exposed "clean energy" as an oxymoron. Adding to public fears of the peaceful atom were belated revelations of deliberate and accidental "releases" at weapons plants exposing workers and nearby residents to dangerous radiation as well as contaminating soil and groundwater. Despite different goals and processes, warhead manufacture and power generation contribute huge amounts of "high-level" waste to an escalating disposal problem for which federal energy officials seek a common solution. Equally dicey is the much larger volume of low-level radioactive waste generated by hospitals, research laboratories, and (principally) nuclear plants. Although nuclear power advocates claim that radioactive waste is largely a political problem, horror stories of radionuclides "migrating off site" reflect the vulnerability to human error and incompetent management of an enormously potent yet inherently dangerous technology. Politics might constrain the solution but it's hardly the cause.

Do I sound pronuclear or antinuclear? I've been called both by true believers on opposite sides of the issue—pronuclear for believing that nuclear power (including waste disposal) can be made to work well and antinuclear for criticizing the industry and its regulators. Alarmed by the incompetent use of geographic analysis in siting nuclear waste dumps, I seldom side with the advocates.[1] Equally alarming, though, is the American consumer's prodigious appetite for energy and the nation's growing reliance on uncertain imports from oil-rich developing nations. As I see it, whatever atomic power foes can do to promote the safe exploitation of nuclear energy serves the long-term national interest. These biases underlie what follows.

Because fear is so much a part of the nuclear power/nuclear waste controversy, I begin this chapter with an examination of how attitudes toward risk and uncertainty explain the public's uneasy tolerance of nuclear technology. To show that these fears are not wholly irrational, I describe two devastating nuclear accidents in the former Soviet Union—the massive airborne release of radiation in the Urals during the winter of 1957–58, an event largely ignored by the media, and the 1986 Chernobyl disaster, which spread fallout and panic across much of Europe. Next, an examination of emergency planning zones surrounding American nuclear plants looks at how federal regulators deal with the possibility of a domestic nuclear disaster, whether arising from flawed design and construction, faulty operation, or terrorist attack. Later sections address evacuation maps and flawed efforts to find safe storage and disposal sites for radioactive waste.

Fear of things nuclear is best understood by comparing nuclear weapons, nuclear reactors, and nuclear waste with other hazards involving machines or substances. Paul Slovic, a psychologist interested in risk perception, devised a concise, revealing comparison by asking groups of lay people (college students, League of Women Voters members) to rate eighty-one hazards on eighteen risk characteristics.[2] In addition to several nuclear categories, the hazards listed on his survey ranged from aspirin, power mowers, and elevators to handguns, nerve gas accidents, and DNA technology. Each characteristic was a word pair, such as *controllable/uncontrollable*, *voluntary/involuntary*, or *effect delayed/effect immediate*. Because of similarities in the responses, Slovic and his colleagues were able to collapse their eighteen-dimensional space to the two-dimensional "cognitive map" in Figure 10.1. For clarity, I've labeled only twenty-six of the eighty-one hazards on Slovic's original drawing, which was much larger.

Although different in appearance from geographic maps showing roads and boundaries, cognitive maps are easy to read because proximity represents similarity. For example, *caffeine* and *aspirin*, not far apart on the cognitive map, have similar perceived risks. In contrast, lay people regard *nuclear weapons (war)* as far more ominous than *caffeine*, which lies at the opposite end of a coordinate axis contrasting catastrophic, dreaded, involuntary hazards holding high risks for future generations with individual, mildly threatening, voluntary hazards imposing low risks on our great-grandchildren. Notably, the other nuclear hazards (*nuclear reactor accidents*, *nuclear weapons fallout*, and *radioactive waste*) also score highly on Slovic's dreaded-risk scale, for which only one other hazard (*nerve gas accidents*) has a comparably high score. The second axis, which contrasts new, unknown hazards having delayed effects with old, well-known hazards having immediate effects, reflects radically different perceptions of *laetrile* (a drug touted to cure

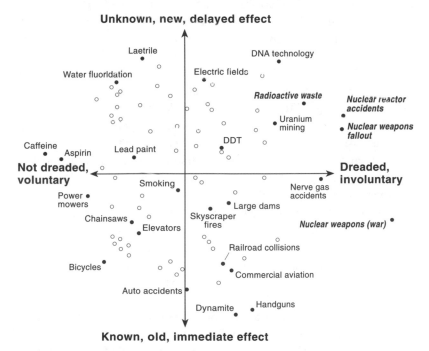

Figure 10.1. Cognitive map of risk perception. Condensed and redrawn from Paul Slovic, "Perception of Risk," *Science* 236 (1987): 280–5; fig. 1 on p. 282.

cancer) and *dynamite*. In general, chemicals and pharmaceuticals collected high scores on the unknown-risk dimension. Among the nuclear hazards, *radioactive waste*, *reactor accidents*, and *fallout* have higher unknown-risk scores than *nuclear weapons (war)*, which are rated similar to *chainsaws*.

Slovic's most revealing finding was a distinctly different cognitive map for technical experts (engineers, scientists) who rated the same hazards. Trained to use and value quantitative measurements, these experts hold a largely one-dimensional view of risk reflecting expected annual mortality rates. Nuclear hazards thus placed well below motor vehicle accidents, smoking, alcoholic beverages, and handguns on their cognitive map—around train wrecks and food coloring, but above skiing and power mowers. No wonder, then, that the technological intelligentsia are often at odds with less (or more broadly) educated folk unwilling to extrapolate an impressive but comparatively short nuclear safety record into the uncertain future. Less impressed by meager body counts, thoughtful nonexperts interpret events such as the 1976 Three Mile Island nuclear accident, which seems not to have killed anyone, as signal reminders of plausible disasters of unprecedented impact.[3] According to Slovic, accidents involving hazards perceived to be above average on both dread and unknown risk (that is, hazards in the upper-right quadrant of Figure 10.1) have "high signal potential." Society acts swiftly and firmly to take control and impose added restrictions in the aftermath of a Chernobyl, a Love Canal, or a TMI.

Besides having a different concept of risk, the public harbors a profound suspicion of nuclear experts. Historian Spencer Weart traced this mistrust to medieval notions of *transmutation*, the elusive goal of sorcerers and alchemists seeking wealth and power.[4] As a technology capable of altering matter, radiation inherited an image of death rays and mad scientists shortly after its discovery in the late nineteenth century. By the 1930s the possibility of nuclear weapons was apparent, and by the 1960s Hiroshima and the cold war had demolished hopes for the primarily peaceful use of atomic energy. Moreover, unfavorable stereotypes of military officers and other authorities suggested, in Weart's words, that "the dreaded energy lurked not only within matter but within the human soul."[5]

Surveys indicate that dread of atomic energy differs little from anxiety over atomic weapons: despite lack of a mushroom cloud, popular images of an accident at a nuclear power plant include fallout and radiation poisoning similar to nuclear war.[6] Moreover, efforts to placate these fears with quantitative risk assessments—numbers that say what most people believe will happen won't happen—only further erode public confidence in nuclear authorities.[7] Once fixed in the national psyche, distrust can be an insurmountable political obstacle that dissipates slowly, if at all.

Fueling public anxiety in the 1960s and 1970s was an emerging realization that the experts had understated the dangers of nuclear technology. Accidents not only happened but the consequences were often bizarre. In 1961, for instance, a maintenance technician at the Atomic Energy Commission's (AEC) National Reactor Testing Station, in Arco, Idaho, near Idaho Falls, was impaled on the ceiling after he and two coworkers attempted to dislodge a stuck control rod by pulling on it.[8] When the stuck control rod suddenly jerked upward eighteen—not four—inches, the reactor "went supercritical," shot the control rod through a hapless worker's groin and into the roof, contaminated the containment (reactor building) with radioactive steam, and killed the other two technicians.[9] But the most gruesome consequence occurred several days later, when a decontamination team placed the clothed parts of the three workers' bodies in lead-lined coffins after severing their highly irradiated heads and hands for disposal as high-level radioactive waste.

More ominous was a radiological accident in the southern Urals, in which one or more explosions of high-level radioactive waste during the winter of 1957–58 contaminated an area of 1,000 km² (400 mi²). Unlike the Idaho incident, which American newspapers reported two days later, the Urals catastrophe was covered up for nearly two decades. The story broke in 1976, when the *New Scientist*, a British popular science magazine, published an article on dissident Soviet scientists by Zhores Medvedev, a Russian geneticist living in London. Unaware that the Urals tragedy was not widely known, Medvedev briefly described the appalling devastation.[10]

> For many years nuclear reactor waste had been buried in a deserted area not more than a dozen miles from the Urals town of Blagoveshensk. The waste was not buried very deep. . . . Suddenly there was an enormous explosion, like a violent volcano. The nuclear reactions led to heating in the underground burial grounds. The explosion poured radioactive dust and materials high up into the sky. . . . It was difficult to gauge the extent of the disaster immediately, and no evacuation plan was put into operation right away. Many villages and towns were only ordered to evacuate when the symptoms of radiation sickness were already quite appar-

ent. Tens of thousands of people were affected, hundreds dying, though the real figures have never been made public.[11]

Although Soviet authorities had kept the explosion and evacuation out of the mass media, Russian wildlife biology journals reported extraordinarily high levels of irradiation in the area. As Medvedev noted, the Soviets had established several biological research stations on the fringe of the contaminated area to study the effects on plants and animals.

Nuclear power advocates like Sir John Hill, head of Britain's Atomic Energy Authority, attacked Medvedev's reports as "science fiction" and "rubbish."[12] But corroboration emerged in a report released by the U.S. Central Intelligence Agency as well as in a letter to the *Jerusalem Post* by Professor Lev Tumerman, a Soviet émigré who described a 1960 visit to the region between Sverdlovsk (now Yekaterinburg) and Chelyabinsk.[13]

> About 100 kilometers [60 miles] from Sverdlovsk a road sign warned drivers not to stop for the next 30 kilometers and to drive through at maximum speed.
>
> On both sides of the road as far as one could see the land was "dead": no villages, no towns, only the chimneys of destroyed houses, no cultivated fields or pastures, no herds, no people . . . nothing.
>
> The whole country around Sverdlovsk was exceedingly "hot." An enormous area, some hundreds of square kilometers, had been laid waste, rendered useless for a very long time, tens or perhaps hundreds of years.
>
> I was later told that this was the site of the famous "Kyshtim catastrophe" in which many hundreds of people had been killed or disabled.[14]

As Medvedev pointed out in a more detailed account the *New Scientist* published several months later, the structural damage Tumerman observed was not a direct result of the blast but a strategy to discourage evacuees from returning.[15]

Scientists at Oak Ridge National Laboratory also confirmed the catastrophe.[16] A systematic analysis of the Soviet ecological and nuclear literature as well as topographic maps published between 1936 and the mid-1970s verified widespread radiation contamination and population resettlement near Kyshtim. A map in the Oak Ridge report (Figure 10.2) described a large L-shaped area where thirty small communities had disappeared, lakes were not restocked with fish, and new canals diverted rivers and streams around contaminated lakes and reservoirs. Although a serious nuclear accident or series of accidents was obvious, the Oak Ridge experts were skeptical of the vague casu-

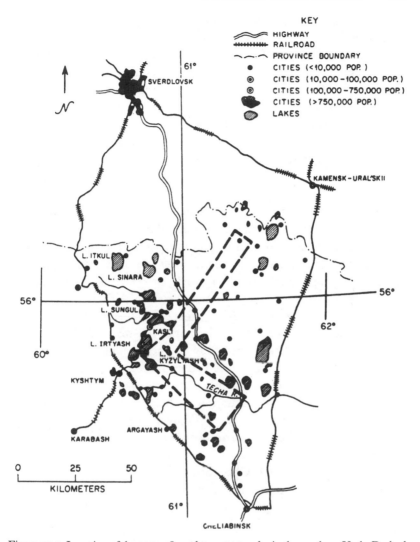

Figure 10.2. Location of the 1957–58 nuclear catastrophe in the southern Urals. Dashed line identifies area where thirty towns and villages disappeared from Soviet topographic maps. Reprinted from J. R. Trabalka, L. D. Eyman, and S. I. Auerbach, *Analysis of the 1957–1958 Soviet Nuclear Accident*, ORNL report no. 5613 (Oak Ridge, Tenn.: Oak Ridge National Laboratory, 1979), p. 2.

alty reports and Medvedev's suggestion that plutonium-rich liquid waste had dried out and gone critical.[17] A devastating explosion did not require nuclear criticality, they argued, and numerous other scenarios could account for a catastrophic dispersal of radioactive waste. After all, high-level waste was often thermally hot as well as radioactive, and a conventional explosion was possible if an airtight storage tank's cooling system failed. Americans had little to worry about, though, because most scenarios plausible for the Soviet nuclear program "would not be credible in the light of U.S. practice or experience."[18]

However diligent the Soviets' effort to keep their nuclear mishaps out of the news, nothing could hide the monstrous radioactive cloud spreading westward across Europe after a three-year-old reactor, unit 4 at the Chernobyl Nuclear Power Station, eighty miles north of Kiev, exploded shortly after midnight on April 26, 1986.[19] By the morning of April 28, easterly winds had carried the fallout through northeast Poland, across the Baltic Sea, and into Sweden, where day-shift workers arriving at the nuclear plant in Forsmark, north of Stockholm, set off alarms when they walked through portal monitors with radioactive dust on their shoes. After similar reports from other nuclear plants, defense officials collected air samples over the Baltic, meteorologists traced the plume back to Chernobyl, and Swedish diplomats in Moscow pressed the Soviets for an explanation. Although analyses of air and dust betrayed the release of most of the fission products of the fuel core (the reactor vessel), Soviet authorities conceded only that a reactor was on fire.[20] Not until May 6, when fire fighters had nearly extinguished the intense graphite fire in the reactor core, did *Pravda* provide a detailed account praising the heroism of the Chernobyl fire brigade and the efficient evacuation of the nearby town of Pripyat.[21] Irked that the Western media were "enjoying the USSR's misery," the Soviet press ignored the government's failure to warn nations in the path of the fallout.

How serious was the Chernobyl catastrophe? The short-term effects, put together over the few weeks following the accident, are grim—31 known dead, 300 persons hospitalized for radiation exposure, more than 135,000 residents evacuated from an eighteen-mile zone around the plant, and the contamination of thousands of square miles of agricultural land—yet less calamitous than the 1984 Union Carbide chemical-plant disaster in Bhopal, India, where more than 2,500 died.[22] Nonetheless, Chernobyl is clearly the worst civilian nuclear accident on record.

The long-term consequences are equally daunting. Burning for ten days, the reactor spewed out fifty tons of evaporated nuclear fuel, the equivalent in long-lived radionuclides of ten atomic bombs of the size dropped on Hiroshima.[23] Four years later, after systematic soil testing and mapping revealed dangerous levels of contamination well beyond the eighteen-mile radius, authorities announced plans to evacuate thousands of additional people in Russia, Ukraine, and what is now Belarus.[24] And because the effects of radiation exposure are cumulative, Chernobyl's fallout will have lingering health effects throughout Europe.[25] Although mortality projections are uncertain and contentious—remember the debate over low-exposure effects of indoor radon?—epidemiologists expect between several thousand and a half million excess deaths over the next half century from cancer, thyroid tumors, and impaired immunity.[26] Even without the characteristic mushroom cloud, a worst-case nuclear accident can be as devastating as a small-scale nuclear attack.

Ukraine's misfortune is not a cause for smug complacency. Despite the U.S. nuclear industry's good, perhaps excellent, safety record, sporadic news reports of lax security make a domestic Chernobyl-style disaster the plausible result of terrorist sabotage, similar in intent to the World Trade Center bombing of 1993.[27] More troubling, America's nuclear plants have had their own close calls, notably the 1979 accident in which equipment failure and operator error led to a partial core meltdown at the Three Mile Island Unit 2 nuclear power plant, in Middletown, Pennsylvania, about 11 miles from Harrisburg, the state capital (Figure 10.3).[28] Despite a lack of identifiable casualties, the TMI-2 incident underscored the fragility of nuclear power and the need for emergency planning.

The incident at Three Mile Island began around 4 A.M. on Wednesday, March 28, when maintenance workers cleaning a clogged pipe accidentally shut off water to the reactor's cooling system.[29] Emergency pumps started automatically, but as water surged into the already overheating reactor, a pressure relief valve popped. Instead of closing a few seconds later, the valve jammed open, draining water from the reactor. A back-up system with its own pumps and water supply couldn't help—someone had disconnected it a few days earlier. Another safety system with its own pumps kicked in, but because instruments erroneously suggested the reactor was getting too much water, control-room operators shut off its pumps and opened a drain line. The reactor grew hotter, vaporized its remaining water, and grew hotter still—well above 4,000°F in the upper part of the core—cracking fuel rods and releasing radioactive material into the containment. Another safety system then pumped thousands of gallons of radioactive water into an adjacent building, insuring a lengthy, costly cleanup.

At 7:24 A.M., after hasty telephone conferences with company officials and the Pennsylvania Emergency Management Agency, plant managers declared a "general emergency"—the first ever at a U.S. commercial nuclear-power plant—and officials of the Nuclear Regulatory Commission (NRC) rushed to the scene. With limited guidance from monitoring systems not designed for this kind of accident, nuclear experts lacked a clear picture of conditions inside the containment. Anxious uncertainty gave way to impending doom on Friday morning, March 30, when NRC staff believed a "hydrogen bubble" growing near the top of the containment might explode, expose the core, and release a radioactive cloud. At 12:30 P.M., Pennsylvania governor Richard Thornburgh recommended voluntary evacuation by pregnant women and preschool children living within 5 miles of the plant. Officials advised everyone else within 10 miles to remain indoors. By then, though, many residents had already left or were leaving, in a voluntary exodus marred only by understandable tensions at hotels, shelters, and private homes at which friends and relatives arrived unexpectedly for an uninvited stay of unknown duration. Spontaneous evacuation is a typical response to portentous industrial accidents, and surveys indicate 60 percent of people living in a 5-mile zone around TMI-2 evacuated, with 50 and 30 percent of residents, respectively, vacating concentric rings 5–10 and 10–15 miles away.[30]

Reconstruction of the accident revealed that an evacuation would have been far more timely two days earlier, when the reactor was thirty to sixty minutes from a full meltdown.[31] Had the NRC and state emergency management officials realized what was happening inside the core, Thornburgh would probably have ordered a demographically broader evacuation on the morning of March 28. Whether his order would have been geographically wider is dubious, though, because existing plans covered only a 5-mile circle around the plant.[32] Yet as the crisis escalated on Friday morning, the NRC first advised state officials to prepare evacuation plans for a radius of 10 miles, and later enlarged the zone to 20 miles (Figure 10.3).[33] By Sunday, though—April Fools' Day—nuclear experts realized that the hydrogen bubble was much smaller than originally thought and was unlikely to explode. Two days later the NRC declared the situation stable, and by April 5 most evacuees had returned. On the eighth Thornburgh withdrew his evacuation advisory and reopened local schools.

Few issues raised by TMI-2 were as contentious as the coordination of emergency planning and the size of the evacuation zone. The high-level commission of politicians and scientists appointed by President Carter to investigate the accident "found an almost total lack of detailed plans in the local communities around Three Mile Island"[34] and blamed a "low priority" for emergency planning on the NRC's confidence in technical safeguards and "desire to avoid raising public concern about safety."[35] In the 1960s and 1970s, when most nuclear power plants were built, licensing standards required off-site emergency planning for a "low population zone" (LPZ) with a radius of only 2 to 3 miles, depending on the design of the reactor. According to federal rules, the "design-base accident" for the TMI-2 plant called for an emergency-planning radius of only 2 miles, but Pennsylvania regulations had increased the zone to 5 miles.

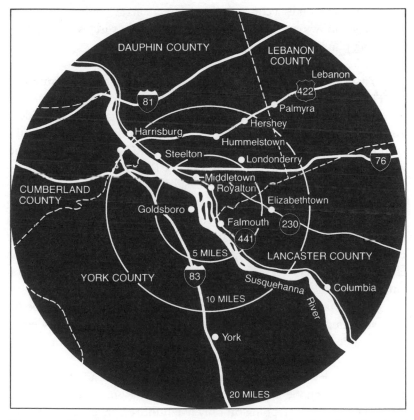

Figure 10.3. Evacuation zones around the Three Mile Island nuclear plant. Reprinted from John G. Kemeny and others, *Report of the President's Commission on the Accident at Three Mile Island: The Need for Change—the Legacy of TMI* (Washington, D.C., 1979), p. 142.

Need for a much larger radius was apparent on March 30, 1979, when NRC experts pondering various accident scenarios considered evacuating areas beyond 20 miles.[36] Focusing attention on a narrow area directly downwind from the plant might seem logical, but shifting winds and uncertainty about time of release dictate circular planning areas. Even in ad hoc planning, no one knows precisely when a release will occur and where the plume will drift. At Three Mile Island, for instance, between noon and early Friday evening, wind speed increased from 6 to 8 mph as movement shifted from west-to-east to southeast-to-northwest—directly toward Harrisburg.[37] Other weather elements affect the extent and shape of the fallout zone. In comparatively stagnant air or a steady downpour most of the radiation might settle out within a few miles of the plant, but if the wind becomes stronger and assumes a constant direction, a narrow, mile-wide plume might lash out with deadly radioactivity as far as 75 miles.[38] In general, a long plume is a narrow plume.

Washington's decision to expand emergency planning around all U.S. nuclear plants was more than expedient politics—the Nuclear Regulatory Commission was well aware that a nuclear accident could have devastating effects far beyond 2 or 3 miles. Three months before TMI-2, in December 1978, a joint NRC-EPA task force had called for two emergency planning zones (EPZs): a 10-mile *plume exposure pathway*, in which residents not evacuating promptly might receive deadly doses of cesium 137 and iodine 131, and a 50-mile *ingestion exposure pathway*, in which a radioactive cloud could contaminate milk, field crops, and soil.[39] Timely evacuation of all or part of the 10-mile zone required coordination of federal, state, and local governments as well as sirens, shelters, detailed maps, periodic drills, and a public information campaign. Although evacuation beyond 10 miles might be necessary, the 50-mile planning zone was largely an agricultural quarantine, to keep strontium 90 out of the food chain. Because dumping milk seemed less complicated than getting people to move out, not freak out, debate focused on the adequacy of the new 10-mile EPZ.

Why 10 miles? If we counted on twelve fingers, instead of ten, would nuclear accidents be any more devastating? If Roman surveyors had gauged distance by stepping off only five hundred paces, instead of an even thousand, would the task force have proposed a 20-mile EPZ? Had the U.S. joined the rest of the developed world in adopting the metric system, would a 10- or 25-kilometer zone (with radii of 6.2 and

15.5 miles, respectively) have been politically acceptable? Why not a larger, safer round-number radius, such as 20, 25 or 50 miles? Bigger is safer, right?

Perhaps, but how much safer and with what trade-offs? Doubling the planning radius would quadruple the planning area, thereby increasing the cost of planning and spreading limited resources more thinly. But would the added security of preparing for a comparatively rare accident warrant extending emergency planning out to 20 miles? The answer, if there is one, is not simple. Even experts who understand how nuclear technology is supposed to work have trouble agreeing on how reactors might fail. Yet as Richard Pollack, director of the Critical Mass Energy Group, told a May 1979 congressional hearing, uncertainty need not mean a stalemate.

> Members of the NRC/EPA task force . . . were divided. Some said, "Perhaps 5 or 8 miles would be sufficient." Others said 25 miles. So, they compromised at an even number of 10 miles. I think it is interesting that this was not the basis of any type of statistical review, but rather this was negotiations between different members about what kinds of policies should be undertaken.[40]

Neither too high, nor too low, 10 miles offered a middle course for reassuring the public without raising anxiety—and in the case of Three Mile Island, reaching toward but not engulfing Harrisburg.

Another factor was money. According to Pollack, the task force chose to "look at what, financially, one would have to invest in order to truly protect the public, and some [members] have concluded it would be such an economic albatross around the neck of the nuclear industry as to not be a viable form."[41] What's more, a radius larger than 10 miles would further antagonize industry groups alarmed that the new regulations, by mandating cooperation of local governments within the planning zone, might grant obstreperous municipalities a de facto veto over the continued operation of existing plants.[42]

However tainted by economic necessity and faith in safeguards, the 10-mile radius also reflects a systematic attempt to identify specific accident sequences, estimate relative likelihood, and establish a "design basis accident." Each accident sequence has a *source term*, the nuclear risk-assessment jargon for "the quantity, timing, and characteristics of the release of radioactive material to the environment following a core melt accident."[43] In a typical accident, an "initiating event" such as a steam-line break or fire triggers a chain of failures of plant personnel, safety systems, or structures. To cope with the huge number of unique sequences, risk assessors plot an "event tree," in which multiple branches diverging from each new event answer the question What can go wrong next? Estimating the chance of rare events requires tenuous assumptions, but once estimated, individual probabilities enable experts to compute odds for specific accident sequences. To estimate off-site consequences, the analyst then considers the amount, half-life, and vaporizability of fission products available for release as well as meteorological data and the probable condition of the reactor vessel and containment.[44]

Each of several reactor designs has its own event tree and design basis accident—a catastrophic accident, to be sure, but not the very worst possible sequence.[45] This numerically based educated guessing has guided a variety of design, siting, licensing, and policy decisions affecting nuclear power, including the older 2- to 3-mile low population zones.[46] In 1975 the Rasmussen Commission issued refined source terms, which not only allowed the NRC to claim that motor vehicles are 100,000 times more deadly than nuclear reactors but also helped the NRC-EPA emergency planning task force compromise on a 10-mile radius.[47] Paradoxically, the Rasmussen study team assumed a maximum plume length of 15 miles—a figure deemed *too low* by nuclear engineer Richard Webb, who criticized the commission for discounting comparatively rare yet nonetheless plausible combinations of severe weather.[48]

Skeptics who consider prevailing source terms incomplete and simplistic question the reliability and alleged objectivity of nuclear risk assessment. The Union of Concerned Scientists, which challenged a 1985 NRC attempt to alleviate earlier shortcomings, compiled a lengthy list of conceptual and procedural flaws, including inadequate treatment of seismic hazards at individual plants and failure to consider deliberate destruction by disaffected employees or terrorists— what better way to get back at the company, get into the history books, or make a memorable political statement.[49] Particularly troubling is the possibility that vandals or saboteurs with a vehicle bomb and knowledge of plant operations could initiate a full core meltdown inside a breached containment.[50]

Not even the 1986 Chernobyl accident, with an evacuation radius nearly twice the current, post-TMI U.S. standard, could persuade the Nuclear Regulatory Commission that 10 miles might be either too short or too shortsighted. The Soviet catastrophe could not be ignored, of course, and emergency planning was one of several safety and preparedness issues the commission addressed in a report on the implications of Chernobyl.[51] Acknowledging that the "most severe and most unlikely accidents" used to justify the 10-mile radius "involve releases of radioactivity comparable in magnitude to that which was actually released at Chernobyl," the NRC disputed the need to base emergency planning "solely on a single highly unlikely event, such as the worst case."[52] After endorsing earlier source-term studies, the report wrapped up discussion of the emergency planning radius with a bureaucratically concise disclaimer: "Consequently, a release magnitude similar to the one associated with Chernobyl and the possibility that ad hoc actions beyond the planning zone boundaries might be needed for very unlikely events were considered and have been factored into the development of U.S. requirements, including the sizes of the EPZs."[53] However arrogant and defensive the NRC position, the politically inspired, scientifically informed muddling-through of nuclear risk assessment is little different from the compromise inherent in planning for earthquakes, coastal storms, and other natural hazards. Nuclear accidents, a new and rare disaster genre, simply allow less fine-tuning than earthquakes or flooding.

Like maps in general, evacuation plans are at least partly rhetorical—authoritative statements that government is prepared and concerned, and that if an emergency occurs, residents must do as they're told. And because preparedness demands accurate, accessible information, the cartographic rhetoric must be clear and coherent as well as convincing. After all, an expanded, 10-mile EPZ is meaningless if area residents cannot quickly figure out who should leave, where they should go, and by what route. Equally important is periodic dissemination within and perhaps a bit beyond the emergency planning zone in a format not readily discarded.

FEMA, not the NRC, oversees off-site emergency planning for nuclear plants.[54] Oversight includes testing alert systems, reviewing detailed evacuation plans, and making certain the plant operator sends residents of the plume exposure EPZ up-to-date emergency information at least once a year. According to FEMA guidelines, these yearly communications should discuss the possible effects of a radiological accident, explain the local warning system (sirens, Emergency Broadcast System radio and TV stations), describe precautionary responses (remaining indoors, evacuation) and special provisions for schoolchildren and the handicapped, and map the evacuation zone and routes to relocation centers.[55] But aside from a few rules reflecting conventional cartographic wisdom—for example, using legible type, showing direction with a compass rose or north arrow, and explaining unfamiliar symbols in a key—the guidelines say little about design and even less about content.[56]

With substantial leeway in preparing emergency information, the utility firm usually hires a corporate communications consultant. Experienced in public relations and graphic arts—but rarely in cartography—the consultant creates a spiffy brochure that tries to tell the public what to do in case of an accident without inciting fears of nuclear power. No wonder, then, that evacuation maps often have the simplistic, glossy look of shareholder newsletters and annual reports. For example, the Vermont Yankee Nuclear Power Corporation's evacuation zone map (Figure 10.4) uses black and tan inks to show the circular EPZ, town boundaries, and principal highways leading to designated "reception centers"—a euphemism for "relocation centers," a FEMA term evoking images of internment camps for refugees. A circle highlights the evacuation zone, a tiny containment icon at the center pinpoints the plant (labeled VY), and different shades of brown partition the circle among Vermont, New Hampshire, and Massachusetts. East of Interstate 91 and in Massachusetts, the map identifies secondary roads by name, not number, and at a larger scale, separate detail maps show neighborhood streets surrounding the relocation centers. The map occupies one page of an attractive, medium-size wall calendar distributed each year within the EPZ as an insert in the *Town Crier*, a free weekly newspaper.[57] Two other Vermont Yankee publications use the same map: a brochure distributed to local motels contains a smaller version, and a poster distributed by mail to businesses uses a larger version.

Vermont Yankee's maps exemplify flaws widespread among nuclear response materials.[58] In particular, most plants ignore federal recommendations to adjust the generic, circular EPZ for local terrain, wind patterns, settlement, and transport routes.[59] Moreover, nuclear evacuation maps typically omit hospitals, pharmacies, and other locations stockpiling thyroid-blocking agents, which can reduce the effects of inhaled radionuclides. In addition, most public-information maps have vague or incomplete titles, lack a north arrow or other direction indicator, and ignore evacuation routes selected by emergency management officials to minimize traffic delays at narrow bridges and other congested areas.

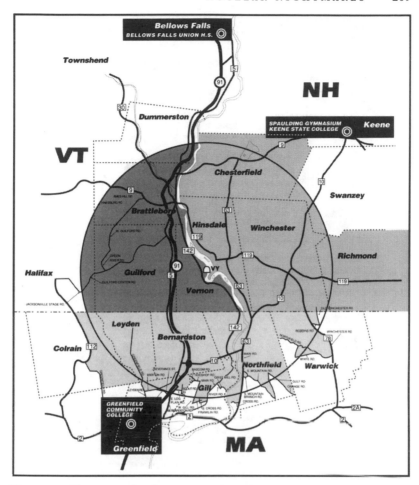

Figure 10.4. Emergency planning zone map from the Vermont Yankee Nuclear Power Corporation's 1995 Emergency Public Information Calendar. Map in calendar is 6.2 inches wide. Courtesy of the Vermont Yankee Nuclear Power Corporation.

To better inform Vermont Yankee's most immediate neighbors, civil defense officials in Vernon, Vermont, developed their own map. As the much reduced black-and-white facsimile in Figure 10.5 illustrates, wide color-coded bands overprinted on a detailed highway map differentiate recommended evacuation routes—not a bad idea given the plant's troubled history. According to an Atomic Energy Commission report, on November 7, 1973, the reactor experienced an "inadvertent criticality incident" after day workers installing a closed-circuit television system cut off power to a primary safety system and failed to

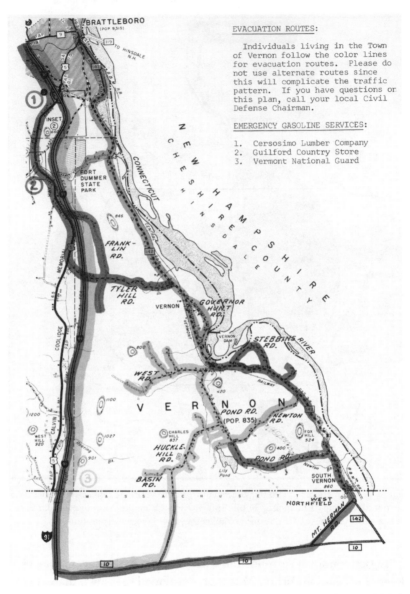

EVACUATION ROUTES:

Individuals living in the Town
of Vernon follow the color lines
for evacuation routes. Please do
not use alternate routes since
this will complicate the traffic
pattern. If you have questions on
this plan, call your local Civil
Defense Chairman.

EMERGENCY GASOLINE SERVICES:

1. Cersosimo Lumber Company
2. Guilford Country Store
3. Vermont National Guard

Figure 10.5. Evacuation-route map for Town of Vernon, Vermont. Hand-lettered labels added to a county highway map identify local roads, and thick lines overprinted in transparent magenta, green, orange, or tan ink identify four evacuation routes leading to Vermont relocation center in Bellows Falls, fifteen miles north of Brattleboro. Reprinted, at a reduced size, from a page-size flyer prepared by the Vernon Department of Public Safety, Civil Defense Division; original map is 6.8 inches wide.

inform the night shift.[60] Because of quick action by the control room staff and an electronic monitoring system, no radiation was released and no workers were exposed. Vermont Yankee has registered other, more recent violations, as in October 1992, when operators failed to shut down the plant when an array of control rods did not move sufficiently rapidly into the reactor core.[61]

FEMA's oversight of dissemination seems almost as loose as its laissez faire attitude toward map design. Although the government requires annual notification of EPZ residents, licensees may select one or more recommended formats: brochures or calendars distributed directly to residents, ads in telephone books, inserts in utility bills, and posters displayed in public places.[62] Even so, many people either don't get the message or forget it too easily. According to a 1986 FEMA survey, only 71 percent of EPZ residents could recall receiving emergency response materials.[63] More ominous is the respondents' wide variation in recall, which ranged from 31 to 84 percent among the fifty-seven nuclear plant sites surveyed. To improve preparedness, FEMA reviewed the public education programs of thirty-five utilities and recommended voluntary improvements to better adapt the emergency maps to the needs of licensees and their neighbors.

The single best vehicle, I am convinced, is the local telephone directory—a format far less likely than a brochure or calendar to be discarded, mislaid, or out-of-date. Most households throw out or recycle their obsolete directories, new residents receive them shortly after moving in, and transients have one in their motel rooms. Even though an emergency broadcast announcer might need to remind residents to look in the phone book, an evacuation-route map like Figure 10.6 is handy and practical. Rochester Gas and Electric Company, which operates the Ginna Nuclear Power Plant on the south shore of Lake Ontario, includes this map in a full-page public service announcement near the back of the Wayne County, New York, telephone directory. Below the bold title "Do you know what to do in case of a nuclear power plant accident?" the ad warns that the steady, three- to five-minute siren blast means turn on the radio, not hit the road. In addition to evacuation routes, the map delineates Emergency Response Planning Areas (ERPAs) local emergency management officials might use for a phased, carefully orchestrated evacuation tailored to a radioactive plume some residents might best avoid by closing windows and staying put for several hours. The ad also encourages residents to

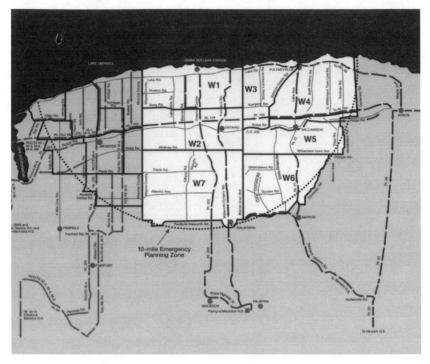

Figure 10.6. Neighbors of the Ginna Nuclear Power Plant in western Wayne County, New York, can find this map of evacuation areas at the back of their local telephone directory. The Rochester telephone book contains a similar map describing evacuation areas and exit routes for the western portion of the ten-mile zone. Courtesy of Rochester Gas and Electric Company.

call or write for RG&E's free, more detailed guide "To Our Neighbors in Wayne County," a large wall calendar distributed annually within the EPZ.

Encouraging residents to stay indoors is an enormous challenge. Most people think invisible radiation quickly blankets the entire area around the plant, rather than moving downwind in a typically narrow or wedge-shaped plume. Neither periodic planning exercises nor emergency information publications adequately address widespread ignorance about radioactive plumes and the folly of simultaneous evacuation.[64] Ex-soldiers and emergency workers conditioned by mandatory drills will probably do as they're told, but civilians are eager to leave, even at the risk of increased exposure. Evacuation maps without response planning areas send a grim warning: Get the hell out, no matter where you live!

In much the same way that a small-scale map of floodplains vague-ly describes the hazards of flooding, the map of commercial nuclear power plants (Figure 10.7) is a generalized portrait of radiological hazards.[65] The picture is incomplete, though, because the country's sixty-seven radiological emergency planning zones ignore military and research reactors, stockpiles of nuclear weapons, repositories of nuclear waste, and radioactive tailings around former uranium mines. Hazardous in various ways, these omissions are not equally risky. After all, research reactors are smaller than power-plant reactors and typically store less nuclear fuel and nuclear waste on site. And even though the 1957–58 catastrophe in the Urals demonstrated that mis-managed high-level nuclear waste can explode or catch fire, launch a radioactive cloud, and devastate a vast area, the principal off-site haz-ard of radioactive waste is contaminated groundwater.[66] Another noteworthy omission is the preferred transport routes for nuclear fuel rods and nuclear waste.[67] Although shipments are infrequent and firms specializing in radioactive transport have an excellent safety record, an accident or terrorist attack in a heavily populated area could be more disastrous than a radiological mishap at a compara-tively remote fixed site.[68]

Emergency Planning Zones at Nuclear Power Plants

Figure 10.7. Commercial nuclear power plants in the United States. Locations repre-sent single reactors or clusters of two or three reactors with a single emergency planning zone. Compiled from data in U.S. Nuclear Regulatory Commission, *Information Digest, 1994 Edition*, NRC publication no. 1350, vol. 6 (1994).

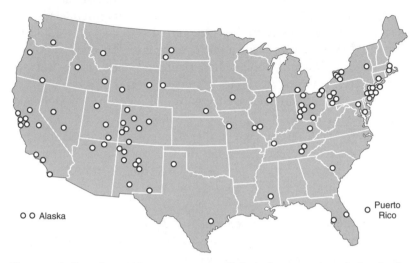

O O Alaska

O Puerto Rico

Figure 10.8. Locations with one or more radiologically contaminated sites in the Department of Energy's thirty-year cleanup program. Redrawn from U.S. Department of Energy, *Environmental Restoration and Waste Management: An Introduction, Student Edition,* DoE publication no. EM-0104 (1993), p. 5.

The geography of nuclear waste in the United States is actually two geographies, one managed by the Department of Energy (DoE) and the other overseen by the Department of Defense. Established in 1977 to consolidate federal agencies concerned with energy regulation, high-tech research, power generation, and nuclear weapons, the Department of Energy now devotes 37 percent of its budget to cleaning up radiologically contaminated sites at more than a hundred locations (Figure 10.8) ranging from national laboratories and obsolete weapons plants to minor facilities once operated by a DoE contractor.[69] In contrast to commercial nuclear plants, most of which lie east of the 100th meridian, DoE nuclear waste sites are largely in the Southwest and the Northeast.

Much less is known about nuclear waste at defense sites. Exempt from AEC and NRC scrutiny, military bases often treated low-level radioactive waste with an arrogance accorded other hazardous materials: bury it and forget about it.[70] In the long run, poor records might prove more detrimental than improper disposal. As a recent General Accounting Office study concluded, "outdated, inaccurate, and incomplete" data are a serious impediment to both the monitoring and the cleanup of military radwaste sites.[71]

A third geography of nuclear waste is emerging: on-site storage of spent fuel rods and low-level radioactive waste at the commercial nuclear power plants in Figure 10.7. With neither a permanent home nor a politically acceptable parking place, high-level waste seems likely to remain where it is, even if nuclear plants must build aboveground vaults when their underground storage pools run out of room for spent fuel rods.[72] No one wants radwaste in his or her home state, it seems, not even on a large, heavily contaminated federal tract like the Nevada Test Site or on a remote Indian reservation.[73] Nevada residents and environmentalists fiercely object to burying high-level radioactive waste under Yucca Mountain, and New Mexico officials adamantly oppose radwaste shipments to a temporary repository approved by the entrepreneurial chiefs of the Mescalero Apache Nation.[74] High-level nuclear waste is not the only radiological pariah. Because of fierce local opposition to new regional disposal sites for low-level waste, hospitals and research laboratories might start storing their contaminated gloves and petri dishes at nearby nuclear plants.[75] And even nuclear plants decommissioned or entombed after their operating license expires might remain on the hazards map as threats to groundwater and nearby streams.

Does a map of commercial nuclear plants belong in a book on hazard zones? Most definitely. Roughly four million Americans live within ten miles of a nuclear power station, and they and other slightly more distant nuclear neighbors run risks, however small, not shared by other citizens.[76] Simply put, a radiological emergency planning zone is as much a hazard zone as a floodplain, a fault, or an active volcano. The principal difference is that aside from rare, unimaginably catastrophic events such as the earth's collision with an asteroid, natural phenomena have a markedly richer record that allows scientists to map risk as well as point out hazards. For example, even though a river may not overflow its banks every year, data on annual peak discharge enable hydrologists to delineate 100-year and 500-year flood zones. Similarly, offsets along faults, ancient ash deposits, radiocarbon dating, and other evidence promote probability-based risk mapping for earthquakes and volcanic eruptions. By contrast, nuclear power plants are too new and poorly understood to warrant statistical strategies, and their hazard boundaries are inherently less certain. For now, the ten-mile compromise with a cartographic focus on selective, phased evacuation provides a faintly rational, politically expedient solution.

Chapter Eleven

Imagining Vulnerability

My grandmother, who came from Massachusetts, enjoyed telling of two famous New England disasters. One was the Great Boston Molasses Flood of 1919, when the rupture of a 58-foot-tall tank released 2.5 million gallons of the sugary goo, which defied its reputation for slowness by promptly smashing nearby buildings and wagons, crushing or smothering twenty-one people and numerous horses, and precipitating 125 lawsuits.[1] The other was the Providence Tidal Wave of 1938—not a tsunami but a real tidal wave, which occurred when the surge from a severe hurricane coincided with an unusually high tide, rushed up Narragansett Bay, and abruptly filled downtown Providence, Rhode Island, with up to 12 feet of water, drowning pedestrians and people in their cars, damaging facades, flushing shoppers out of stores, and providing an entrée for looters.[2] Although officials in Boston and Providence bore no blame for these tragedies, the startling variety of "low-probability, rapid-onset, high-consequence events" (as emergency preparedness experts call them) has forced public administrators to treat every factory, waterfront, and transport line as a potential source of danger.

216

This chapter examines the use of paper maps and electronic geographic information systems in emergency management. Although the focus is upon technological disasters, emergency preparedness must also address natural calamities, like earthquakes and floods, which can rupture pipelines, cause explosions, and trigger the release of toxic materials. Discussion begins with the Emergency Planning and Community Right-to-Know Act, which required industries to reveal hazardous materials and established Local Emergency Planning Committees (LEPCs) to develop emergency response plans for chemical accidents. The chapter then looks at the adoption of electronic mapping and plume modeling by LEPCs and at computer techniques that promote understanding of vulnerability by linking plausible plumes with population data.

By any standard, the 1984 chemical disaster in Bhopal, India, was horrendous, but after a day or so the loss of over 2,500 lives meant less to the American media than the culpability of Union Carbide Corporation, which owned the pesticide plant.[3] Although the workers and managers were Indian, Union Carbide engineers at the firm's Danbury, Connecticut, headquarters had designed the production process, and safety experts from Danbury had visited the plant in 1982 and warned of a "runaway reaction."[4] Less than a year after the catastrophe, Union Carbide was in the news again, when toxic chemicals leaking from its plant at Institute, West Virginia, sent 135 people to the hospital for nausea, breathlessness, and other symptoms. Although no one died, the Institute accident underscored the hazard of living near a chemical plant.

Catastrophes like Bhopal are hardly surprising. After all, a chemical plant is an integrated complex of tanks and pipes designed specifically to bring ingredients together—including substances better kept apart. Plumbing can leak or rupture, and opening or closing the wrong valve can create a toxic mixture or set off an explosion. In the case of Bhopal, the immediate cause was the unexplained addition of large amounts of water to a tank of methyl isocyanate (MIC), which reacted by releasing heat, which in turn vaporized much of the deadly MIC.[5] Distracted by what turned out to be a hasty, futile effort to neutralize the escaping poison, plant employees failed to warn nearby residents. Union Carbide's neighbors at Institute were comparatively lucky: although the facility manufactured MIC, none of the twenty-three chemicals that escaped were as deadly.[6]

To calm fears aroused by Bhopal and Institute, Congress passed the Emergency Planning and Community Right-to-Know Act (EPCRA) in late 1986.[7] Concerned primarily with hazardous chemicals used in manufacturing, lawmakers settled on a mix of prevention and preparedness, and focused on localities, rather than states or federal agencies. Local governments, the bill's sponsors argued, were typically the first to suspect a hazard or respond to a crisis, and thus were better situated to involve citizens and work with plant managers in reducing risk. To provide data essential for planning, the law made manufacturers responsible for an up-to-date inventory of the locations and amounts of all hazardous chemicals, the immediate reporting of all dangerous spills, and an annual summary of all toxic releases, including air emissions, landfill disposal, and discharges to streams and lakes. In making these data accessible to the public, leaders hoped a better informed citizenry would assume a fuller role in hazard mitigation and emergency planning.

In addition to new information, EPCRA established two new institutions. The basic unit is the Local Emergency Planning Committee, established to develop an emergency response plan addressing the more than 350 highly hazardous chemicals in the right-to-know inventory as well as other locally significant natural and technological threats. To promote community participation, Congress called for a broad, diverse membership that would include elected officials; law enforcement, civil defense, fire-fighting, environmental protection, and planning personnel; hospital officials; emergency health care and rescue workers; manufacturers; representatives of the media and relevant community groups; and other interested citizens. The law also instituted State Emergency Response Commissions (SERCs) to partition each state among local emergency planning districts and review each local emergency response plan at least once a year. Fifty-six SERCs (including those in the District of Columbia and various territories) oversee and coordinate four thousand LEPCs—slightly smaller on average than the nation's more than three thousand counties. According to a survey by the International City/County Management Association, single-county LEPCs are the rule, with notable exceptions in New England (where towns are strong and counties are little more than statistical units), California (which has six mutual-aid planning districts), and Oregon (where a single LEPC serves the entire state).[8]

In 1986 many counties already had both an emergency plan and an emergency manager. Younger managers tended to have diplomas in public administration or fire safety, whereas older ones were often veterans of civil defense and the cold war era of the late 1940s and 1950s, when public bomb shelters stockpiling food and water seemed a practicable preparation for a nuclear attack from Russia.[9] Nuclear preparedness was the predominant concern in the 1970s, when emergency management officials (as they prefer to call themselves) began to address a broader range of natural and technological hazards. As a response to Bhopal, EPCRA shifted the emphasis even more by providing a wealth of data on hazardous chemicals in the hope that knowledgeable citizens would make planning more effective.

Congressional hopes for citizen participation remain largely unfulfilled.[10] According to political scientist Susan Hadden, "few LEPCs have managed to increase the level of awareness in the broader community or to involve citizens in a meaningful way."[11] Even though deliberate efforts to exclude amateurs are rare, emergency professionals dominate most LEPCs, and turf battles disguised in jargon repel ordinary citizens and elected officials. Moreover, industry tends to be overrepresented, and some LEPCs, out of deference to important local employers, have declined to release or publish data clearly within the public's "right to know." The SERCs could correct this imbalance, but few (if any) have. And Congress, having neglected to appropriate new funding for these new institutions, has few grounds for complaint.

Perhaps the best (if not the only) study of maps used by LEPCs is geographer Ute Dymon's concise, tellingly titled essay "Mapping—The Missing Link in Reducing Risk under SARA III."[12] A cartographer focusing on the design and use of environmental maps, Dymon identified two types of maps essential to the LEPC mission: *risk maps*, which point out areas of high risk for natural and technological hazards, and *response maps*, which describe evacuation routes, shelters, monitoring points and other spatial information useful immediately after a disaster. LEPCs can prepare both types of maps before an emergency, but the typical strategy is to collect existing maps and hope the information is adequate for a crisis. Under EPCRA, local committees can request a comprehensive federal review by the interagency National Response Team, but few do: of the one thousand LEPCs in the six northeastern states, only ten committees in eight years had submitted plans for review. And none of these ten plans included risk and response maps satisfying rather minimal FEMA guidelines on cartographic content.[13]

A near calamity in Marlborough, Massachusetts, illustrates the value of local hazard and evacuation maps.[14] On January 6, 1991, during a Sunday night delivery to a downtown gas station, a gasoline tank truck's hose ruptured, and a hundred gallons of gasoline spilled into a storm drain, flowed away from the site, and started to vaporize. The driver notified authorities, who prepared for several days of chaos, as sporadic explosions damaged buildings and hurled manhole covers through the air. The town's emergency plan placed the fire chief in charge, and after notifying other emergency personnel, he ordered an evacuation of houses along the storm drain. Although storm sewer maps identified areas at risk, officials worried that the fuel or its fumes might enter an older, abandoned storm drain network not shown on the town's hazard maps. Because of this uncertainty, the chief evacuated a sizable area that the deserted storm sewers might have served. No map of the old system existed, but a search of files at the public works department turned up obsolete engineering drawings, from which officials compiled a map eight feet long describing the abandoned drainage lines. This ad hoc crisis map let the chief focus on areas at risk and enabled many residents to return home with minimal inconvenience. In narrowing the crisis zone and informing the public through the news media, local officials won the cooperation needed to avert disaster.

Marlborough was fortunate that the town's emergency program manager was a fire department employee who appreciated maps. In addition to collecting existing maps and training himself and coworkers in emergency preparedness, this conscientious fire fighter had compiled detailed emergency planning and evacuation maps in accord with FEMA guidelines. Drawn in color on large-scale base maps from the town engineering department, the maps identified firms reporting dangerous chemicals as well as routes for transporting hazardous materials. To represent risk, he drew circles with a one-mile radius around each storage site and delineated buffer strips along the transport routes. Had he included natural hazards and estimated risk probabilities for hazard zones, the maps would have met FEMA standards for an emergency operations plan.

Evacuation planning maps were especially helpful during the Marlborough gasoline spill. A critical facilities map identified hospitals, fire and police stations, and potential shelters such as churches and schools. A plastic overlay not only protected the map in the field (and from coffee drinkers at the firehouse) but allowed officials to add new information relevant to the crisis. Additional maps identified apartment buildings, nursing homes, day care centers, schools, and other sites with special needs. Lists keyed to the maps pointed out elderly and handicapped citizens requiring individual notification and assistance.

In locating and delineating buffers around chemical hazards, Marlborough's LEPC had begun to exploit the new EPCRA data. But the town's emergency response maps still reflected commonsense notions of simplicity and flexibility succinctly summarized in a 1972 disaster handbook prepared for local governments by the U.S. Defense Civil Preparedness Agency. According to the handbook, maps of the local operations area were the first of six key types of supporting information. But equally essential was a work area for displaying and marking the maps:

> County and town road maps (as well as other maps) may be tacked to sheets of wallboard. Grease pencils and colored pins may be used on clear plastic overlay to depict emergency situations, and to show the locations of available manpower and equipment. The use of overlays to visualize the situation has proven successful not only to illustrate what's happening, but also to help make decisions on emergency actions to be taken. Simple magnetic maps may be made by placing magnets on maps that have metal screening underneath. Regardless of the method used, the information should be kept simple, with color coding used as much as possible.[15]

Marlborough had computer software for mapping emergency data, but its LEPC had no money to implement it and no one knew how to use it.[16] The software was CAMEO (for Computer-Aided Management of Emergency Operations), a menu-driven system for building a local inventory of hazardous chemicals stored at various sites, displaying maps, and generating dispersion plumes for a variety of weather conditions.[17] Developed by NOAA and redesigned with the support of the EPA, CAMEO contains response information for over 2,600 hazardous substances, but users must develop their local data base.[18] Despite free distribution by the EPA, CAMEO was useless if the LEPC lacked either a computer or the staff expertise to set up the data base.

To understand better how local emergency management agencies use maps and mapping software, I visited Joseph (Joe) Falge (pronounced *fallj*) in the subbasement of the Onondaga County office building, in a bunker designed to withstand atomic attack.[19] Falge directs the county's Office of Emergency Management, which coordinates emergency planning and response between county departments, New York State, the city of Syracuse, and the county's nineteen towns. A political appointee who had worked in the Department of Social Services, Falge became county disaster preparedness officer in 1983. His outgoing personality seems well suited to the role of liaison between local police and fire chiefs and county and state officials. Not an emergency professional, he's the occasional target of our local newspapers, which criticized his ill-timed vacation during the 1993 spring floods and his close ties to the county executive, for whom he coordinates an annual golf tournament/fund-raiser.[20]

Except for two part-time deputies, who work in another building, and a secretary, Joe Falge *is* the Office of Emergency Management. Like most county emergency preparedness officials, he has no enforcement powers and gets involved only when requested by the city mayor or a town supervisor. The initial response is usually handled by the Emergency 911 dispatch office or the county fire coordinator. For a transportation accident with a toxic release, a call to 911 would result in automatic notification of local police and fire officials, the fire coordinator, and emergency management—that is, Joe Falge, who would serve only as an information resource unless the town or city relinquished control to the county.

A rigid command-and-control structure lets officials focus on helping victims, not defending turf. If a toxic release occurs in a village, for instance, the village mayor is in charge. If the village can't handle it, the mayor cedes control to the town. If the town can't handle it, the town supervisor puts the county in charge. And if the disaster overwhelms the county, the county executive will ask the state to step in. The elected official typically delegates authority to the fire chief, the police chief, the coroner, or someone else with appropriate expertise, as outlined in the plan for various stages of a disaster. The first question every emergency response plan must address is, Who's in charge?

In Onondaga County, each town has its own emergency plan, which Falge reviews for feasibility and completeness. Among other duties, Falge conducts or participates in drills and training programs, maintains a good working relationship with Conrail and local manufacturers, and chairs the county's LEPC. Formed in the early 1990s, the LEPC meets every two months to review the county's emergency plan. Appointed by the county executive, the twenty-two-member LEPC includes a variety of relevant state and county officials, a radiologist at the SUNY Upstate Health Sciences Center, two people from local consulting engineering firms, and representatives from the print and broadcast media. Its meetings are public yet not widely advertised.

Maps are important, but not as important as I had thought. When asked to name the map he used most frequently, Falge gave me a copy of the 1:100,000 county highway map, which identifies roads by name and municipal turf by color.[21] Towns are either yellow, green, orange, or pink, with no two adjoining units sharing the same hue. Rounding out this jurisdictional collage, the city of Syracuse is gray, the villages (which are never adjacent) are purple, the Onondaga Indian Reservation is a reddish orange, state-owned land is a green tree-like pattern, and neighboring counties are identified by name but otherwise left blank. Among other cartographic references on hand, a set of New York Department of Transportation 1:24,000 maps, reformatted by town, provides greater spatial detail (but no terrain contours), and a Syracuse/Central New York street atlas shows fire district boundaries, which often deviate from town boundaries.[22] No maps hang in Falge's office, but in another part of the subbasement, the county's emergency operations center has large wall-mounted maps, covered with write-on plastic, for both the city and the county. County officials required to assemble here during an emergency include the heads of the planning and transportation departments, in whose offices more detailed maps are readily available.

Does Onondaga County have any risk maps? I asked. No need, I was surprised to hear: our area's not *that* dangerous. With his hands, Falge placed two imaginary circles, each about ten miles in diameter, on the highway map. One was centered just east of Syracuse and the other near the city's western edge. These slightly overlapping circles, he pointed out, cover most of the county's industries as well as Conrail's east-west mainline from Boston to Chicago, the New York Thruway, and Interstate 690, which runs through downtown. Any serious industrial or transportation accident would almost certainly occur within this dumbbell-shaped hazard zone, which includes the gasoline storage tanks in "Oil City" and Bristol Laboratories' pharmaceutical plant in East Syracuse. Maybe. Maybe not.

I should not have been surprised: emergency planning is not like land-use planning, where making maps supersedes making lists. By contrast, emergency managers are obsessed with lists of whom to notify, what to fear, how to respond, when to act, where to seek shelter, which supplies to send, where supplies are kept, which neighboring jurisdictions to call for help, which neighbors to depend on for back-up while helping another neighbor, when to hold training sessions, what to cover, whom to invite—and even how to update all the other lists.[23] Especially important are lists of hazardous materials at each storage site, lists of the dangers of each toxic substance, and lists of indicators reflecting the nature and severity of a spill. And when a disaster occurs, effective communication depends on lists of government officials, media contacts, relief agencies, and hospitals—lists like the hand-lettered roster of local officials and other contacts dominating a wall of Joe Falge's office. Next to each name is a job title and telephone number. Organized alphabetically, the list includes at least two officials for each of the county's nineteen towns and fifteen villages. Location is an unavoidable element in emergency management, but response planning focuses on lists and guidelines, not maps.

Despite having few paper maps on file, Falge has an electronic information system, which not only retrieves and integrates lists of toxic chemicals and hazardous sites but makes maps. The software is Emergency Information System, better known by its acronym EIS. In 1993 New York's State Emergency Management Office (SEMO) purchased EIS for all counties that didn't have it but were willing to commit a computer and staff time.[24] Onondaga County made the commitment by buying a computer, recruiting a graduate student intern from the SUNY College of Environmental Science and Forestry to install the software and set up the data, and designating a county data processing employee to receive training and operate the system during a disaster. Neither Falge nor other LEPC members use the software directly, but the emergency management office conveniently shares the subbasement with the data processing division.

SEMO chose EIS for several reasons: the software's straightforward operation, with menus and documentation based on the jargon of emergency managers, and an on-line *event log* for chronicling the locality's response to an emergency.[25] Helpful after an incident in evaluating preparedness and performance, the event log automatically records the operator's use of EIS as well as notes, notifications, and descriptions of actions entered manually during the crisis. A common, statewide system has other advantages: a substantial discount on the single-copy price, ready exchange of data over telephone lines among neighboring counties and the state operations center in Albany, and cost-effective training and support through state-sponsored workshops and regional user groups. User meetings every two months proved especially helpful in SEMO's "Lakes District," between Syracuse and Rochester, where ideas and solutions exchanged among county emergency officers led to several highly effective implementations of EIS.

Like other software packages designed for emergency response, EIS links an electronic map of local streets, boundaries, and landmarks with an air-dispersion model so that officials can identify and evacuate areas endangered by a release.[26] Each site storing highly hazardous chemicals is an icon on the map: a tiny black building with a smokestack represents a factory, while a red skull and crossbones signifies a pesticide service, a water treatment plant with a huge chlorine tank, or a farm supply store with anhydrous ammonia. By pointing to an icon with the computer's mouse and clicking on the mouse button, the user retrieves information about dangerous chemicals stored at the site as well as the names and phone numbers of the firm and its emergency contact person. To estimate the impact of a release, the user indicates the toxin and computes a chemical plume by entering information about wind speed, wind direction, precipitation, temperature, atmospheric stability, and other weather conditions, as well as surface roughness and time of day.

EIS buyers have a choice of two dispersion models: ALOHA (for Areal Locations of Hazardous Atmospheres) can estimate plumes for nearly one thousand chemicals, whereas CHARM (for Complex Hazardous Air Release Model) provides a more accurate treatment of dense chemical gases such as chlorine and anhydrous ammonia.[27] An option requiring the more expensive Windows version of EIS, CHARM can plot the plume for any time after the start of a release, whereas ALOHA, developed by the EPA and provided free to EIS users, merely projects the maximum extent of the plume. To save money as well as avoid needless sophistication, SEMO bought the basic DOS-ALOHA version.

Emergency coordinators use expected concentration to divide the plume into two parts: an inner zone, where the toxin is an immediate danger to life and health, and an outer zone, where the concentration would exceed the "level of concern" for evacuation, typically a half or a tenth the danger level. For the plume in the lower-right quadrant of Figure 11.1, isolines representing 500 and 250 parts per million (ppm) of anhydrous ammonia bound the two zones.[28] Although both zones require evacuation, the inner zone poses a more immediate threat to residents as well as a greater danger to emergency personnel. When

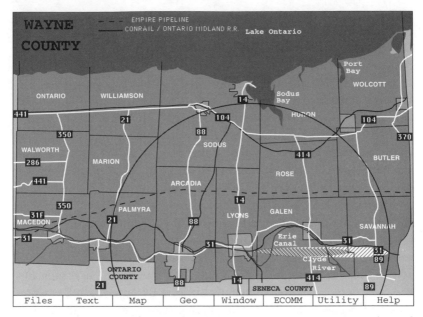

Figure 11.1. Plume (lower right) representing the worst-case instantaneous release of seventy-five tons of anhydrous ammonia from a fertilizer plant in Lyons, New York. Inner zone shows areas where the concentration would exceed 500 parts per million, the level considered immediately dangerous to life and health. Emergency officials would also evacuate the outer zone, with concentrations more than half the danger level. A circular arc defines a zone of vulnerability around the plant. Courtesy of Rich Cobb.

evaluating community vulnerability to specific hazards, analysts sometimes plot a third, smaller area: the "lethal zone," where higher concentrations bring instant death.

Emergency officials can prepare for an accident by developing and storing for each highly hazardous chemical at every site a generic plume based on a worst-case spill of the maximum amount of material likely to be stored. Should a release occur, the operator quickly retrieves the generic plume, adjusts the model for current weather and the estimated amount of hazardous material in the tank or warehouse, and overlays the plume on an electronic map of the area. Around each plume, the system then constructs an evacuation zone, flared outward to allow for a possible shift in wind direction. Figure 11.2 compares the narrow 40-degree evacuation zone used for steady winds with the broader 120-degree zone for variable winds.

Generic plumes can be the basis of a crude vulnerability map portraying each hazardous chemical site as a circular hazard zone.[29] As

Figure 11.2 illustrates, the maximum reach of the plume's outer isoline, representing concentrations exceeding the "level of concern" (LOC), provides the radius for a vulnerability zone. A countywide plot of these circles—a few big, most small—reveals at a glance areas requiring greater surveillance. To identify areas with higher levels of risk, emergency planners can plot the markedly smaller circles for the isolines reflecting concentrations deemed "immediately dangerous to life and health" (IDLH) and lethal. Because plume length varies with wind speed and atmospheric stability, worst-case vulnerability analysis would reflect the longer plumes of strong, steady winds.[30] The plume in Figure 11.1, for instance, represents a 10 mph wind under stable nighttime conditions.

Because enlarging part of a map can be sluggish on older computers, EIS encourages users to compose and store all base maps for instant retrieval. When a map is needed in a hurry, a "click and zoom" feature lets the operator move rapidly from a view of the entire county to progressively more detailed maps showing a section of the county, an individual town, and an enlarged portion of the town. In addition, EIS supports timely decision making by storing a variety of important lists under names such as "resources" (for highway con-

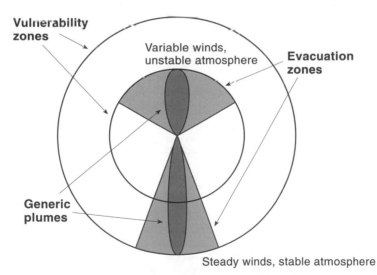

Figure 11.2. Relationship between the lengths of generic plumes for steady and variable winds and their evacuation zones and vulnerability zones. Steady winds yield a longer plume, narrower evacuation zone, and larger vulnerability zone than variable winds.

struction and fire equipment), "needs" (for nursing homes, hospitals, day-care centers and other special needs locations), and "shelters" (for the capacities of schools and churches).

Plume modeling can be useful in training fire fighters, police, and persons in charge of special needs facilities to deal appropriately with an emergency.[31] Tom Bowman, emergency management director in Jefferson County, New York (and a former Roman Catholic priest), showed me a plume model developed for a paper mill in downtown Watertown, New York, a block away from an apartment building for senior citizens.[32] Using ALOHA to model a worst-case release of chlorine—the paper industry uses large amounts of chlorine as a bleaching agent—he generated a two-zone map of the plume's "foot-print" and a plot of chlorine concentration at the apartment building against time. On Figure 11.3, a black-and-white version of the plot, the solid line represents outdoor concentration, which quickly reaches the dangerous level. In contrast, the dashed line represents indoor concentration, which is consistently much lower. From these estimates Bowman concluded that an evacuation—which responders are conditioned to execute—would be impractical unless the tank were to empty over a full hour, say, rather than in five or ten minutes. Because chlorine is two and a half times as dense as air (and even heavier at the outset of a leak in its chilled, near-liquid form), a rapid leak from a nearby tank might easily asphyxiate evacuees just leaving the building.

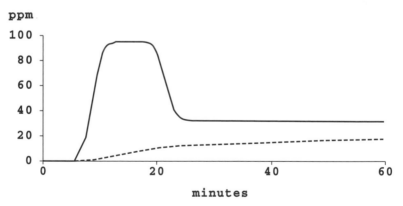

Figure 11.3. ALOHA plot showing temporal change in concentration of chlorine in parts per million (ppm) at a point 200 yards downwind from the release site. Outdoor concentration (solid line) reflects the dense toxic cloud early in the scenario, when the release rate approaches 86 pounds per minute. Indoor concentration (dashed line), which reflects the slow seepage of gas into the building, does not reach the 30 ppm level considered immediately dangerous to life and health. Courtesy of Tom Bowman.

A safer strategy, the plume model suggested, is for residents to close doors and windows and move to the building's upper floors. In contrast, a release of ammonia, lighter than air once vaporized, might be avoided by evacuating to the ground floor and leaving the building if concentrations were low and unlikely to rise. However flat and homogeneous their cartographic footprints, toxic clouds are three-dimensional phenomena perhaps best evaded by jumping or ducking, rather than running.

Bowman is quick to emphasize the model's limitations: the terrain information is imprecise, and ALOHA doesn't work well with stable air, low wind speed, or situations in which the terrain might divert or "steer" the plume. "The model has to be taken as a tool," he notes, "not as what's going to happen."[33] Even so, conscientiously applied modeling encourages emergency managers to ponder a broad range of plausible events and ask pointed questions. For the senior citizens apartment building, for instance, can the air circulation system be shut down on a moment's notice? If so, by whom? Who is the contact person during the day? At night? Are the building's managers aware of the danger? Do they know how to respond? Are the residents aware of the danger? Would they know how to respond? Modeling is an excellent tool for devising realistic drills and demonstrating the need for preparedness.

Data entry can consume weeks of work, especially when maps must be composed manually. Fortunately for New York officials, EIS readily accepts TIGER/Line files, an extensive geographic data base developed by the U.S. Bureau of the Census and the U.S. Geological Survey. An acronym for Topologically Integrated Geographically Encoded Referencing, TIGER is a series of carefully structured lists of node points (mostly street intersections), lines (street segments between consecutive intersections as well as portions of railways, boundaries, streams, and shorelines), and areas (blocks bounded by a closed chain of street segments).[34] An electronic street map of sorts, TIGER/Line files integrate street addresses and census areas with linear features such as political boundaries, roads, railways, pipelines, streams, and shorelines recorded with sufficient detail for display at 1:100,000.

TIGER data are neither complete nor error-free. In Wayne County, New York, for example, emergency operations officer Rich Cobb discovered that the TIGER maps furnished by SEMO included obso-

lete features (abandoned or dismantled railways), omitted a recently
completed pipeline, failed to identify landmarks and other reference
features, and used the same color for country lanes, superhighways,
and all other roads.[35] To make the maps more informative, Cobb
exported relevant TIGER features from EIS to two illustration packages (PC PaintBox and Dr. Halo); assigned colors differentiating private, town, county, and state roads; and added highway numbers and
other appropriate labels. Lacking a digitizer or scanner, he transferred
the pipeline by eye from maps provided by the pipeline company. In
addition to cleaning up the TIGER files, Cobb composed maps of
varying detail for quarter sections of the county, each of the county's
fifteen towns, and quarter portions of each town. And for the sixty-two
sites storing highly hazardous chemicals, he developed detailed site
maps or floor plans showing tanks and other storage facilities as well
as buildings, access roads, and property lines. Although Cobb's system

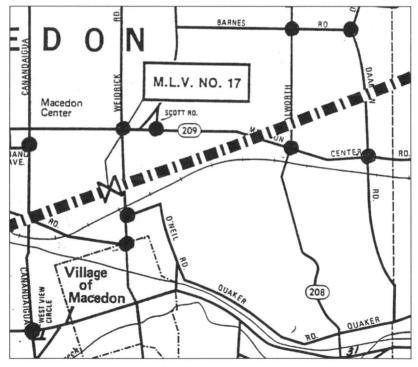

Figure 11.4. Excerpt from map for the Town of Macedon, Wayne County, New York,
showing traffic control points at least a quarter mile distant from a natural-gas pipeline.
"M.L.V." refers to a main line valve, which automatically shuts down the system should
a rupture occur. Courtesy of Rich Cobb.

doesn't provide printouts or hard-copy plots, a lap-top computer with identical maps and plumes is available for use in the field.

Complex software, scarce computers, and limited data are persistent impediments to the wider use of electronic emergency mapping. However rapid and helpful a computer-based information system, town fire fighters and other highly local responders rely largely on paper maps and printed lists of hazardous chemical sites. Because the learning curve is steep and the vast majority of volunteer fire fighters and local police are not yet computer literate, the photocopy machine remains the most reliable technology for disseminating cartographic information. That's clearly evident even in Wayne County, where Rich Cobb has not only mastered EIS but built an impressive emergency response data base. When I visited his office, Cobb had recently developed town-level maps of traffic control points set back a mile from railways and a quarter mile from the pipeline. As the excerpt in Figure 11.4 demonstrates for the pipeline, he merely marked ideal roadblock locations as large black dots on maps provided by the pipeline company. For the county's two railroads, he drew a milewide buffer by hand on a highway map and used an office photocopier to make enlarged town-format, page-size sections for distribution to local officials. Although Cobb will enter these roadblocks into EIS later, when time permits, this "quick and dirty" black-and-white version filled a more pressing need.

Electronic systems for emergency planning and response must deal with three locational formats: street addresses, plane coordinates, and areal units such as census blocks, towns, and fire districts. Street addresses are particularly important in the built-up parts of a community, where most residents and businesses identify location by number and street. *Come quickly—our kitchen's on fire! We're at 415 Elm Street.* Road names are also useful, especially in parts of the county without numerical addresses. *Block off MacDougall Drive between Highway 14 and Townline Road.* Plane coordinates afford a more direct, computer-friendly description of point locations as well as roads and other line features represented by a list of points. Moreover, *eastings* (Xs) and *northings* (Ys) measured along perpendicular axes promote the rapid calculation of straight-line distance between points as well as the length of curved routes approximated by chains of straight-line segments. *Go out Route 12, 1.4 miles past River Road.* Emergency responders must also relate points and addresses to political and

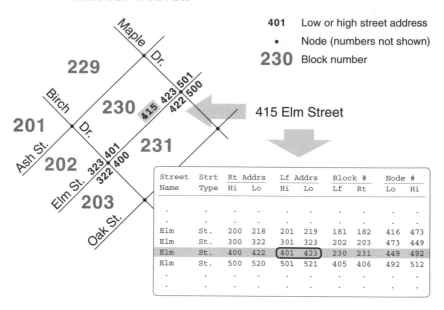

Figure 11.5. Street-segment record for the 400 block of Elm Street relates the residence at 415 Elm to census block 230.

administrative subdivisions. *That's in the South Madison fire district. Also get the Waterloo police and the ambulance corps from Wilson Corners.* Census areas, particularly *blocks* in cities and suburbs and *block numbering areas* in rural areas, enable geographic information systems to estimate the population at risk. *Get the high school in Bernardsville ready, and the community hall too. In half an hour, we'll have 850 people out of their homes.* Population counts are especially valuable to emergency planners, who must calculate needs, designate shelters, and arrange supplies before—not during—a disaster.

TIGER/Line files help the computer match addresses with areal units.[36] A separate record in the file represents each street segment between consecutive intersections. As Figure 11.5 describes, TIGER records relate three types of information: the two nodes (intersections) bounding the segment, the two blocks "cobounding" the segment, and dual address ranges on the odd and even sides of the street. Because the street segment file is sorted alphabetically, the computer searching for 415 Elm Street quickly identifies the handful of records for street segments named Elm, eliminates those for Elm Circle and Elm Drive, and after identifying 415 as an odd number, focuses on the record for odd-number addresses between 401 and 423. The TIGER file thus links 415 Elm Street not only to block 230 but also to whatever jurisdictions (town, police patrol zone, fire district, school attendance area, voting precinct, etc.) the local system administrator has assigned to its block or block-face. In this example, the address range 401–423 constitutes the street segment's left-hand block-face, whereas the even-number addresses 400 through 422 comprise the right-hand block-face. (By convention, "left" and "right" refer to the direction from the low-address end of the street to the high-address end.) Although the block is the Census Bureau's smallest reporting unit, the block-face is important to municipalities that split blocks in order to place opposite sides of the street in the same jurisdiction.

If all lots are equally wide and have consecutive addresses, TIGER/Line files allow a nearly precise estimate of an address's plane coordinates. Among the twelve lots along the odd-numbered side of Elm Street in Figure 11.5, for instance, the home at 415 is only four lots away from the corner house at 423, so that its coordinates are closer to the easting and northing of the block-face's high-address intersection.[37] In contrast, the house at 403 Elm, next to the corner lot at 401, has plane coordinates even closer to the easting and northing of the low-address intersection. If lot width varies and house numbers are not consecutive, plane coordinates will be less precise but still adequate for a moderate-resolution computer monitor. Coordinates estimated for winding streets are generally satisfactory because TIGER uses "shape points" between intersections to treat curved features as chains of straight-line segments.

Distances computed with plane coordinates can reveal instantly whether a nursing home or school is within a circular vulnerability zone around a leaking tank car or burning warehouse. Precise, automatic identification is especially important at the onset of an emergency, when an auto-dialing telephone notification system can prepare residents for evacuation or in-home sheltering. And flashing map symbols can point out the most threatened facilities. For circular hazard zones, automatic threat detection is a straightforward process of calculating the facility's distance from the hazard and comparing this distance with the radius of the zone. But irregularly shaped vulnerability zones require the more sophisticated overlay functions available in a geographic information system (GIS).

The simplest and most rapid approach to automatic overlay is a rectangular grid of square cells, which cartographers call *raster data*. Straightforward formulas for converting plane coordinates to grid coordinates assign every point to a cell, every line to a chain of cells, and every area to a cluster of cells. The GIS treats each type of feature as a separate layer, or *coverage*, and assigns 1 to cells with the feature and 0 to all other cells. Overlay is a straightforward process of generating a new coverage by multiplying corresponding cells in coverages A and B; as Figure 11.6 illustrates, only cells with 1 for both A and B will contain a 1 in the overlay coverage. Precise analysis requires a fine-grained grid of comparatively small cells, which can increase processing time as well as consume enormous amounts of electronic memory. But because most cells in most coverages are empty, the GIS can take advantage of efficient data-compression and raster processing techniques.[38]

Planners concerned with emergency preparedness rely heavily on another GIS-generated coverage, the buffer representing locations within a given distance of a point, line, or area feature. Circular buffers

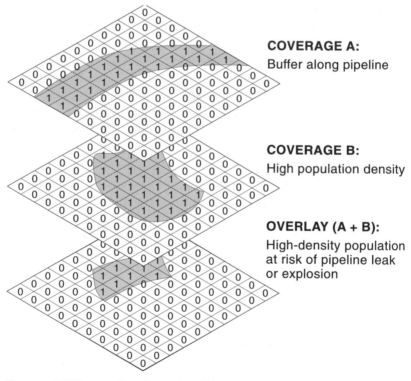

COVERAGE A:
Buffer along pipeline

COVERAGE B:
High population density

OVERLAY (A + B):
High-density population at risk of pipeline leak or explosion

Figure 11.6. With raster data, the overlay of binary coverages A and B, in which 1 represents the presence of a feature or characteristic and 0 represents absence, is a simple process of multiplication.

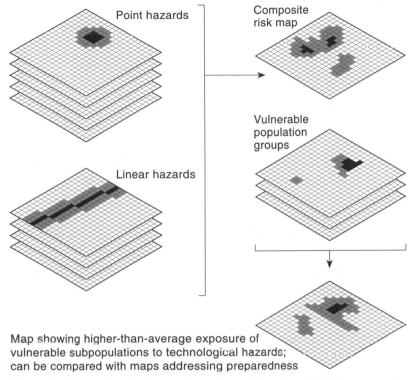

Figure 11.7. Summation of risk scores for point and linear hazards yields a composite risk map, which can be overlaid with coverages for vulnerable subpopulations to identify areas most in need of vigilance, mitigation, and preparedness.

describe vulnerability zones around chemical storage sites; linear buffers represent vulnerability along pipelines, roads, and railways used to transport dangerous materials; and area buffers depict hazardous fringe zones around large, irregularly shaped tank farms or manufacturing sites. The GIS constructs buffers with a *proximity* or *spread* function, in which a sophisticated algorithm propagates a wave outward from irregular lines or boundaries.[39] Once delineated, the buffer can be overlaid on coverages for special needs facilities or census blocks to identify populations at risk. When used to examine population along transportation corridors, buffers help planners reroute shipments of hazardous materials away from heavily populated areas.[40]

Buffers and overlays can help policymakers assess the combined effect of multiple hazards on neighborhoods.[41] Figure 11.7 illustrates how a GIS can summarize risk by adding overlays representing vulnerability zones around point and linear hazards. Higher scores reflect higher risk for cells near the center of circular vulnerability zones around chemical storage sites, and lower scores represent lower risk near the perimeter. Similarly, belts with progressively lower scores toward the outer edge reflect declining risk with increased distance from a pipeline, railway, or road used regularly for hazardous shipments.[42] In addition to plumes for gaseous spills, individual hazard zones might reflect the influence of terrain on releases of toxic or flammable liquids. The analyst can produce a composite risk map by assigning each coverage a weight for relative hazardousness and adding together the weighted coverages. Integrating the resulting map with coverages for vulnerable subpopulations (the very young, the old, the poor) and disadvantaged ethnic or racial minorities offers insights for addressing environmental justice, land-use restrictions, and the locations of firehouses and emergency response teams.[43] Municipal administrators must ensure congruence between the maps of risk and the maps of preparedness.[44]

Community right-to-know laws have made available an enormous amount of data, but most citizens have little appreciation of chemical hazards hidden behind factory walls or inside enormous high-pressure gas tanks. And those who bother to obtain SARA III data seldom succeed in assessing a hazard's geographic dimensions. Maps could help—if the right maps existed—but they would have limited value unless the local emergency planning committee included citizens eager to understand plume modeling and electronic mapping. But why only LEPC members? Why not high school science classes, volunteer fire departments, patrons of local public libraries, and anyone with a computer, a modem, and a phone line?[45] Though widely touted as "the information superhighway," the Internet might prove equally valuable as the "information country lane" that links remote places with the rest of the world.

Citizens who explore their communities with plume-modeling software are in for some surprises. Modeling, they often discover, is never as straightforward as the software suggests. Different models yield different plumes, and different sources recommend different worst-case scenarios. For example, Bill Stone, who chairs the LEPC in Schenectady County, New York, found that for a hypothetical release of chlorine, three different plume models produced circular vulnerability zones ranging in area from 7 to 922 square miles.[46] Because ALOHA yielded generally smaller zones, Stone wondered whether the county might be underestimating vulnerability.

Another discovery was the huge population threatened by a local sewage treatment plant.[47] In his analysis Stone overlaid a population map based on census blocks and a plume representing the release of 1,000 pounds of chlorine. According to ALOHA, within two minutes the release would expose everyone in a 0.02-square-mile circle around the plant to lethal levels (100 ppm or higher) of chlorine. The release would also subject 5,900 people in a 0.84-square-mile area to concentrations greater than 30 ppm, which kills healthy adults in a half hour. What's more, a circular vulnerability zone based on a short-time general exposure limit of 3 ppm included 36,000 people in a 5.04-square-mile area—a population much too large to evacuate in half an hour. However unlikely a disaster with an enormous circular footprint, the LEPC recommended that the plant either find a safer alternative to chlorine or stop using one-ton containers, build a secondary containment structure, and improve its leak-detection/warning system. Because of the inevitable delay in verifying a leak and activating the emergency broadcast network, Schenectady—and thousands of other cities with a wastewater treatment plant near a residential neighborhood—could become a mini-Bhopal.

An obvious solution is to declare a "dead zone" around any plant that cannot reduce its use of hazardous materials to an acceptably safe level. Zone width might depend on safety inspections, mitigation measures, and warning systems as well as on the amounts and hazardousness of chemicals stored on site. A firm investing in a less risky process would thus have a narrower dead zone than a competitor who rejects mitigation. A persistent menace must move, shut down its plant, face substantially increased premiums for liability insurance, or pay a sizable "toxic death risk index tax" to subsidize residential relocation and compensate victims of technological accidents.[48] Although heavy-handed tax and zoning laws might be the only way to motivate some contumacious corporations, lawsuits claiming damages for impaired health and diminished property values have encouraged chemical manufacturers in Louisiana to create their own safety zones by buying up and tearing down nearby homes.[49] However willingly a company might purchase its lethal zone, the cost of shrinking or depopulating its much larger IDLH and LOC buffers would be enormous. Even so, a firm might recover some of the cost by donating the buffer to an environmental land trust. Ironically, where the dead zone is dedicated to agriculture or wildlife, a hazardous industry could prove an ecological asset.

Like floodplains and earthquake fault zones, hazardous industrial sites are inherently contentious. However clear and compelling the rationale for establishing dead zones, estimating risk and drawing perimeters is neither exact nor objective. But unlike natural hazards, chemical storage sites offer society the opportunity to negotiate the nature and riskiness of individual threats as well as their cartographic expression.

Chapter Twelve

Crimescapes

My father lives outside Baltimore and talks frequently of crime, which seems rampant in the city. There's much less of it, though, in his suburban neighborhood of narrow streets and large shade trees. My mother considered Larchmont "safely outside the city"—until their next-door neighbor was murdered. At least the crime met Maryland's legal definition of murder, which is sufficiently broad to include a fatal heart attack during a robbery and felony assault. Gregarious and living alone, the retired construction superintendent had befriended some area teenagers, who stopped by for a visit. While the girls distracted him, the boys went upstairs looking for money. When the man became suspicious and got up, apparently to get his handgun, one of the boys jumped on his back. The man collapsed and died; the kids panicked and fled. Dad saw them run across the lawn, went over to investigate, discovered the body, and testified at the trial. Deeply affected by the sudden, tragic loss of a friend, he still hires neighborhood children to cut grass and rake leaves. But he stays in at night, refuses to open his door to strangers, and avoids the city.

My father's fears, like those of many senior citizens, reflect news reports, not personal experience. Newspapers, TV newscasts, and radio talk shows thrive on murders, rapes, armed robberies, and brutal assaults. Although incidents involving celebrities, multiple victims, very young perpetrators, or unusual brutality receive the most attention, a barrage of brief accounts of less sensational crimes foster the perception of crime as a hazard. Awareness, if not fear, of crime is difficult to avoid in big cities, where running tallies for serious crimes are as prominent as box scores for major league sports. In Syracuse, where I live now, the local media count killings but mention the city's year-to-date total only when announcing a new murder. With fewer than twenty homicides last year, we're reminded only once or twice a month, on average. In contrast, with well over three hundred homicides a year, Baltimoreans cannot easily escape the message that theirs is a dangerous city. But how dangerous is it? And how uniformly dangerous?

This chapter treats crime (except the white-collar variety) as a geographic hazard, that is, as a harmful phenomenon with a pattern of risk worth mapping. Neither a natural hazard like earthquakes and flooding nor a technological hazard like radiological and chemical accidents, crime is a social and behavioral hazard, usually perpetrated by one person against another. I could try to define crime and explore its causes, but I won't. As a social construct with a geographic dimension, crime is satisfactorily defined, at least for our purposes, by the news media and publications such as the *Places Rated Almanac*, which rely heavily on data compiled by the Federal Bureau of Investigation. After discussing the use and limitations of FBI crime rates, the chapter examines broad national trends based on climate and population density. Turning to streets and neighborhoods, I then explore variations within metropolitan areas and how words and maps in the media portray areas as unsafe. Discussion concludes with a look at law enforcement's use of geographic information systems as a crime-fighting tool.

Geographic comparisons of crime in America depend upon the Uniform Crime Reporting (UCR) Program, initiated in 1930 by the FBI and the International Association of Chiefs of Police to collect standardized national data measuring fluctuations in crime.[1] UCR staff receive monthly reports from more than sixteen thousand sheriffs and police departments. Participation is voluntary, with some law enforcement agencies reporting directly and others funneling their counts through a state-level clearinghouse. Because state laws often describe offenses differently than does the FBI, coders at the local level must consult the precisely worded definitions in the *Uniform Crime Reporting Handbook*, an eighty-eight-page guide to describing, classifying, and counting crimes.[2] In addition to publishing an annual statistical summary, *Crime in the United States*, the FBI uses UCR information for special reports on selected aspects of criminal behavior, victimization, and law enforcement.[3]

To monitor national trends, the FBI collects data for eight "index crimes," computes separate indexes for violent crimes and property crimes, and adds these two indexes together to construct an overall, total Crime Index. The four violent index crimes are murder, forcible rape, robbery, and aggravated assault, which the *Handbook* defines as an attack or threat that places the victim at greater risk of bodily injury or death than simple assault.[4] The four property index crimes are burglary, larceny, motor vehicle theft, and arson. Congress made arson an index crime in 1978, but because of "insufficient data"—many jurisdictions don't report suspicious fires, which can be difficult to prosecute—the Bureau ignores arson in computing crime indexes.[5]

Critics have numerous complaints about Uniform Crime Reporting. Among objections concerning definitions, the criteria differentiating aggravated and simple assault are complex and subjective, and the motor vehicle theft category equates a car taken for a joyride and recovered intact with a stolen vehicle that is stripped, wrecked, or never recovered.[6] Other objections concern the program's reliance on voluntary, unaudited reports, easily corrupted by lazy cops— some police apparently hate writing reports so much they'll ignore an incident to avoid the paperwork—or by an unscrupulous police chief who exaggerates need or performance by reporting more or fewer crimes.[7] More troublesome, though, is the inherent omission of crimes that citizens don't report or the police don't notice. In general, crime counts are low where victims either mistrust authority or have little confidence that the police can protect them or recover their property. According to the National Crime Victimization Survey, citizens report less than 40 percent of all crimes to the police.[8] Reporting varies, though, with the seriousness of the offense: in 1992, for instance, victims or their acquaintances reported only 15 percent of thefts less than $50 but 88 percent of thefts involving more than $1,000.[9] Surprisingly, the survey found that poor households report household crime only slightly less frequently than more affluent households.[10] Although Hispanics have lower reporting rates than non-Hispanics, black households report slightly more frequently than white households.[11]

Crime rates are partly an artifact of the number, deployment, and attitudes of police.[12] For instance, the city that beefs up patrols increases the likelihood that its officers will discover additional crime themselves, without information from citizens. An enhanced police presence can also make the public more willing to report crime, especially if local government projects a concern for minority victims. But policing strategy is far more than delineating and staffing patrol zones. According to political scientist James Wilson, "policing style," which can vary by city, zone commander, or individual officer, also affects reporting.[13] For example, crime rates would generally be higher where police adopt a zero-tolerance, law-and-order "legalistic style" than in areas with the "watchman style" of preserving order without reacting to all illegal actions with equal vigor. A third approach, which Wilson labeled the "service style," encourages the police to see themselves as both helpers and enforcers. Despite comparatively accurate counts of burglaries and larcenies committed by outsiders, service-oriented police who mollycoddle middle- and upper-income residents can seriously underreport domestic violence and other crimes by locals.

To complement the FBI's data, the Bureau of Justice Statistics, another branch of the Department of Justice, annually surveys approximately 110,000 persons age twelve and older in 66,000 households. Respondents meet with interviewers for the National Crime Victimization Survey every six months over a period of three years. Using a short screening questionnaire, the interviewer asks questions such as "During the last six months, did anyone break into or somehow illegally get into your (apartment/home), garage, or another building on your property?"[14] Once a crime is identified, the interviewer asks detailed questions about the location, time of day, weapon, and perpetrator; the victim's injuries or loss; the offender's relationship (if any) to the victim; whether the crime was reported, and if not, why not; and for a crime that was reported, the response by the police. A substantial improvement over UCR data in its portrait of victims and their losses, the victimization survey offers very little geographic detail: annual summary statistics break the country into four large regions but provide no information about states, metropolitan areas, counties, or cities.[15] Despite many shortcomings, Uniform Crime Reporting offers the most useful (if not the only) national picture of the geography of crime.

Because participation is voluntary, some parts of the picture are fuzzier than others. In 1993, for instance, law enforcement agencies serving approximately 5 percent of the population did not participate.

To provide more reliable counts for states without full participation, the FBI uses crime rates for the areas that do report to estimate numbers of offenses for those that don't. Because crime rates tend to be higher in big cities, the UCR staff adjusts the data separately for each state's metropolitan, nonmetropolitan-urban, and rural populations. (Social scientists call this stratification.) The following table illustrates this process for forcible rape in Vermont, where the UCR participation rate varied from 95.4 percent for rural areas to only 22.1 percent for cities outside metropolitan areas.[16]

	POPULATION	FORCIBLE RAPE
Metropolitan areas	113,634	
Area actually reporting	62.3%	34
Rate per 100,000 inhabitants		48.0
Estimated totals	100.0%	55
Cities outside metropolitan areas	199,095	
Area actually reporting	22.1%	25
Rate per 100,000 inhabitants		56.8
Estimated totals	100.0%	113
Rural areas	263,271	
Area actually reporting	95.4%	58
Rate per 100,000 inhabitants		23.1
Estimated totals	100.0%	61
State Total	576,000	229
Rate per 100,000 inhabitants		39.8

As the upper part of the table shows, the 34 rapes reported for metropolitan areas represent jurisdictions containing only 70,794 people (62.3% of the state's metropolitan population of 113,634). Applying the resulting rate of 48.0 rapes per 100,000 to the entire metropolitan population yields an adjusted estimate of 55 rapes. Similar adjustments add 88 rapes to the count for nonmetropolitan cities, where UCR participation is very low (22.1%) and the crime rate somewhat higher (56.8 rapes per 100,000), but only 3 rapes to the count for rural areas, for which participation is high (95.4%) but the rate much lower (23.1 rapes per 100,000). Despite this appropriate and reasonable adjustment of the data, nonparticipation also adds considerable uncertainty to crime rates estimated for Indiana, Mississippi, Ohio, and Tennessee, which in 1993 had participation rates well below 90 percent.

Although *Crime in the United States* provides considerable geographic detail, the only map in the four-hundred-page report is a

garishly colored, not very informative four-region comparison of the
FBI's violent and property crime indexes.[17] To examine regional differences more fully, I constructed state-unit maps (as in Figure 12.1)
for violent crimes, property crimes, the combined Crime Index, and
individual index crimes. To make Rhode Island and Delaware visible,
I used the Visibility Base Map, and to avoid the distraction of an anomalous, highly troubled city, I amalgamated the District of Columbia
with Maryland and recomputed the state's rates and indexes. To promote simple, straightforward comparisons among maps, I grouped the
fifty states on each display into five equal-size categories of ten states—
with one exception: the murder map has eleven states in its top category because two states tied for tenth place. This "quintile" (or equal-
fifths) classification obviates frequent references to the map key with a
straightforward, easy-to-learn graphic code whereby more ink means
greater danger: solid black identifies high-crime states in the top fifth;
white points out low-crime states in the bottom fifth; and three clear-

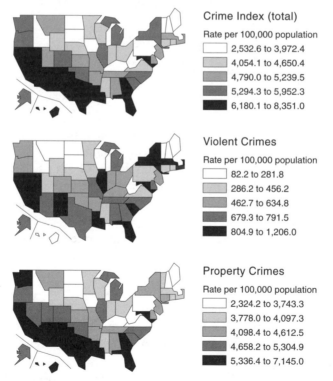

Figure 12.1. State-level maps of the UCR Crime Index and the summary indexes for violent and property crimes. Data from *Crime in the United States, 1993*, pp. 60–6.

ly differentiated shades of gray promote comparisons among the intermediate categories.[18]

Organized cartographically, FBI crime data reveal a latitudinal pattern of lower rates in the North and higher rates in the South. For example, in the uppermost map in Figure 12.1, the ten states with the lowest overall crime rates are all north of Tennessee, whereas the ten highest states form a crime belt stretching from California through Georgia and interrupted only by lower rates for Mississippi and Alabama. The lowermost map portrays a nearly identical north-south trend for property crimes—not surprising, really, because over 85 percent of Crime Index incidents are property crimes. In contrast, the middle map reflects the associations of violent crime with latitude and large metropolitan centers. Although well north of the California-to-Georgia crime belt, distinctly higher rates in Illinois, Michigan, Maryland, New York State, and Massachusetts reflect big city lawlessness in Chicago, Detroit, Washington, Baltimore, New York City, and Boston.

Explanations are straightforward, if a bit simplistic: thermal stress promotes aggressive behavior in the North in the summer but in the South throughout much of the year; burglars, robbers, and pickpockets prefer warmer climates, which attract well-heeled, ready-to-pluck tourists; and big cities offer numerous opportunities for escape to comparatively young, often transient populations, stressed by alienation and overcrowding.[19] Because most perpetrators are males under twenty-five, crime rates are especially high in large cities with high birthrates as well as bright lights that attract young migrants from other regions. And because back alleys and freeways offer efficient escape routes, late-night "convenience stores" serving urban and suburban neighborhoods become convenient targets for robbers, shoplifters, and burglars.

With a few noteworthy exceptions, individual maps for the four violent index crimes (Figure 12.2) also reflect the combined effects of warmer climate and large, densely settled populations. More significant, these influences are most apparent for murder and nonnegligent manslaughter, the violent crime with (we hope) the least underreporting. Illinois, Maryland, and New York are among the top eleven murder states, and all "Sunbelt" states except New Mexico are in the top twenty. In contrast, the bottom ten states are not only in the North but are comparatively rural. Minnesota might seem anomalous at first glance because Minneapolis-St. Paul (where crime is indeed higher than most places in the state) is the nation's sixteenth largest metro-

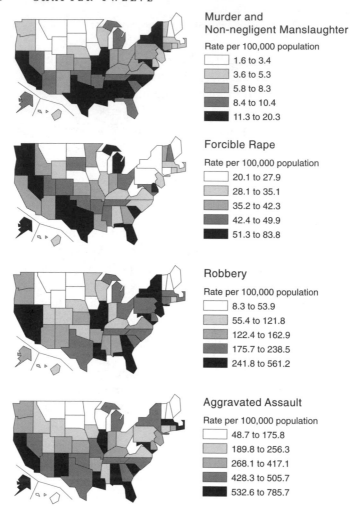

Figure 12.2. State-level maps of the four UCR violent crimes. Data from *Crime in the United States, 1993*, pp. 60–6.

politan center, but cooler weather, comparative prosperity, and substantial rural and small-city populations foster low statewide rates for both murder and aggravated assault.

Forcible rape has a geography markedly different from those of murder, robbery, and assault. Although the ten lowest-rate states on the rape map are all in the northern half of the map, the group includes three comparatively metropolitan states—New York, Pennsylvania, and Connecticut—with higher categories on the other three maps. Of these, New York, which registered the fourth highest state-level violent crime rate (1,073.5 per 100,000) in 1993, is the most puzzling. I found no single explanation (underreporting by emotionally crushed victims? better policing? proportionately fewer young males? preference for other forms of aggression? more street-smart females?) yet am certain that the answer (if there is one) is to be found in New York City, where most of the state's crimes occur. Geographer Barbara Shortridge suggests an intriguing hypothesis: Roman Catholics, who are comparatively numerous in the urban Northeast, are exposed since childhood to church dogma condemning sexual deviancy.[20] The maps reveal other anomalies: strongly Roman Catholic Illinois, which falls in the top fifth for all violent crimes, is in the next-to-lowest category for rape, whereas Oregon and Washington, which place in the next-to-lowest category for violent crimes, are in the upper fifth for rape. Although underreporting may play a role, the National Crime Victimization Survey suggests that the reporting rate for rape (52% in 1992), although low, is not a great deal lower than for robbery (62%) or aggravated

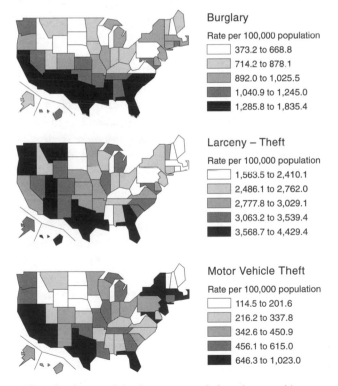

Figure 12.3. State-level maps of the three property index crimes used in computing the UCR Crime Index. Data from *Crime in the United States, 1993*, pp. 60–6.

assault (68%).[21] Whatever the explanation, UCR data for other recent years confirm that the pattern of rape in Figure 12.2 is not a fluke.

Property crimes have their own eccentric geographies. Although climate and large-scale urbanization are important factors, the three maps in Figure 12.3 reveal substantial differences among burglary, larceny, and motor vehicle theft. Big cities offer many attractive targets for breaking and entering, but the uppermost map suggests strongly that burglars don't appreciate cool or freezing temperatures, even in cities. The burglary belt, which extends south from North Carolina into Florida and west to California, turns north along the Pacific Coast to include Oregon and Washington, where most residents live sufficiently close to the ocean to enjoy pleasant summers, mild winters, and nearly year-round housebreaking. Month-by-month crime data demonstrate that burglary, like aggravated assault, is a fair-weather crime.[22] By contrast, larceny (which includes pocket picking and other forms of theft not involving violence or fraud) and motor vehicle theft are much less typical of the Sunbelt. As examples, Mississippi reports one of the lowest larceny rates, while the Northeast, except for northern New England, seems a risky place to park a car on the street.

A fourth set of state-level maps (Figure 12.4) questions the meaningfulness of the Crime Index, which the FBI constructs by adding the rates for violent and property crimes. How useful is an index that equates murder with shoplifting? asked David Savageau and Richard Boyer, authors of the *Places Rated Almanac*. To provide a more reliable measure of overall risk, they computed their own "crime score" by dividing the property crime rate by 10 and adding the result to the violent crime rate.[23] A denominator of 10 seemed reasonable because property crime is about ten times as common as violent crime, and Savageau and Boyer wanted to equate the two broad categories, not their individual incidents. But even though violent crimes account for roughly 10 percent of all index crimes, the middle map shows the violence component varying widely from less than 3 percent in Vermont and North Dakota to over 19 percent in (gulp) New York. And as the lower map shows, by counting only a tenth of all property crimes, the *Places Rated* crime score paints a more dismal picture of New York and Pennsylvania (lower map) than the government's official Crime Index (upper map).

Places Rated Almanac took two other precautions: averaging together five years of crime data to dampen insignificant fluctuations and

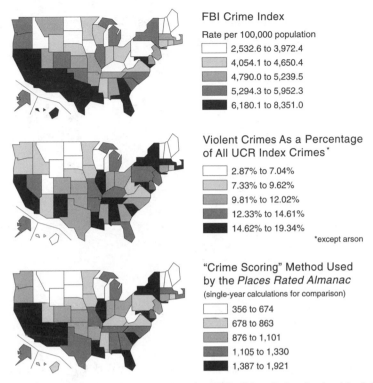

FBI Crime Index

Rate per 100,000 population

☐ 2,532.6 to 3,972.4
▨ 4,054.1 to 4,650.4
▨ 4,790.0 to 5,239.5
▨ 5,294.3 to 5,952.3
■ 6,180.1 to 8,351.0

Violent Crimes As a Percentage
of All UCR Index Crimes*

☐ 2.87% to 7.04%
▨ 7.33% to 9.62%
▨ 9.81% to 12.02%
▨ 12.33% to 14.61%
■ 14.62% to 19.34%

*except arson

"Crime Scoring" Method Used
by the *Places Rated Almanac*
(single-year calculations for comparison)

☐ 356 to 674
▨ 678 to 863
▨ 876 to 1,101
▨ 1,105 to 1,330
■ 1,387 to 1,921

Figure 12.4. State-level maps comparing the FBI's Crime Index (top) with violent crimes as a percentage of all index crimes except arson (middle) and the crime scoring method of the *Places Rated Almanac* (bottom).

focusing on metropolitan areas, which are more meaningful than entire states to retirees seeking a new community. Although tables compare statewide statutes addressing drunk driving, handgun control, and capital punishment, the centerpiece of the *Almanac*'s crime chapter is an eight-page "place profiles" table of crime rates for the nation's 318 metropolitan statistical areas and their 25 Canadian counterparts. This alphabetical listing includes individual rates for seven index crimes (all but arson), separate summary indexes for violent and property crimes, the combined *Places Rated* crime score, and a trend symbol (Δ, ∇, or –) indicating rising, falling, or stable crime rates. To promote comparison, the last column reports a rank based on the crime score. Johnstown, Pennsylvania, captured first place by beating out Wheeling, West Virginia, 284 to 303 for the lowest crime score in North America. At the other extreme, Miami, Florida (crime score = 2,821), demonstrated that "Miami Vice" was not just a catchy name for a television series

Metropolitan Areas with 35 Lowest Crime Scores
Places Rated Almanac, 1993

Figure 12.5. Low-crime metropolitan areas. Data from *Places Rated Almanac, 1993*, pp. 217–18.

about hip cops by snatching rank 343 from New York City (crime score = 2,264). The *Almanac*'s numbers also confirm my perception that Baltimore (crime score = 1,430, rank = 314) has much more crime than Syracuse (crime score = 594, rank = 66).

Places Rated Almanac displays crime rankings in three other ways: check marks before names point out the 35 safest metro areas in the alphabetical listing; a two-page table lists places and their crime scores in rank order; and the crime chapter's sole map ("Crime: Winners & Losers") identifies the five lowest- and five highest-ranking metropolitan areas. Although the pagewide map has adequate space for large black thumbs-down symbols marking losers (Miami, New York, Tallahassee, Baton Rouge, and Jacksonville, Florida), it requires an enlarged, three-state inset for the quintet of winners (Johnstown, Wheeling, Parkersburg-Marietta [West Virginia-Ohio], Williamsport [Pennsylvania], and Altoona [Pennsylvania]).[24] More geographically concentrated than the high-crime losers, these relatively safe areas have consistently smaller populations as well.

To explore geographic patterns in *Places Rated* crime scores, I plotted separate dot maps for the 35 safest and the 35 most crime-ridden metro areas.[25] As Figure 12.5 illustrates, low-crime areas are heavily concentrated in the northern Appalachians, the Ohio Valley, and central Wisconsin.[26] The sole southern outpost, Punta Gorda (crime score = 445, rank = 22), on Florida's Gulf coast, is a prosperous Anglo-American enclave with many retirees and few tourists.[27] Not surprisingly, most of these 35 safest places have comparatively small populations, and many became metropolitan statistical areas only recently, after 1983, when the Office of Management and Budget dropped its rigid population-size requirement for central cities and allowed small cities like State College, Pennsylvania (crime score = 416, rank = 16), to count residents of surrounding townships.[28] The most populous area on the map is Worcester-Fitchburg-Leominster, Massachusetts (crime score = 467, rank = 27), largely an outer suburb of Boston. Although the Census Bureau counted 710,000 people here in 1990, most of the metropolitan areas in Figure 12.5 have fewer than 150,000 residents.

Some of these 35 crime-safe areas are affluent and growing. State College, Pennsylvania, for example, added 11,000 residents during the 1980s, for a solid but unspectacular 10 percent increase. But most are economic and demographic backwaters with comparatively older, law-abiding populations and few attractions for young, lawless in-migrants. Don't get me wrong: these are nice places to raise children or enjoy a placid (if not chilly) retirement. Utica, N.Y. (crime score = 392, rank = 11), which we visit three or four times a year, has pleasant suburbs, several good ethnic restaurants, and the best art museum in central New York—but comparatively few attractions for thugs and thieves.

The 35 high-crime metropolitan areas in Figure 12.6 are more geographically dispersed than their low-crime counterparts, and their pattern is distinctly southern, stretching from North Carolina to California. Some have large populations, others are medium-size, and a few are comparatively small. The group includes the nation's three largest metropolitan areas: New York (crime score = 2,264, rank = 342), Los Angeles-Long Beach (crime score = 1,838, rank = 337), and Chicago (crime score = 1,566, rank = 330). Other massive, well-known metro regions with a serious crime problem are Albuquerque, Atlanta, Baltimore, Dallas, Fort Worth, Houston, New Orleans, Orlando, and San Antonio—stable if not growing cities attractive to tourists, transients, drug dealers, and organized crime bosses. Although several high-crime metropolitan areas registered populations under 200,000 in the last census, these smaller centers are anomalous. For example, Yuma, Arizona (crime score = 1,506, rank = 323), the smallest of the high-crime

Figure 12.6. High-crime metropolitan areas. Data from *Places Rated Almanac, 1993*, pp. 217–18.

group, is a newly anointed metro area near the Mexican border and a smaller, marginally less turbulent version of El Paso (crime score = 1,514, rank = 324). Overall, Figure 12.6 reinforces earlier observations of above-average crime rates in warm climates, big cities, and other areas with high concentrations of young adults and strangers.

As any police officer or news reporter knows, maps of states and data listings of metropolitan areas offer few insights to local patterns of crime and fear. Bad neighborhoods, crime spots, and "no-go" areas are a reality in most cities, and longtime residents develop mental maps of dangerous places from personal experience, friends, newspapers, and television. Newcomers, of course, must rely more heavily on the media, especially in high-crime cities. Quite pointedly, *Places Rated Almanac* warns new migrants to the Sunbelt to pay attention to "crime's share in the local evening news."[29]

How accurately do news gatherers report their local geography of crime? Not very well, judging from spatial biases identified by media critics. In her book *Making Local News*, communications scholar Phyllis Kaniss dissected the professional preference for city news, which journalists consider more important than suburban news.[30] This bias leads to unflattering stories of crime, racial unrest, and corruption, which portray cities as much riskier than their suburbs. In a similar vein, geographer Susan Smith blamed the press for linking race with crime and promoting interracial fear.[31] And sociologist Roy Edward Lotz identified another racial facet of news distortion: suppression of stories about black victims and black neighborhoods, which some editors consider unnewsworthy.[32] Lotz used the term "redlining" to compare editorial and financial neglect of African American neighborhoods. Although the banker who discriminates is almost always racially biased, the editor who underplays crime in minority neighborhoods could be trying to promote racial harmony.

A favorite target of media critics is the manufactured crime wave that uses fear to boost newspaper sales and TV ratings.[33] With nearly predictable regularity, print and electronic news providers suddenly and relentlessly turn their attention to drugs, spouse abuse, date rape, serial homicide, guns in schools, or some other newly discovered social epidemic, only to drop the subject several weeks later. Although crime (or at least incidents known to police) has its ups and downs, news editors eagerly extrapolate a 15 percent one-year upswing in the crime rate into an apocalyptic 75 percent increase four years hence. Alarmist reporting has multiple roots: the necessity of selective coverage in large cities with an abundance of serious crime, the rewards (individual and corporate) of scooping the competition, and the natural tendency of events to cluster in time, thereby inviting discovery, intrigue, and hype. Despite equally alarmist criticism, sensational crime journalism is not new: according to media historian Mitchell Stephens, the most rabid present-day tabloid publisher would blanch at the more lurid stories found in sixteenth-century newsbooks.[34]

Are the mass media to blame? Not always, say scholars who have studied the public's ravenous appetite for crime news. After all, evil is fascinating, observed sociologist Kai Erikson, who suggested that stories about deviant behavior seem to "satisfy a number of psychological perversities among the mass audience." Erikson further opined that "newspapers (and now radio and television) offer their readers the same kind of entertainment once supplied by public hangings or the use of stocks and pillories."[35] (A similar ghoulish, morbid curiosity also underlies public interest in a wide range of natural and technological disasters.) Despite the obvious link between crime and media coverage, a cause-and-effect relationship remains contentious:

although some researchers attribute public anxiety to watching and reading about crime, others either find no relationship between fear and media exposure or reverse the argument by ascribing a strong interest in crime news to preexisting apprehension.[36] Fear varies geographically, though, from suburban residents uneasy about going downtown to city residents afraid to step out the front door.

City size has a marked effect on crime coverage. Because Baltimore's media can describe only a tiny fraction of the thousands of offenses its citizens report each day, my father receives a more sensational and socially deviant picture of his city than I do of mine—for information about neighborhood burglaries and simple assaults, Dad must rely on the free suburban "shopper" tossed onto his porch each week. By contrast, in a medium-size city like Syracuse every murder and armed robbery has news value, as do many far less serious offenses. In still smaller markets, editors glean what they can from police radios, court records, and other government sources. One of my strongest memories of four years of graduate school in State College, Pennsylvania, is the late evening TV news from Altoona—one of *Places Rated Almanac*'s crime-safe areas—where the lead story seemed to alternate between truck accidents and plant closings and where the rare murder might remain front-page news for days to come. In contrast, the *New York Times* rarely reports local homicides in comparable detail, much less on page one.

Because crime stories affect perceptions and property values, residents complain when the locations reported are vague or wrong. Inaccuracies are especially troublesome in large metropolitan areas with identical street names in different neighborhoods or municipalities. To avoid ambiguity, the *St. Louis Post-Dispatch* now mentions by name both the street and the neighborhood.[37] Editors also consult a large map showing the names and boundaries of the city's seventy-four neighborhoods, as delineated by the Community Development Agency. In a column explaining the paper's new policy, "reader's advocate" Sue Ann Wood conceded that the agency's map includes several new, sanitized neighborhood names unfamiliar to longtime residents. Particularly puzzling to her were repeated references to the "Clayton-Tamm" area, traditionally known as "Dogtown."

Although crime stories almost always name streets or neighborhoods, newspapers differ from television and radio by occasionally including a map. In addition to catching readers' attention and

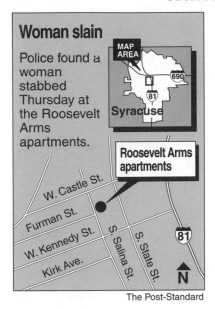

Woman slain

Police found a woman stabbed Thursday at the Roosevelt Arms apartments.

MAP AREA

690

81

Syracuse

Roosevelt Arms apartments

W. Castle St.

Furman St.

W. Kennedy St.

Kirk Ave.

S. Salina St.

S. State St.

81

N

The Post-Standard

Figure 12.7. Locator map used to illustrate a crime story. Reprinted, with permission, from the *Syracuse Post-Standard*, 2 June 1995, p. A-1.

providing a design element for page layout, a map can enhance a crime story by showing how a specific crime happened, where it happened, or where crimes of a particular type are happening.[38]

An explanatory map depicting events in different parts of a building or along the route of a chase is usually reserved for the most heinous or spatially complex crimes, such as the senseless mass murder in a chain restaurant, shopping mall, post office, or commuter train. Unless describing a chase or multistate rampage, the explanatory crime map is usually a large-scale site plan or floor plan. The artist places events in chronological order with prominent sequence numbers (1, 2, 3, . . .), each accompanied by a small block of text.[39] Because of its large scale and need for detail, the cartographic crime-scene narrative is often less a map than an illustration.

A simple map showing only location is the most common newspaper crime map. Figure 12.7, from the front page of Syracuse's morning newspaper, describes the site cartographically and summarizes the crime with words.[40] A concise title in bold type ("Woman slain") catches the reader's attention, while a one-sentence subtitle ("Police found a woman stabbed Thursday at the Roosevelt Arms apartments") reveals a few key facts. A balloon label ("Roosevelt Arms apartments") pointing to a prominent black dot links the crime scene to a portion of the street network, and a smaller inset map relates the neighborhood ("Map Area") to the city boundary and Interstate highways 81 and 690. Look! the map tells readers: more violence on the Near West Side.

Integrating over time all crime locations shown or mentioned in news stories is hardly an efficient or reliable way to construct a mental map. Aside from the cognitive challenge of summarizing months or years of newspaper articles, people miss issues while on vacation and crimes revealed on slow news days receive more prominent coverage than those forced by happenstance to compete with more newsworthy events. Because of these limitations, the most valuable newspaper crime map is the dot map summarizing a year of homicide reports. Even the *New York Times*, which seldom stoops to the sensationalism of the *Post* and *Daily News*, is not averse to an annual dot map of murders.[41] Used to illustrate stories about local crime patterns, dot-distribution maps can help readers form accurate mental maps of other serious crimes.

A good example is the distribution map in Figure 12.8, created by Steve Segal—a former student of mine—who handles information graphics for the *Tribune-Review*, a thriving metropolitan daily published in Greensburg, Pennsylvania, thirty miles southeast of Pittsburgh. In late June 1993, after a rash of car thefts by armed robbers in various parts of the country focused media attention on a frightening and trendy crime dubbed carjacking, the *Tribune-Review* ran a special report, "Carjacking: When Thieves Become Bold."[42] One of several

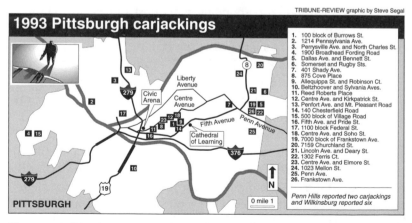

Figure 12.8. Distribution map summarizing locations of carjackings in Pittsburgh. Reprinted, with permission, from *Tribune-Review*, 27 June 1995, p. A10.

illustrations for the feature, the map provides a concise, geographically organized summary of twenty-six armed auto robberies committed
within the city limits since the beginning of the year. (A sidebar
described eight suburban carjackings.) Although Pittsburgh is far less
dangerous than Chicago, Miami, or San Juan, local carjackers had
injured several victims. The article did not discuss Steve's map directly, but a detailed examination of precautionary and defensive strategies
advised readers to know where they're going and avoid strange neighborhoods.

Citizens seldom see them, but maps are an essential tool of law
enforcement, particularly in large cities, where record keeping
and data analysis are an enormous challenge.[43] Almost all police and
court records contain locational information, and decision makers at
different levels rely on maps for a concise overview as well as for specific facts about particular places. At the citywide level, for instance,
maps help the chief's staff monitor workload and redraw beat boundaries or assign temporary reinforcements as needed.[44] In precincts or
divisional headquarters, maps help shift supervisors not only identify
and respond to hot spots but communicate these concerns to patrol
officers. And in specific investigations, maps help detectives locate suspects, recognize *modus operandi* (MOs), and comprehend the hunting
patterns of serial rapists and other predators. By reconstructing a
felon's decision making, the police have a better chance of apprehend-
ing him before he claims his next victim.[45]

Computers have made mapping easier, more timely, and more
revealing. Before interactive computing, investigators filed three-by-
five-inch index cards by street address and inserted small colored pins
in a street map mounted on a cork bulletin board. Different colors represented different crimes, the homes of suspects or known offenders, or
places where stolen property was fenced or abandoned. According to
the FBI's *Manual of Law Enforcement Records*, "spot maps" could reveal
meaningful patterns for burglaries, purse snatchings, robberies, motor
vehicle thefts and recoveries, and the homes of suspects.[46] A hypothetical example compared the spatial patterns of automobiles stolen and
recovered. More clustered than car thefts, recoveries were strikingly
similar in geography to a second map, showing residences of youthful
offenders. "If regularly maintained," the *Manual* observed, maps could
be very helpful for crime analysis. Even so, yesterday's spot maps,
tediously constructed with paper street guides and metal pins, were less

effective than electronic displays based on TIGER data and address matching.

Like other fields with a cartographic approach to hazard analysis, law enforcement employs specialized software. In addition to a variety of commercial mapping packages and geographic information systems, a growing number of police departments rely on the Spatial and Temporal Analysis of Crime (STAC) package developed by the Illinois Criminal Justice Information Authority. Distributed free by the Authority as well as by the Bureau of Justice Statistics in the U.S. Department of Justice, STAC is a toolbox of statistical and cartographic techniques organized as two modules: a Space Program, which not only provides electronic pin maps but automatically identifies concentrations of crime, and a Time Analyzer, which helps police recognize when specific types of crime are most likely to occur.[47] Not a complete cartographic system, STAC provides statistical summaries, which the user transfers to a mapping package for display.[48]

STAC evolved because existing software, although useful in law enforcement, didn't provide a reliable picture of hot spots. Computer maps of urban information usually portray rates and counts compiled for city blocks, census tracts, and other small areas. Appropriate for census data and other information aggregated areally to protect privacy, a rigid network of predetermined territories would sacrifice much of the richness of crime data reported by address. An analyst would want to know, for instance, that police made four separate arrests for identical offenses on the four corners of a single street intersection—information blurred if not lost altogether by a block-level map. An electronic pin map might avoid the blurring, but clusters of tiny dots on small computer screens with modest resolution are difficult to compare, especially when multiple incidents at the same or neighboring locations require plotting one dot on top of another, or nearly so. Efficient, dependable crime analysis requires unambiguous map symbols that point out significant clusters.

If a statistical test confirms that crimes are more clustered than random—which they usually are—STAC uses the iterative search strategy described in Figure 12.9 to pinpoint hot spots.[49] The search is based on a grid of evenly spaced parallel lines and a set of uniform overlapping circles. Each grid intersection is the center of a circle, and the spacing of grid lines is equal to the radius. The analyst specifies the radius and selects a grid of either squares or triangles (not shown), defined by two or three sets of parallel lines, respectively.[50] STAC

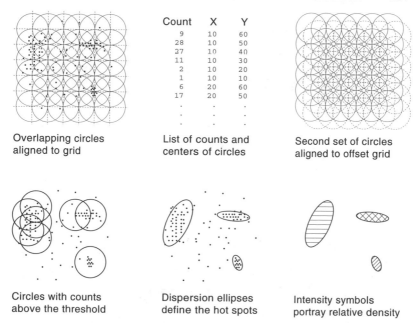

Count	X	Y
9	10	60
28	10	50
27	10	40
11	10	30
2	10	20
1	10	10
6	20	60
17	20	50
.	.	.
.	.	.

Overlapping circles
aligned to grid

List of counts and
centers of circles

Second set of circles
aligned to offset grid

Circles with counts
above the threshold

Dispersion ellipses
define the hot spots

Intensity symbols
portray relative density

Figure 12.9. The STAC strategy for identifying and portraying hot spots.

counts the number of incidents within each circle and adds these counts to a list together with the coordinates for the center. Because the origin of the grid affects the counts, the computer compiles another list of incident counts based on a second grid, with lines midway between those of the first grid. STAC merges the two lists and identifies circles with counts exceeding a specified threshold. Single, isolated circles that exceed the threshold define small clusters, and groups of overlapping circles with above-threshold counts define large clusters. Because incidents are rarely symmetrical or evenly spaced within a cluster, STAC describes each hot spot with a dispersion ellipse representing the compactness, elongation, and orientation of its incident locations.[51] Because a big cluster need not be a dense cluster, the interior of each ellipse is an intensity symbol representing relative concentration.

Two closely related examples demonstrate that STAC maps are easy to read. In Figure 12.10, dots depict individual incidents of street gang-motivated nonlethal violence, while thick dispersion ellipses point out hot spots, where these incidents are highly clustered. The ellipses appear again in Figure 12.11, on which ellipses with distinctly thinner lines portray hot spots for gang-related drug crimes. On this

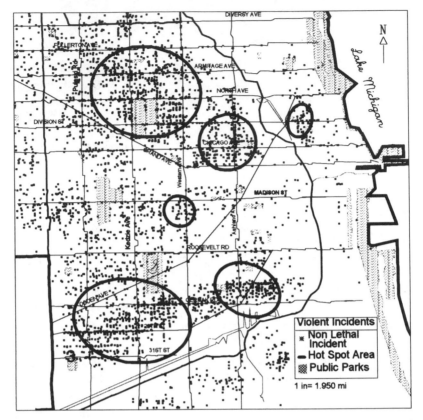

Figure 12.10. Street gang–motivated nonlethal violence in Chicago Area 4: incidents and hot spots, 1991–94. Data source: Chicago Police Department. Data for 1994 are preliminary. Courtesy of the Illinois Criminal Justice Information Authority.

second map, dots represent instances of street gang–related homicide, which often occurs in hot spots for nonlethal violence. These high-assault, high-murder areas mark the overlap of territories of rival gangs.[52] In contrast, the geographic pattern of drug-crime hot spots— generally located in comparatively "safe" areas, well away from disputed turf—contradicts a popular assumption that street-gang violence is drug-related.

Hot spots for other types of crime are equally revealing when compared cartographically to the locations of liquor stores, bars, transportation stops, public phones, schools, city parks, and other locations that attract criminals or their victims. STAC analyses have shown, for example, that robbery-assaults around "el-stops" (stations on Chicago's elevated transit system) increase steadily up to a block and a half

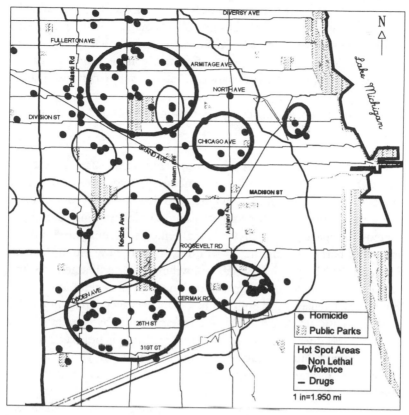

Figure 12.11. Street gang–motivated crimes in Chicago Area 4: incidents of homicide (dots) compared with hot spots of drug crimes (thin ellipses) and nonlethal violence (thick ellipses), 1991–94. Data source: Chicago Police Department. Data for 1994 are preliminary. Courtesy of the Illinois Criminal Justice Information Authority.

away and then fall.[53] Crime reflects surveillance, which is highest in the immediate vicinity of the station, as well as opportunity, which declines with increasing distance. Simply put, muggers pounce in low-vigilance areas where their prey is not yet too dispersed.

According to Carolyn Rebecca Block, a senior research analyst at the Illinois Criminal Justice Information Authority, STAC is both a statistical toolbox and a "network of innovative people" eager to use geographic information to understand and reduce crime.[54] Users not only participate actively in developing and testing the software but share experiences at conferences and through newsletters. Block, who oversees STAC development for the Authority, and her husband, Loyola University sociologist Richard Block, are also active STAC users.

In addition to insightful joint research on violence and drug dealing by Chicago street gangs, the Blocks use STAC maps to help community groups ease residents' fears.[55] Knowing where crime is helps citizens avoid trouble spots as well as lobby for more conscientious regulation of liquor outlets and other nuisances. Because public communication has an important role in law enforcement, the police can increase their effectiveness by presenting crime maps at neighborhood meetings and in the media.[56]

Law enforcement has a unique opportunity: by sharing its maps with the public, the police can encourage crime prevention on a variety of fronts, from home owners who install improved locks and join neighborhood watch groups to city administrations that create "defensible spaces" with surveillance cameras, reconfigured streets, and other barriers to flight.[57] I wonder, though, how much data the cops and their bosses at city hall will willingly share. Trained in semi-military, command-and-control traditions, the police tend to be tight-lipped and defensive, and politicians too are well aware that knowledge is power. Officials might argue, for instance, that by revealing areas requiring less intensive patrol, the crime map points out fruitful territory for enterprising burglars and drug dealers. And by showing where government has failed to protect citizens, the maps can contribute to accelerated decay, despair, frustration, and more crime. Both arguments have merit: as the physicist-philosopher Werner Heisenberg noted, observation affects the observed phenomenon and injects uncertainty. Even so, in an age of Internet, talk radio, and electronic rumor, crime maps are not the only views offered—yet are quite possibly the least biased. Moreover, if government acts responsibly, the effects of knowing can be more positive than negative.[58] Because the democratic process can mitigate hazards, suppressing information about crime locations is no more enlightened than ignoring earthquakes and coastal storms.

Chapter Thirteen

John Snow's Legacy

W hen asked about disease maps, most epidemiologists and geographers think immediately of John Snow's 1854 map of cholera deaths and the Broad Street Pump. A London physician who attributed the epidemic to drinking water contaminated by sewage, Snow plotted the homes of victims on a map together with the locations of public pumps, which local residents visited daily with buckets to collect water.[1] As he suspected, cholera deaths were clustered around a single pump, on Broad Street in Soho. Rather than debate his findings with skeptical colleagues, Snow persuaded local officials to remove the pump's handle and cut off the infectious water. As he predicted, new cases of the disease dropped precipitously.[2] The John Snow Pub, on the site of the infamous pump, commemorates his insight and assertiveness.

Real epidemiology isn't like that, at least not in late-twentieth-century America. Cholera is rare, if not extinct, and contagious diseases like pneumonia and influenza are far less troublesome than heart disease, cancer, stroke, and numerous degenerative ailments once ascribed to old age. Occasional outbreaks of salmonella or hepatitis might be traced to a church picnic or contagious restaurant worker, but advances in sanitary engineering and public health have reduced the impact of infectious diseases transmitted through food and drinking water. Location is important, but increased mobility and delayed diagnosis becloud the effects of geography. AIDS victims, for instance, are typically unaware of their disease until years after a single fatal contact, which few can pinpoint with certainty in time or space. Even so, health agencies must be alert to highly infectious tropical viruses as well as slower acting environmental poisons or parasites disseminated locally by air, soil, water, insects, or other animals. Because a cluster of otherwise uninteresting isolated cases might exhibit a suggestive or revealing pattern, mapping is as important in public health as it is in law enforcement.[3]

Despite a shared need to identify hot spots, the disease police and the crime police have different approaches to data collection and mapping. Epidemiologists interested in environmental poisons and other etiologic factors want to know where victims lived, not where they died. To avoid misleading maps that largely reflect the distribution of hospitals, trauma centers, and nursing homes, health officials painstakingly compile death statistics by place of residence, not place of occurrence. Another difference is completeness: mortality counts are far more accurate than crime counts, but a second data-collection network, used to count cases (fatal or not) of highly contagious diseases, is often less thorough than the FBI's Uniform Crime Reporting. A third difference is the level at which public health and law enforcement officials look for geographic trends. As this chapter illustrates, systematic cartographic monitoring of disease is more common at the broad, nationwide level than at the state or local level. By contrast, detailed local investigations typically reflect the predilections of researchers at a nearby health science school or an apparent local hot spot, brought to light more often by concerned citizens than by systematic scrutiny of national data. After reviewing mortality atlases and computer visualization as broad-brush tools for studying cancers and other chronic diseases, the chapter looks at how state health departments use tumor registries and case-cluster analysis to screen for cancer hot spots and at how local health agencies have begun to use geographic information systems for modeling hazards and monitoring environmental health.

Like crime reports, mortality counts begin at the local level. The physician or coroner who certifies cause of death typically fills out the death certificate jointly with an undertaker or hospital clerk, who forwards the form to the city or county health department for verification.[4] The local office then transmits the information to the state health department, which codes and collates the data and reports each month to the National Center for Health Statistics (NCHS), in Hyattsville, Maryland. In addition to aggregating the data by state and county of residence as well as by cause of death, the NCHS computes rates and publishes an annual mortality report as part of the *Vital Statistics of the United States*.[5] Provisional counts reach public health officials and research libraries five months later via the *Monthly Vital Statistics Report*.[6]

Death reporting is nearly universal—or as close as Mafia hits and other suspicious disappearances might allow. All states have provided death registration since 1933, coverage has grown to include all counties, and as the confidently optimistic National Center for Health Statistics cautiously asserts, "It is believed that more than 99 percent of births and deaths occurring in this country are registered."[7] As late as the 1950s, though, maps of county-level death rates portrayed spurious cool spots in the southern and central Appalachians, where perfunctory record keeping apparently fostered burial without registration.[8]

D eath counts have two uses: one demographic and the other epidemiological. The demographer combines census and vital registration data to estimate net migration for the period between two censuses. For a state, county, or other areal unit, the demographic equation

$$\Delta P = \Delta M + (B - D)$$

treats population change ΔP as the sum of net migration ΔM and natural increase, computed by subtracting the number of deaths D for the same period from the number of births B. Because the demographer's intent is to solve for ΔM, the only unknown in the equation, how the people died and at what age is not important. By contrast, the epidemiologist is highly interested in cause of death as well as age, race, and sex. Decedents' income, education, ethnicity, and social standing might prove equally informative, but death certificates have ignored these often subjective, hard-to-verify traits.[9]

Although focused on disease and vulnerability, the epidemiologist relies heavily on the demographer's data. At the most basic level, the crude death rate (CDR), expressed in deaths per thousand people

$$CDR = (D/P) \times 1,000$$

adjusts for the obvious expectation that a county with a million residents will register many more deaths than a county with only ten thousand inhabitants. "Crude" applies because the rate ignores the equally obvious effect of age differences among areas. For example, Alaska, which appeals to young migrants seeking outdoor adventures and well-paying jobs, has a much younger population than Florida, which offers retirees warm weather and year-round golf. The Sunshine State thus has a much greater share of citizens 65 and older: 18 percent in 1990, in contrast to only 4 percent for Alaska. Because Floridians are thus more likely to die than Alaskans, the epidemiologist is hardly impressed by a map showing crude death rates that are high in Florida and low in Alaska.

To create a meaningful cartographic pattern, the epidemiologist computes an age-adjusted death rate (AADR) for each area on the map.[10] Adjustment is based on a reference population, typically the entire nation for a specific reference year, and a set of age groups, or *cohorts*, such as the eighteen five-year age groups (0–4, 5–9, . . . , 80–84, 85 and older) used in census publications. As described in the formula

$$AADR = (actual\ deaths\ /\ expected\ deaths)\ x\ CDR_{ref}$$

the crude death rate for reference population (CDR_{ref}) is multiplied by a ratio of actual to expected deaths for the area in question. This ratio relates the number of deaths registered to the number of deaths expected if each of the area's cohorts experienced the same mortality as the nation as a whole. In other words, the calculation considers whether the area's five-year age groups were dying more or less rapidly than similar cohorts in the reference population. An area registering more deaths than expected yields a ratio greater than one and receives an age-adjusted rate higher than that of the reference population, whereas an area with fewer deaths than expected receives a lower rate.

A hypothetical example based on three broad cohorts (0–19, 20–64, and 65 and older) explains the rationale and process of age adjustment. In this example, areas A and B differ markedly from each other as well as from the reference population. A table describing age structure and mortality indicates that residents of A are younger on average than the reference population and much younger generally than inhabitants of B, which has a higher death rate.

COHORT	REFERENCE POPULATION	AREA A	AREA B
0–19	25,000	3,000	1,000
20–64	60,000	6,000	6,000
65 and older	15,000	1,000	3,000
Total	100,000	10,000	10,000
Deaths	1,000	80	150
Crude Death Rate	10.0	8.0	15.0

The higher crude rate for B is no surprise to the epidemiologist, who uses age adjustment to answer a thornier question: Given the disparity in age structure, are 150 deaths in B really as worrisome as 80 deaths in A?

A second table describes the calculation of the reference population's age-specific death rates, used for age adjustment. As appropriate, the Grim Reaper claims a disproportionately large share of the oldest cohort.

COHORT	DEATHS	POPULATION	RATE/1,000
0–19	50	25,000	2.0
20–64	300	60,000	5.0
65 and older	650	15,000	43.3
Overall	1,000	100,000	10.0

These age-specific rates reappear in the next table, which describes the calculation of 79.3 expected deaths for area A.

COHORT	RATE/1,000	POPULATION	EXPECTED DEATHS
0–19	2.0	3,000	6.0
20–64	5.0	6,000	30.0
65 and older	43.3	1,000	43.3
Total		10,000	79.3

AADR = (80 / 79.3) x 10.0 = 10.1 per 1,000

When compared to A's 80 actual deaths, the result is an age-adjusted death rate of 10.1 per 1,000—well above A's unadjusted rate of 8.0 and a shade worse than the reference population's crude rate. In contrast, the age structure of area B yields 162 expected deaths and an age-adjusted death rate of 9.3 per 1,000, markedly lower than its unadjusted rate of 15.0.

COHORT	RATE/1,000	POPULATION	EXPECTED DEATHS
0–19	2.0	1,000	2.0
20–64	5.0	6,000	30.0
65 and older	43.3	3,000	130.0
Total		10,000	162.0

AADR = (150 / 162.0) x 10.0 = 9.3 per 1,000

Age structure and its expectations thus posit a paradox: area B is not only healthier than the reference population but healthier than area A—which has the lower unadjusted death rate. This kind of reversal is not only common but easy to account for if area B, like Florida, attracts affluent and comparatively healthy senior citizens.

Does age adjustment significantly alter the map of death rates? Most certainly, according to the state-level maps in Figure 13.1, which compares crude and age-adjusted death rates.[11] The maps portray male deaths from heart disease for 1989 through 1991, a three-year period that conveniently straddles the decennial census enumeration. Division by 3 converts both sets of rates to annual averages, and a five-category equal-size (quintile) classification promotes comparison.[12] Both maps show higher rates in the East and lower rates in the West, but careful inspection reveals several noteworthy differences. In the left-hand map, for example, the top fifth includes Florida, Iowa, and South Dakota, where generally older populations yield above-average unadjusted death rates. Yet the right-hand map places Florida in the next-to-lowest category and assigns Iowa and South Dakota to the middle fifth. Apparently Florida's retiree-migrants were healthier on average than most older Americans, while residents of the two Midwestern states succumbed to heart disease less readily than age structure would suggest. By contrast, age adjustment detects higher-than-expected death rates in Georgia, Louisiana, and South Carolina. Comparison of the two maps indicates that crude death rates underestimate the severity of heart disease in the Southeast—except for Florida—while exaggerating its effects on the somewhat older populations of New England and the Midwest. Race accounts for much of this pattern because African Americans, who have a high incidence of hypertension, are more concentrated in both the rural South and the large cities of the Northeast and the Mississippi valley.[13]

Comparison of the map keys also reveals higher age-adjusted rates for most states—could this be a mistake? No, merely a reflection that

Diseases of the heart, for males, 1989–91

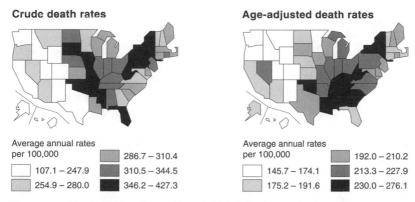

Crude death rates   **Age-adjusted death rates**

Average annual rates
per 100,000   286.7 – 310.4

107.1 – 247.9   310.5 – 344.5

254.9 – 280.0   346.2 – 427.3

Average annual rates
per 100,000   192.0 – 210.2

145.7 – 174.1   213.3 – 227.9

175.2 – 191.6   230.0 – 276.1

Figure 13.1. Crude (left) and age-adjusted (right) death rates for heart disease among males, 1989–91. Data from *Monthly Vital Statistics Report* 43 (24 October 1994): 8–9.

the maps address only male mortality, which for heart disease is much higher than female mortality. Georgia, for example, has unadjusted and adjusted rates of 264.2 and 235.4 per 100,000 for males, both well above the corresponding rates of 246.8 and 123.2 per 100,000 for females. Although heart disease claimed almost equal numbers of the state's males (24,921) and females (24,690) during the three-year period, age adjustment reflects shorter life spans and earlier fatal heart attacks for men, usually attributed to higher levels of job-related stress, hostility, smoking, and obesity.[14]

In search of potentially important physiological and lifestyle factors, epidemiologists typically prepare separate maps for males and females. Dissimilar maps might reflect hazards specific to male-dominated occupations such as mining or agriculture, whereas similar maps could suggest causal factors affecting both sexes, such as race, environment, and access to health care. None of this works for heart disease, though, as the similar yet environmentally uninformative sex-specific maps in Figure 13.2 attest. Their lopsided geography correlates with the distribution of African Americans, but lack of a clear urban-rural division suggests an insignificant role for stress and other occupational factors, at least at the state level. However enigmatic, this pattern is no fluke: data for the early 1970s and mid-1980s yield a similar east-west dichotomy, as do separate maps for African Americans and whites.[15] Medical geographers Gary Shannon and Gerald Pyle, who observed a steady overall decline in fatal heart disease since 1950, considered hygiene and health care plausible factors yet hardly conclusive.

Diseases of the heart, age-adjusted death rates, 1989–91

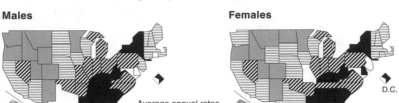

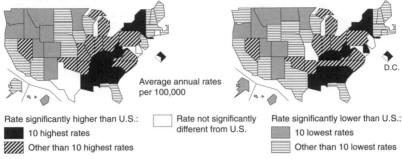

Figure 13.2. Age-adjusted death rates for heart disease among males (left) and females (right). Redrawn from maps in *Monthly Vital Statistics Report* 43 (24 October 1994): 8–9.

> [T]hese strong state and regional coronary heart disease mortality differences add interest to the question of unknown factors. Do the populations with high rates continue to practice poor health habits and thus have higher death rates? Or is the care of the heart attack victim in these areas such that survival rates are not as high as in other states? The answers to these questions are that we do not know which factor or combination of factors is responsible.[16]

As this example illustrates, in real epidemiology, maps often raise more questions than they answer.

The map key in Figure 13.2 reflects the epidemiologist's fondness for the statistician's notion of significance. The pair of maps appeared on separate pages in the *Monthly Vital Statistics Report*; in combining them at a much smaller size, I substituted my Visibility Base Map and a more efficient set of area symbols but retained the categories and wording of their original keys—except for fine print relating "significantly" to what statisticians call the 5 percent level. In statistical parlance, 5 percent significance refers to a probability model describing a hypothetical experiment in which Ping-Pong balls stamped with rates are plucked at random from a huge bowl.[17] The mean of all the rates in the bowl equals the overall nationwide rate, and rates close to the mean rate are much more likely than comparatively extreme rates, indicated by the tapered ends of the smoothed frequency graph in Figure 13.3. Five percent is the odds of drawing a rate either higher than R_{high} or lower than R_{low}, and according to the model, 95 percent of all randomly drawn rates will fall between these two markers. Because extreme rates are divided among separate regions, one with high values and the other with low ones, statisticians call this a "two-tailed test" for significance. The epidemiologist who applies the test to California pronounces the Golden State's actual rate "not significantly different" from the national average if it falls in the middle range. Otherwise, a rate above R_{high} is considered "significantly high," whereas a rate below R_{low} is "significantly low." Because the rates portrayed in Figure 13.2 reflect actual differences among substantial populations, rather than random events, few states fall in the middle, "not significantly different" category for either map.

My simplified explanation of the statistician's test for significance requires an important elaboration: the range of rates deemed not significant varies with the number of deaths reported.[18] Statisticians call this range the "confidence interval." For states like California and Texas, with large populations and many deaths, the confidence interval is comparatively narrow, so that a rate not too different from the national mean will be statistically significant at the 5 percent level. Less populous states, like Wyoming and Vermont, have many fewer deaths and a wider, more tolerant confidence interval, so that a rate deemed significant (but barely so) for California might prove insignificant for Vermont. Because the requirement for significance becomes more

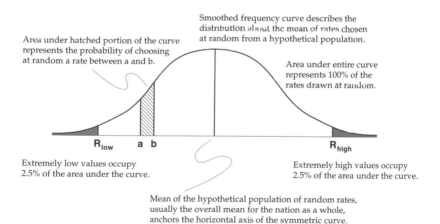

Figure 13.3. Mathematical curve describes the variation of randomly chosen rates in a hypothetical statistical experiment. Area under the curve represents relative frequency; 5 percent of this area lies in a zone of extremely high or extremely low rates and 95 percent falls in a middle zone representing values "not significantly different" from the mean.

stringent as the number of deaths declines, a rate that is statistically significant for Vermont might be insignificant for one of Vermont's counties or towns. Moreover, a city block, where only one person—or perhaps no one—dies of heart disease in a typical year might not achieve statistical significance unless several people die, yielding a rate several times the national rate. In statistical jargon, small areas with few cases typically lack the "power" to generate rates significantly different from the overall mean.

Because size affects significance, there's a troublesome trade-off between statistical reliability and geographic detail. A map based on subcounty units such as towns or census tracts might provide the detail needed to identify the effects of toxic waste sites, industrial pollution, or radon, but most—perhaps all—of its above-average rates could prove statistically insignificant. Moreover, rates that pass the significance test might be spurious. For example, a suburban neighborhood of tract houses built three or four decades ago might have appealed largely to young families similar in age. Back then the death rate was low, but now, with their children grown and gone, the parents are beginning to die in substantial numbers, largely of natural causes. And if some of these healthier and now more affluent folk move to the Sunbelt and leave behind a less healthy contingent of aging seniors, even age adjustment might not prevent an above-average death rate.

Uncertainty reflecting small numbers becomes even more problematic when the epidemiologist not only subdivides the population by race and sex but makes separate maps for specific diseases. The goal is clear: a more precisely defined disease, such as cancer of the lip, with a narrower range of risk factors than for cancer in general, is more likely to yield meaningful, readily understandable hot spots. Unfortunately, this greater specificity reduces the number of deaths and raises the threshold for statistical significance. To obtain more reliable rates, the epidemiologist might resort to larger areal units or a longer period of time, but larger areas dilute the effects of point sources of pollution, whereas a longer time span (one or two decades instead of three years, say) not only increases the proportion of decedents who spent much of their lives elsewhere but embraces a broader range of local exposures. An intensive local survey, with clinical examinations, interviews of relatives, and careful analysis of death certificates, could remove much of the noise, but these investigations are slow and expensive. Besides, an important role of disease mapping is screening for areas where a clinical or interview study might be useful.

The most conspicuous and persistent target of epidemiological cartography is cancer, a poorly understood collection of diseases, each identified with a particular part of the body, or site.[19] Oncologists theorize that each site reflects exposure to one or more specific carcinogens. Mouth cancer, for instance, is strongly correlated with pipe smoking and snuff dipping, whereas lung cancer is related not only to cigarette smoking but also to residential radon and occupational contact with asbestos fibers. Interaction among risk factors is also important, as demonstrated by a high incidence among smokers exposed to elevated concentrations of radon. Because cancer sites are numerous, the search for revealing hot spots requires not just a handful of maps but an entire atlas. The National Cancer Institute, which has published several cancer atlases since 1975, sees cancer mapping as a long shot with a substantial payoff for medical science and health policy.[20] The U.S. Environmental Protection Agency, which produced its own cancer atlas in 1987, considered the maps a contribution to both hypothesis generation and sound regulatory policy.[21]

The typical map in a U.S. cancer atlas addresses a single cancer site at the county-unit level with males and females covered on separate maps, usually on facing pages. The first atlas, published in 1975, covered the twenty-year period 1950–1969 but only for the white population. To avoid predominantly insignificant rates for less common cancer sites, the authors based thirty-eight of the sixty-five cancer maps on State Economic Areas (SEAs), which are individual counties or groups of similar counties, nearby if not contiguous. Each Standard Metropolitan Area is a separate SEA, and according to the Census Bureau's typology for counties, an SEA consists of either metropolitan or nonmetropolitan counties, never a mixture. A 1987 atlas addressing the white population relied entirely on SEAs, as did a companion volume published in 1990 for the nonwhite population.[22] Both atlases adopted a two-page layout juxtaposing separate maps for the periods 1950–59, 1960–69, and 1970–80 with an overall trend map identifying areas where rates generally increased (or decreased) between 1950 and 1980 and where change was significantly more (or less) rapid than for the country as a whole.[23]

Most cancer atlases use a classification similar in principle if not wording and format to that in Figure 13.2. The 1975 atlas, for instance, treated statistical significance and membership in the highest decile (top 10 percent) separately in allocating three of its five categories to above-average rates:

Significantly high, in highest decile
Significantly high, not in highest decile
In highest decile, not significant
Not significantly different from U.S.
Significantly lower than U.S.

Separate color schemes distinguished county maps from SEA maps, and hues focused on significant hot spots. A curious mix of unambiguous contrast and blatant ugliness, the county maps portrayed the first three categories with red, orange, and golden yellow, respectively, and the fourth, nonsignificant category in pea-soup green. The authors made the SEA maps look different by assigning the first four categories hot pink, light purple, dark periwinkle, and sky blue. In both schemes, though, white for the fifth, significantly low category deemphasized cool spots. In contrast, the 1987 atlas, which employed similar categories, adopted a temperature metaphor in assigning white to the fourth, nonsignificant category and allocating warm red and orange to the first two categories, a faint reddish purple to the third, and a cold blue to the last, significantly low group. As a result, hot spots and cool spots were distinct and readily apparent.

I asked Linda Pickle, project leader for the 1987 white and 1990 nonwhite cancer atlases, which maps were most revealing. With no hesitation, she identified temporal sequences of maps depicting lung cancer among white females and prostate cancer among nonwhite males.[24] As the black-and-white excerpts in Figure 13.4 illustrate, a cluster of significantly high death rates for cancers of the trachea, bronchus, lung, and pleura emerged between 1950 and 1980 in California, Nevada, Oregon, and Washington. Already apparent in more heavily urban areas during the 1950s, high mortality spread rapidly to the hinterland during the 1960s. A similar ominous upward trend in female lung cancer can be seen in other parts of the country, but the effect is most obvious in the leading-edge states of the West Coast. Because fatal lung cancer usually reflects a long period of exposure, Figure 13.4 suggests increased smoking inspired less by women's lib and Virginia Slims ads than by earlier propaganda touting tobacco's effectiveness as an appetite suppressant—my mother, a light smoker who died from lung cancer in 1989 in Maryland, well remembered commercials with the slogan "Reach for a Lucky, not a sweet!" Male lung cancer has a very different geography, less indicative of West Coast sophistication than the use of hand-rolled cigarettes in the lower

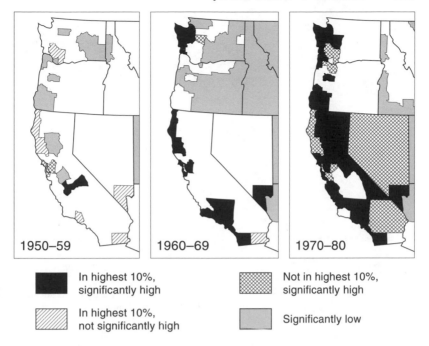

1950–59   1960–69   1970–80

■ In highest 10%, significantly high

▨ Not in highest 10%, significantly high

▨ In highest 10%, not significantly high

▢ Significantly low

Figure 13.4. Age-adjusted mortality rates among white females for cancers of the trachea, bronchus, lung, and pleura, by decade and State Economic Areas, 1950 to 1980. Excerpt redrawn from color maps in Linda Williams Pickle and others, *Atlas of U.S. Cancer Mortality among Whites: 1950–1980*, pp. 78–9.

Mississippi valley and occupational exposure to asbestos in shipyards along the Atlantic and Gulf coasts.[25]

More enigmatic is the recent increase in prostate cancer among nonwhites in the Southeast.[26] Figure 13.5 shows a remarkable expansion during the 1970s, when significantly high rates occupied a two-pronged belt extending from southeastern Virginia through western Georgia, with a substantial outlier across central Florida. Large numbers of African Americans in the Southeast helps account for the statistical significance, but why so many SEAs have significantly high rates remains a mystery. Because the 1987 atlas shows a very different geography of prostate cancer among whites, the explanation probably lies in either genetics or race-related traditions, including practices affecting environmental exposure.[27] The brief description of geographic patterns at the front of the 1990 nonwhite atlas noted a "dramatic" increase in fatal prostate cancer among blacks, with an average age-adjusted rate twice that for whites.[28] Aware of the problem in the

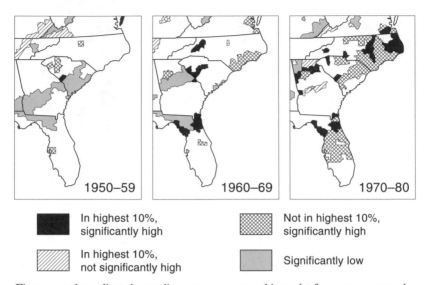

In highest 10%, significantly high

Not in highest 10%, significantly high

In highest 10%, not significantly high

Significantly low

Figure 13.5. Age-adjusted mortality rates among nonwhite males for prostate cancer, by decade and State Economic Areas, 1950 to 1980. Excerpt redrawn from color maps in Linda Williams Pickle and others, *Atlas of U.S. Cancer Mortality among Nonwhites: 1950–1980*, pp. 92–3.

late 1970s, researchers at the National Cancer Institute found no conclusive explanation.[29] There's no shortage of hunches, though, from poor use of preventive health care to ingestion of catfish contaminated with mercury and other heavy metals, which often accumulate in the beds of industrially polluted streams and lakes. However inconclusive, the maps stress the need for periodic prostate examinations and more rigorous testing of bottom-feeding seafood.

Valuable for both discovery and propaganda, disease atlases address a diverse audience of health scientists, policymakers, health-care advocates, and concerned citizens. Concerned that their maps be accessible across a wide range of uses, epidemiologists at the National Center for Health Statistics initiated a promising dialogue with cognitive scientists and cartographic researchers,[30] The first fruits of this collaboration appeared in 1996, with the center's atlas for leading causes of death among the nation's 805 health service areas for 1988–1992.[31] Especially innovative are the atlas's three styles of choropleth map, each with a color scheme tailored to a distinct map-reading task.[32] For example, a sequence of monochrome area symbols, ordered from light to dark, reflects broad regional patterns, whereas a double-ended scheme with distinctly different hues for high and low rates and more saturated colors for the more extreme values promotes comparison from map to map.[33] A third style, based on the traditional significance test, uses a nonordered sequence of highly varied hues to promote quick and accurate matches between color symbols on the map and corresponding samples in the key. (To further promote "rate readout," these maps occupy a full page, whereas maps in the other two styles fill only a quarter of a page.) Each style has a distinctive color scheme and legend, based on extensive testing of human subjects and designed to promote reliable map reading.[34]

I applaud maps tailored to map reading but remain wary of map categories tied to statistical significance and other rigid, standardized schemes. Although statistical significance provides a workable assessment of reliability, which varies with number of deaths, slavish adherence to the customary 5 percent level exiles moderately high rates deemed "not significant" to a visual limbo, which might well mask a weak but nonetheless meaningful geographic pattern evident, for example, at 15 percent but not at 5 percent. Impressive to uninformed lay readers, "significant" and "not significant" are little more than an excuse for not experimenting with cut-points. What's more, statistical testing—developed to help researchers draw reliable inferences about large populations from small samples of independent, random observations—is hardly necessary for death rates based on essentially complete vital registration records, which constitute a population, not a sample, random or otherwise. Inferential statistics, focused on the risks of wrongly rejecting (Type I error) and wrongly accepting (Type II error) a "null hypothesis," ignores the corresponding cartographic risks of "seeing wrongly" and "not seeing."[35]

What's the solution, then, aside from recruiting numerous epidemiologists, medical geographers, and other experts to experiment with cut-points and examine many different views of each map? That's a possibility, and not too farfetched, as suggested by promising experiments exploiting the Internet for computer-moderated collaborative design.[36] What I have in mind, though, is an expert system that seeks out hot spots by automatically experimenting with map categories and time periods while attempting to match maps of mortality data with maps of risk factors.[37] Generating and testing new cut-points for geographic correspondence is a tedious, repetitive task, at which computers excel, but vigilance also requires an exhaustive catalog of potentially meaningful patterns. The catalog would consist of geographic templates, each describing a health-hazard zone such as the neighborhood of a Superfund site, the area downwind or downstream from a firm reporting carcinogens to the Toxics Release Inventory, or a county where the *Census of Manufactures* suggests that workers might have been exposed to asbestos, mercury, lead, or another hazardous substance. Fragments of the catalog exist in data bases scattered throughout federal agencies, which should be encouraged to disseminate the information in a standardized format, as part of what Washington calls the National Spatial Data Infrastructure.[38] Published electronically, the catalog could enhance interactive mapping packages such as Epi Map, which the Centers for Disease Control distributes free over the Internet to encourage geographic surveillance of mortality data.[39] As a visual pump-primer for cartographic analysis, template-based prescreening can increase the social benefit of collecting mortality data as well as reduce the risk of not seeing a serendipitously revealing pattern.

In practice, cancer surveillance relies less on systematic attempts at discovery than on citizens worried about the high number of tumor cases in their neighborhood or town. Cancers clustered in space or time, especially those clustered in both space and time, are terrifying to relatives and neighbors, inherently intriguing to scientists, and a cause of deep concern for politicians, who respond in part by funding tumor registries and cancer surveillance programs.

Not all states have a tumor registry, and cancer surveillance programs are comparatively new.[40] The New York State Health Department, which has maintained a cancer registry since 1940, established a formal surveillance program in 1981 because of increased inquiries from the public in the wake of Love Canal.[41] According to program director Aura Weinstein, New York is atypical in not only responding to all inquiries about perceived excess cancers but conducting an in-depth cancer incidence investigation whenever requested.[42] "We don't turn anybody down," says Weinstein, whose program handles two to three hundred inquiries in an average year and initiates ten to twelve new investigations. First, though, a community liaison representative listens to citizens' concerns and provides information about cancer and its varieties, risk factors, and geographic patterns. If this information is insufficient and the citizen wants a formal study, Weinstein schedules an investigation based on her staff's current workload, the size of the perceived cluster, and the apparent social concern. A cluster of a half dozen cases of childhood leukemia, for instance, would most certainly take precedence over an equal or smaller number of more widely varied adult tumors.

Before determining the time period to be studied and whether to consider all cancers or only those for a particular site or group of sites, Weinstein and the Cancer Surveillance Advisory Committee, a diverse panel of experts within the department, tailor the study area to data requirements as well as the territory of concern to the requestor. An inappropriately large area would be costly and time-consuming to investigate, especially if the study requires interviews with patients or next of kin, whereas a small area might include too few cases or ignore potentially important causal factors. As a practical matter, study area boundaries reflect the availability of population data, required to calculate the expected number of cancers. A largely rural study area might consist of several towns or one or more Zip Code areas, whereas an urban or suburban study area could comprise several census tracts, each a more or less socioeconomically homogeneous area with roughly four thousand residents. When the requestor identifies a specific exposure, known or suspected, officials attempt to match the study area to the zone of most likely exposure.

The investigator culls cases (alive or deceased) from the state cancer registry, which integrates information from hospitals, private physicians, testing laboratories, death certificates, and health departments in neighboring states and is believed to be 95 percent complete. Determining whether a cancer case falls within the study area can be difficult, perhaps hopeless, if the reported address is a post office box or a rural delivery route, but Zip Codes help the researcher select cases to verify with other sources, such as phone directories and motor vehicle records. Voter registration lists, which always list the town or city of residence, are especially useful. An accurate count of confirmed cases is essential because statistical significance depends on a comparison with the expected number of cancer cases, computed from the study area's demographic structure and statewide age-specific incidence rates. In addition, the surveillance investigator looks for geographically significant clusters by plotting each case's residence at the time of diagnosis on an electronic "pin map" and applies one or more standard epidemiological strategies, such as examining whatever clinical data seem relevant, exploring deceased workers' occupational history and likely exposure, or matching each case with a demographically similar person chosen from the general population and known as the "control." Although research methods differ from study to study, each report must address two key questions:

- Do the accumulated statistical, epidemiological, and clino-pathologic data indicate the existence of unusual cancer patterns in the community?
- Do the accumulated data strongly suggest a positive relationship between these unusual cancer patterns and exposure data for any suspect environmental pollutants?[43]

The Cancer Surveillance Advisory Committee, which meets monthly and oversees the investigation, must approve the study report before its release to the requestor, county health officials, and the media. To avoid needless public anxiety, the committee makes certain the writer carefully avoids the inflammatory "C-word" (cluster).

Although tumor patients who perceive a cluster might eagerly identify themselves to the media, incidence investigation reports are deliberately vague about the location and even the number of confirmed cancer cases. Because the surveillance program does not release numbers smaller than six, a small cluster is described as "an unusual incidence of cancer cases." The report usually contains only one map, describing study area boundaries. Privacy is a paramount concern, and the health department neither identifies individual tumor cases by name nor reveals their addresses even vaguely, as with a dot on a map. Program staff and the Advisory Committee examine spatially detailed maps, but because the tumor registry guarantees confidentiality, readers of the report receive only the experts' interpretation. A pity, though, because local newspapers or ad hoc groups of victims' relatives sometimes publish their own maps, based on sources less thorough and reliable than the cancer registry.

Most investigations yield negative or mixed findings, as illustrated by the cancer study for Zip Code 10504 (Armonk), in Westchester County, for the period 1978–87.[44] A citizen had requested the study because of dry-cleaning solvents found in local water supply wells in the late 1970s. The investigation treated males and females separately and considered eight groups of tumor sites as well as cancer in general. A two-page summary preceding the eleven-page report offered a three-sentence conclusion:

> Among males overall, there was no statistically significant difference from the expected number of newly-diagnosed cancer cases (97 cases observed; 94 cases expected). Among females overall, a statistically significant excess of cases was observed for newly-diagnosed cancer cases

(121 cases observed; 86 cases expected). Specific cancer sites where differences were observed are identified below.

Only two specific sites proved statistically significant: male prostate cancer (23 cases observed; 13 cases expected) and female breast cancer (40 cases observed; 25 cases expected). For these sites as well as kidney cancer and leukemia (two more likely consequences of ingesting dry-cleaning fluid), the study found "no obvious spatial clustering or clustering in time." Nor could the investigation pinpoint particular environmental or occupational factors. However inconclusive, these results are nonetheless reassuring. (Although cancer is a common disease, striking one in three individuals and three-quarters of all families, people who choose a healthy lifestyle and periodic checkups have a right to know of deadly carcinogens in air, groundwater, soil, or local produce.) If reported cases were significantly higher for all eight tumor sites, the public would, indeed, have cause for alarm.

The health department's Center for Environmental Health attacks cancer clusters from a different direction.[45] Instead of focusing on places pointed out by concerned citizens, researchers in the Center's Bureau of Environmental and Occupational Epidemiology look first at exposure and then at incidence. In screening for areas with high levels of environmental pollution, the Bureau relies heavily on the EPA's Toxics Release Inventory (TRI), to which manufacturers must report the amounts of hazardous chemicals released to air, water, or soil. Because many of the substances listed are known or suspected carcinogens, the TRI data can suggest places that warrant careful monitoring, as can state and federal inventories of landfills and toxic waste dumps. To estimate potential exposure at a TRI site, an analyst uses a geographic information system to count residents within a one-mile circle. The GIS merely computes the distance between the potential hot spot and the centroids of each census block in the general area, and adds up the populations of all blocks within a mile. Applying address-matching and GIS software to the tumor registry yields a corresponding count of cancer cases, which might warrant further, more detailed investigation.

Public anxiety sometimes requires a multicounty regional analysis, as in the early 1990s, when a high incidence of breast cancer on Long Island suggested a causal link with air pollution. To address these concerns, the Bureau divided Nassau and Suffolk counties, Long Island's

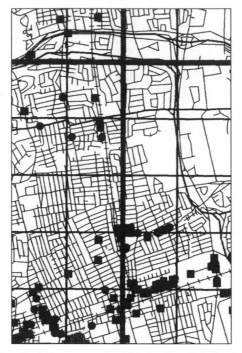

Figure 13.6. Black-and-white excerpt of a GIS plot (in color) showing industrial locations for two different years. Grid lines are one kilometer apart, and lines with short perpendicular cross-ticks represent railroads. Courtesy of the New York Center for Environmental Health.

two outer counties, into 5,809 grid squares, one kilometer on a side.[46] Because lifetime exposure is important, researchers identified 1,420 women who had lived on the Island for at least twenty years and were diagnosed with breast cancer between 1984 and 1986. They matched each case with a control similar in age and county of residence, interviewed the women in the case and control groups, and used address matching to assign participants to grid cells. Because breast cancer typically is not diagnosed until ten to twenty years after whatever exposure might have been critical, they relied on industrial directories for 1965 and 1975, well before the TRI. The directories identified facilities by address, industry, and number of employees. After eliminating facilities unlikely to emit airborne toxins and plants with fewer than twenty workers, the researchers converted addresses to grid coordinates and counted the number of chemical facilities in each cell for each time period. Figure 13.6, a portion of a GIS plot for the study, shows considerable geographic variation in the distribution of sites, which generally clustered near railway lines and older traffic arteries. In addition to comparing number of tumor cases with number of facilities across all grid cells, the researchers assessed the effects of traffic-related air pollution by comparing tumor incidence with traffic counts for twenty-five grid cells.

As with many cancer studies, the results were suggestive but not definitive. The study found no association between breast cancer and traffic volume or, for premenopausal women, between breast cancer and proximity to industry. But postmenopausal women with breast cancer tended to live closer to chemical facilities than women without breast cancer. Moreover, this association was stronger for cases who lived near a chemical plant before 1975, when air pollution regulations were less strict. Association is not causation, though, and other factors, such as income, could account for the relationship.[47] A public information handout summarizing the study and its implications posed a key question and then answered it with clear but classic scientific mugwumpery.

Should women living near a chemical plant on Long Island or elsewhere be concerned about these findings?

Women living near chemical sites on Long Island or elsewhere should not become overly concerned about these preliminary findings. The study results need to be confirmed through other investigations before any firm conclusions can be drawn.

No doubt follow-up studies will rely on a GIS to integrate exposure and incidence data as well as screen patterns for spatial association.[48]

A s social hazards amenable to mapping, crime and disease differ fundamentally in goals and constraints. Crime analysts, for instance, are concerned less with root causes than with effective control. This is understandable, I suppose, given the metaphysical complexity (not to mention ideological controversy) of determining causes of crime—one can map unemployment and poverty, of course, but how does one map meanness? Because citizens, public officials, and the police want effective control more than scholarly debate, the emphasis in crime mapping is on timeliness. If reported incidents are mapped promptly, with trends reliably detected and displayed, law enforcement's response has an excellent chance of being socially efficient, at least in the short run. Disease mapping, by contrast, involves much longer time periods as well as the logic and mores of laboratory science and statistical estimation. Unfortunately, the epidemiologist's

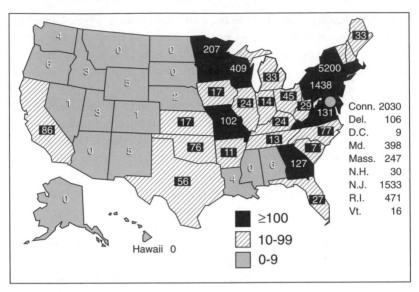

Figure 13.7. Number of reported Lyme disease cases, by state, 1994. Reprinted from *Morbidity and Mortality Weekly Report* 43 (23 June 1995): 460.

data often cannot sustain the level of proof to which medical scientists aspire. We might know a person's address at time of death or diagnosis, for instance, but pinpointing the place of exposure—if indeed there was a unique place of exposure—is usually impossible. The best we can do, it seems, is to try out hunches and search for connections ranging from narrowly focused occupational exposures in mine or factory to broader concerns such as the influence of state or local public policy on access to preventive health care. But because maps of death and disease might prove revealing, epidemiologists, geographers, and spatial statisticians work at refining their methods of aggregating data, screening trends, and simulating hypothetical patterns that provide a foundation for assessing significance. However elusive, discovery has a huge payoff, for the researcher as well as for society at large, that makes the wild goose worth chasing.[49]

Although epidemiological cartography is directed toward discovering linkages between disease and environment, maps of mortality and illness often serve a second goal: public awareness of new or escalating threats such as AIDS and Lyme disease, a bacterial infection carried by deer ticks and first identified in Lyme, Connecticut. Although propaganda plain and simple, these maps are neither devious nor sinister.

For example, the geographer Peter Gould developed a dramatic cartographic narrative describing the spread of AIDS in the United States in the 1980s, when public health officials ignored the diffusion of the disease down the urban hierarchy from big cities to smaller cities as well as outward, into the rural hinterlands.[50] Based on controversial projections from incomplete data, Gould's maps reminded readers of *Forbes*, *Playboy*, and *Time* that HIV is not just a scourge of big cities like New York, San Francisco, and Miami. When presented at the International AIDS conference in 1992, in the Netherlands, the maps were an epiphany to epidemiologists who had viewed the disease as a temporal, not a spatial-temporal, phenomenon. And as an animation on educational television, Gould's maps showed middle and high school students that AIDS is a nationwide threat to which no community, large or small, is immune. With less flair and poorer spatial resolution, the U.S. Public Health Service used a black-and-white state-level map of case counts (Figure 13.7) to advise local health officials that by the end of 1994 Lyme disease had spread to all but seven states.[51] The map also caught the attention of the *New York Times* and other newspapers, which reminded readers of the risks of walking in the woods with arms or legs exposed.[52] Because they relate health hazards to where people live, disease maps are pointedly effective warnings of danger.

Chapter Fourteen

Emerging Cartographies of Danger

A momentous adaptation of electronics and numerical analysis, hazard-zone maps are a new cartographic genre, as distinct in appearance and function as property maps and navigation charts. Equally significant is the likely effect of environmental risk maps on politics and society as the world's burgeoning population and complex economy grow ever more vulnerable to catastrophe. In the same way that boundary symbols and rhumb lines were catalysts for territorial claims and international trade, maps of natural and technological hazards afford pragmatic strategies for regulating land use and dodging disaster. Preventive measures can be painful, though, and cartographic warnings often go unheeded until an unprecedented yet foreseeable catastrophe demonstrates the wisdom of listening.

Hazard-zone mapping is a recent phenomenon, with few representatives of the genre in existence fifty years ago, before digital cartographic data and computer simulation.[1] Whereas primitive precedents merely hinted at risk in their depictions of faults, volcanic deposits, and mortality rates, contemporary delineations of geographic hazards and relative risk are novel in their specificity, variety, and ostensible precision. And while earlier hazard maps offered description and explanation, current efforts focus on forecasting and monitoring. The data might be flawed and the models unrefined, but agencies authoring hazard maps clearly know what they want to show. Although the risk maps for many threats merit refinement, the conceptual revolution—broad recognition that society can and must delineate hazardous geographic environments—is essentially complete.

A consequence of this revolutionary dependence on maps is tolerance and trust. Government and the public are enthusiastic believers, often willing to overlook imprecision and uncertainty when (and where) hazard maps provide the scientific foundation, however tenuous and rhetorical, for much-needed regulation and mitigation. Although atmospheric dispersion models and flood insurance maps have yet to attain the relative certitude of aeronautical charts, elected officials recognize air pollution and flood damage as serious threats that demand analysis and description, however tentative. Necessity begets acceptance, and if control and insurance programs must be based on incomplete data and questionable delineations, so be it. One need not be an avid postmodernist to recognize that hazard-zone maps are social constructions.

Although widely accepted, hazard-zone maps are often controversial. Developers, property owners, and local officials understandably object when higher authorities thwart their plans by invoking maps that are conceptually weak if not obviously flawed. And project opponents lodge equally vehement protests when hazard maps appear to understate perceived threats to the environment, racial minorities, or nearby residents. Because complaints about specific delineations often reflect ignorance of cartographic generalization, regulatory agencies that adopt a less than perfect "official map" must be prepared to defend these representations as appropriate, consistent, and fair. Imprecision and uncertainty are unavoidable, but autocratic pronouncements that "the map is the map" are politically explosive, if not ultimately self-defeating, unless government maintains a high, uniform standard of data quality and provides prompt correction of obvious oversights.

Uniform accuracy is often impossible, if not pragmatically undesirable, because detailed mapping is expensive and slow, especially when reliable estimates require measurements recorded continuously at many sample points over many years. Flood insurance maps are especially difficult to upgrade because hydrologic models based on stream discharge and water elevation data depend upon thirty or more years of record. Indoor radon poses a different yet equally challenging sampling problem: not only can potential risk vary considerably from house to house within a neighborhood, but measurable concentration varies seasonally as well as with the occupants' lifestyle. What's more, radon testing is usually at the discretion of the resident, who can bias the result by manipulating air flow or misstating the placement of canisters within the home; because fudging is so easy, laws requiring radon testing prior to sale offer the home buyer little more than a naïve sense of security. Although environmental scientists address these sources of uncertainty with additional data focusing on radon's "geologic potential," the resulting maps can only hint at a lung-cancer hazard lurking in the basement. What the maps really show are zones where conscientious and widespread testing might be most revealing.

The radon controversy is a classic example of the map's power to validate a dubious hazard. As propaganda to encourage radon testing, the EPA's flashy reddish orange–yellow map is an effective attention-getter. But how medically sound and socially beneficial are its large, ominous red zones, based on a 4 pC/l "action level" that predicts dire consequences for much of the nation? Unless backed by vigorous federal, state, or local efforts to educate home owners and encourage (if not subsidize) repeated testing and appropriate remediation, this colorful abstract portrait of risk is at once alarmist and trivial. Is it reasonable, for instance, for the map to lump together all concentrations greater than 4 pC/l, thereby recognizing no difference between 5, 25, or even 105 pC/l? Would not a higher cut-point, such as Canada's 20 pC/l threshold, target a much smaller area within which a comparatively greater risk warrants a more focused (and presumably more effective) effort to identify "hot zones"? Is this not an attempt to impose certainty where uncertainty should be a central element in the discourse? This missed opportunity calls into question the EPA's motives: Is its three-class radon map largely a ploy for exaggerating the importance (and funding needs) of the agency's radon program?

It's wise to question the map author's motives, especially when a federal agency is pitching Congress for a new program. Fortunately, few hazard maps are as simultaneously threatening and outlandish as the pair of images with which the National Aeronautics and Space Administration promoted a high-tech defense against colliding asteroids such as the large "near earth object," or NEO, 10–15 km (6–9 miles) in diameter that many paleontologists now blame for the rapid demise of

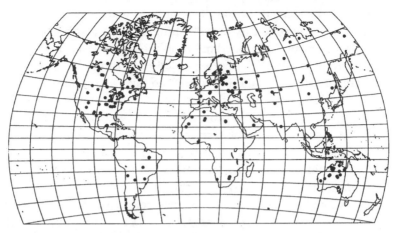

Figure 14.1. The distribution of terrestrial impact craters as a risk map. Reprinted from David Morrison, ed., *The Spaceguard Survey: Report of the NASA International Near-Earth-Object Detection Workshop*, included in U.S. Congress, House Committee on Science, Space, and Technology, *The Threat of Large Earth-Orbit Crossing Asteroids: Hearing before the Subcommittee on Space*, 103d Cong., 1st sess., 1993, p. 121.

the dinosaurs.[2] To convince Washington to appropriate the $50 million start-up cost of the "Spaceguard Survey," a sophisticated network of satellites and early warning telescopes with which astronomers would catalog NEOs and forecast disastrous intersections with the earth's orbit, NASA technocrats documented the threat with a map (Figure 14.1) showing 130 terrestrial impact craters roughly 140 to 200 km (87 to 124 miles) in diameter accumulated over the past two billion years.[3] Doomsday space collisions are not science fiction, the map says: asteroids, comets, and giant meteorites have struck the earth many times in the past, and can strike again with devastating results.

Were the climatic consequences of an asteroid collision not worldwide, the pattern of dots might suggest that Western society is at greater risk than the rest of the planet. Not so, Spaceguard's backers eagerly point out in the map's caption: the greater density of dots in the eastern United States, western Europe, and northwestern Australia merely reflects geographic variation in geologic stability and geophysical exploration.[4] Because erosion and subduction have erased some cosmic craters while others lie undiscovered beneath the sea, polar ice caps, or tropical jungles, the map's 130 dots (in their number if not their size) understate the threat.

Asteroids big enough to create craters are not the only danger. Lest a well-read legislator recall that most meteorites burn up in the atmos-

phere and never reach the ground, the authors juxtaposed simple outline maps of New York City and Washington, D.C., with a similarly
scaled plan view of the 2,000 km² (800 mi²) area devastated by the
comparatively small NEO that exploded over Tunguska, Siberia, in
1908 (Figure 14.2). Looming over the maps, a photograph of
scorched and shattered trees attests to the damage resulting when this
60-meter (200-foot) "cosmic projectile" disintegrated at an altitude of
about 8 km (5 miles) with the force of a 12-megaton bomb.

What makes NASA's maps seem overwrought if not preposterous
is the implication that there's something we earthlings can do—besides
pray, commit mass suicide, or plan the bacchanalian party to end all
parties—should the Spaceguard Survey forecast a fatal terrestrial
impact for, say, April 3, 2049, at 10:32 A.M. EST. It turns out, though,
that Big Science can construct a shield no more outlandish than the
Star Wars defense system touted by the Reagan administration in the
mid-1980s. The principle is much the same: compute the approaching
object's trajectory, launch an unmanned rocket with a massive warhead, and blow up the bugger before it does any harm.[5] Whereas Star
Wars must destroy incoming intercontinental ballistic missiles in mid-
flight, a Spaceguard interceptor would move into a parallel path,
approach to an optimum distance, and detonate a bomb with sufficient power to deflect the asteroid into a slightly different, significantly safer orbit. And by counterattacking millions of miles from earth
and years before the slated collision, this asteroid deflection scenario
avoids fallout and other debris and still leaves time, should the mission
fail, to launch another interceptor. And if hurling atomic missiles into
outer space doesn't work, a mirror six miles wide could vaporize just
enough of the approaching asteroid to alter its orbit.[6] Too bad the
space agency can't claim the film rights.

Even so, aren't the odds of colliding with an asteroid too low to justify the expense of a comprehensive inventory? To address this question
NASA calculated the risk for an average year.[7] The risk assessment
focused on two events: a "globally catastrophic impact" with an object
at least 1 km in diameter and "Tunguska-class impacts" similar to the
devastation portrayed in Figure 14.2. NEOs large enough to threaten
civilization with a worldwide fire storm, persistent crop failures, and
massive mortality have an estimated recurrence interval of 500,000
years, which reflects an annual probability of 1/500,000. If a quarter of
the world's population were to perish, each person's annual probability of death would be 1/2,000,000, which applied to a U.S. population

of about 250 million yields only 125 "annual equivalent deaths"—a bit more than the yearly average of 100 tornado deaths but far below our annual tally of 50,000 traffic fatalities.[8] Worldwide, the average annual threat of a global catastrophe is 2,500 deaths, still a fraction of America's human roadkill. Tunguska-class events prove even less worrisome when progressively more specific recurrence intervals are estimated for the entire globe (300 years), the earth's populated areas (3,000 years), urban areas (100,000 years), and urban areas in the United States (1,000,000 years). The annual probability of death for an individual is 1/30,000,000, which translates into a mere 15 equivalent annual deaths for the nation and only 150 deaths worldwide.

Is it worth $50 million down and $10 million a year for 25 years just to catalog NEOs and forecast asteroid collisions? Absolutely. We spend billions of dollars on Big Science anyway, and $300 million for a better sense of whether and when we might lose the game of cosmic roulette makes the Spaceguard Survey seems a good deal for taxpayers—especially before the Congress starts investing in an Asteroid Wars defense system. We'd gladly pay, I'm sure, an equal amount for an accurate, long-range worldwide forecast of tornado paths. Tornadoes, of course, are a recurrent hazard, and several times a year our

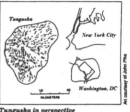

Tunguska in perspective

Figure 2-2. On June 30, 1908, at 7:40 AM, a cosmic projectile exploded in the sky over Siberia. It flattened 2,000 square kilometers of forest in the Tunguska region. If a similar event were to occur over an urban area today, hundreds of thousands of people would be killed, and damage would be measured in hundreds of billions of dollars.

Figure 14.2. Maps and a photograph authenticate the hazards of cosmic impacts. Reprinted from David Morrison, ed., *The Spaceguard Survey: Report of the NASA International Near-Earth-Object Detection Workshop*, included in U.S. Congress, House Committee on Science, Space, and Technology, *The Threat of Large Earth-Orbit Crossing Asteroids: Hearing before the Subcommittee on Space*, 103d Cong., 1st sess., 1993, p. 125.

leaders would have to justify their stinginess to thousands of the victims' friends, relatives, and creditors. Asteroid catastrophes, despite their higher consequences, are far too rare to attract support from a public largely ignorant of mathematical probability and downright skeptical of any project immediately beneficial only to astronomers, space agency employees, and government contractors of the high-technology kind.

There's a caveat, though: Spaceguard cannot reliably detect objects less than 100 m (330 feet) in diameter.[9] Even though the NEO catalog would grow prodigiously in its first decade, a nearly complete inventory of earth-crossing objects in the 100 m range would require several centuries of observation. Recognition depends on size and distance, and despite impressive advances in recent decades, imaging technology does not readily recognize small but deadly objects such as the 60 m asteroid that exploded in the lower atmosphere over Tunguska. Particularly troublesome are long-period comets, which require twenty years or more to orbit the sun. Because Spaceguard focuses on the asteroid belt, a new comet large enough to be locally if not globally catastrophic might become apparent a mere three months before impact. At that range, deflection might be impossible, but there's still time for evacuation—if scientists armed with maps could convince a skeptical public that the sky is falling.

Giant mirrors and interceptor missiles with nuclear warheads create another hazard, warns astronomer Carl Sagan.[10] Although an atomic explosion on or just above the surface of an earth-crossing asteroid would tend to deflect the object away from a fatal impact, a few carefully planned nuclear explosions could deflect an otherwise harmless asteroid onto a collision path with the planet. "There is no other way known," warns Sagan, "in which a few nuclear weapons could by themselves threaten the global civilization." Whatever the motive of his hypothetical mad scientist—on a smaller scale we've seen them all: criminal extortion, lust for power, religious obsession, utter meanness, amoral curiosity—the threat of nuclear leveraging is daunting. What's more, even a well-intended direct hit on an incoming NEO might shatter the target into a cluster of smaller but deadly asteroids, some of which could impact simultaneously on different parts of the planet. This professedly farfetched scenario adds another dimension to the unforeseen vulnerability of elaborate mitigation schemes, such as groins, artificial levees, and asbestos removal.

Most risk maps involve statistical models of some sort for estimating the likelihood of a rare event, such as a volcanic eruption or disastrous flood; for assessing the cumulative impact on humans of small but sustained doses, as for indoor radon; or for describing variation in toxic exposure throughout an area, as for air pollution. Although statistical guesswork and quantitative modeling have become highly sophisticated, climatic change and land development can undermine the most rigorous estimation techniques. Roofs, parking lots, and drained wetlands expedite runoff and increase the chance of flooding, for example, and the poor family's new mobile home provides less protection from tornadic winds than a ramshackle farmhouse or an urban tenement. When forecasting requires a representative record of the hazard's magnitude and variability, an unsteady environment can make any period of observation too short. And comparatively rare hazards, like volcanic eruptions, are inherently uncertain because rational projections based on the most comprehensive data cannot guarantee a future that uniformly replicates the past.

What if there is no past? Should we bother forecasting terrorist attacks on dams and levees, worst-case nuclear meltdowns, and global nuclear warfare? Of course—we have little choice. Uncertainty is no excuse for ignorance, and however unthinkable the catastrophe, forecasting and planning are more rational than blind acceptance of a fate we might at least influence, if not control. Planning promotes vigilance, after all, and vigilance reduces risk. Governments must be vigilant individually and collectively, and emergency managers must prepare for a full range of plausible disasters. Evacuation and rescue are appropriate strategies, which require maps of relative risk. For nuclear plants, for instance, we can plan evacuations based on what we know about plant layout and operation, the amounts of stored fissionable material on site, the surrounding population, wind patterns, the food chain, and of course, the behavior and deadliness of various reaction products. And in pointing out where an orderly evacuation is impossible, maps can show where not to site an atomic power station in the first place. On a global scale, models of "nuclear winter" can identify world regions especially vulnerable to crop failure.[11] In pointing out where survival is questionable, maps can be a powerful argument for nuclear nonproliferation agreements.

That risk maps are rhetorical should surprise no one—persuasion is their prime role, after all, and if they fail to convince or at least command attention, they miss their most important target. There's a danger, though, in maps that under- or overstate risk, thereby either promising false security or raising needless alarm. All risk maps are at least partly rhetorical, if only to enlarge understanding of a hazard by momentarily ignoring the attendant uncertainty. And as social constructions, risk maps can always be manipulated—sometimes a little, sometimes a lot—to serve the map author's agenda. Skepticism (but not cynicism) is always healthy.

For rapid-onset weather hazards like tornadoes and flash floods, monitoring is an essential complement to statistical forecasting. While probabilities reflected in hazard maps safeguard our homes and businesses through building codes, land-use controls, and insurance programs, severe-storm warnings tell us when to evacuate or take shelter. But make no mistake: monitoring is a complement to mapping, not a substitute for it. Warning systems might appear to save more lives, but risk maps promote shelters and safety drills as well as guide the design of optimally efficient monitoring systems. In turn, data collected through monitoring promote cartographic refinement. And for somewhat predictable phenomena like street crime and brush fires, monitoring combined with mapping can help local authorities take corrective measures and issue timely warnings.

Electronic cartography and the World Wide Web, which can link local users to vast data archives and powerful software, promise marked improvements in timeliness and specificity. Essential in emergency management, maps tailored to specific situations are equally valuable in emergency preparedness. As examples in chapter 2 demonstrated, seismic risk maps are especially useful when customized to reflect type of construction or a rupture with a specific magnitude and direction at a specific location along a specific fault. Similarly, maps projecting the consequences of heavy rains or strong winds can reflect season of the year or time of day, both of which influence the size and locations of the aged, the very young, and other vulnerable populations as well as soil moisture, which affects the likelihood of flooding, landslides, and tree fall. By encouraging users to explore the effects of specific circumstances, interactive mapping promotes an understanding of risk that is difficult, if not impossible, through traditional, print-based cartography.

The most telling question is last: Why do we map some hazards but not others? Its answer lies in each hazard's unique geography, the number of people a single disaster can kill or leave homeless, and the ability of potential victims to reduce their individual or collective risk.

We map hazards because the maps afford control, either real or imagined, over fate and nature. For low-probability, high-consequence events like cosmic impactors and reactor meltdowns, mapping can inform and justify a course of action, such as the Safeguard Survey and emergency planning zones around nuclear plants. Local variation in vulnerability makes risk mapping especially worthwhile: whereas a small-scale world map of massive impact craters is blatant propaganda, large-scale flood zone maps provide a rational base for insurance programs and land-use restrictions. Even crime and traffic crashes, widespread if not ubiquitous threats, warrant maps that point out promising surveillance sites. By contrast, lightning strikes and chainsaw mishaps don't inspire hazard maps because random events or personal carelessness make their geographies unimportant.

Equally significant is our ability to observe, measure, and forecast danger. Hazard mapping's technological dependence is especially apparent in seismic-risk maps, which advanced from crude plots of earthquake damage to maps delineating hazard zones around surface faults to multisheet atlases of risk maps customized for various types of construction. In addition to refining risk maps of other natural hazards, technology promises the discovery of new, heretofore unsuspected dangers—like nonexplosive chemicals, indoor radon, and earth-crossing asteroids just a few decades ago. And as nuclear power, air pollution, and electromagnetic fields around power lines testify, technology creates its own hazards, which we must map, if only to determine whether to take the new threats seriously.

Not all dangers merit large-scale maps with precisely delineated hazard zones. Because reliable data are expensive, the social benefits of detailed risk mapping must be apparent through safety education, hazard mitigation, or emergency management. Where cartographically orchestrated risk reduction requires land-use restrictions many citizens consider intrusive, the public benefits must be sufficiently obvious and widespread to offset libertarian complaints and local opposition. Experience in managing the risk of earthquakes, riverine flooding, and coastal storms suggests increasing public approval—propelled by informed fear and a sense of control—of hazard-zone maps. Acceptance of cartographic simulation and risk assessment as normal and necessary may prove as monumental a turning point in public administration as the acceptance centuries ago of boundary maps, without which land ownership, taxation, and zoning would be impossible.

NOTES

CHAPTER ONE

1. Rodger Brown, National Severe Storms Laboratory, telephone conversation with author, 22 April 1994.

2. NOAA's National Weather Service uses this risk-contour map in an eight-page handout on tornado safety.

3. David Franke and Holly Franke, *Safe Places* (New York: Arlington House, 1972); and David Franke and Holly Franke, *Safe Places for the 80s* (New York: Dial Press/Doubleday, 1984).

4. Franke and Franke, *Safe Places*, 20.

5. Franke and Franke, *Safe Places for the 80s*, 1–4.

6. Ibid., xiii.

7. Franke and Franke, *Safe Places*, 121.

8. See, for example, Jeanne B. Perkins, *The San Francisco Bay Area—On Shaky Ground* (Oakland, Calif.: Association of Bay Area Governments, 1987), 6–7.

9. David Savageau and Richard Boyer, *Places Rated Almanac: Your Guide to Finding the Best Places to Live in North America* (Chicago: Rand McNally, 1981, 1985; New York: Prentice Hall Travel, 1989, 1993).

10. Norman Crampton, *The 100 Best Small Towns in America* (New York: Prentice-Hall, 1993).

11. David Franke, *America's 50 Safest Cities* (New Rochelle, N.Y.: Arlington House, 1974).

12. Hugh Bayless, *The Best Towns in America: A Where-to-Go Guide to a Better Life* (Boston: Houghton Mifflin, 1983).

13. Ibid., 17.

14. Benjamin A. Goldman, *The Truth about Where You Live: An Atlas for Action on Toxins and Mortality* (New York: Random House/Times Books, 1991).

15. Federal Emergency Management Agency, *Risks and Hazards: A State by State Guide*, FEMA publication no. 196 (Washington, D.C., 1990).

CHAPTER TWO

General references on earthquake hazards include Philip R. Berke and Timothy Beatley, *Planning for Earthquakes: Risk, Politics, and Policy* (Baltimore: Johns Hopkins University Press, 1992); Bruce A. Bolt, *Earthquakes* (New York: W.H. Freeman and Co., 1978, 1993); James M. Gere and Haresh C. Shah, *Terra Non Firma: Understanding and Preparing for Earthquakes* (New York: W.H. Freeman and Co., 1984); Richard Holden and Charles R. Real, *Seismic Hazard Information Needs of the Insurance Industry, Local Government, and Property Owners in California: An Analysis* (Sacramento: California Division of Mines and Geology, 1990); Karl V. Steinbrugge, *Earthquakes, Volcanoes, and Tsunamis: An Anatomy of Hazards* (New York: Skandia America Group, 1982); and J. I. Ziony, ed., *Evaluating Earthquake Hazards in the Los Angeles Region—An Earth Science Perspective*, U.S. Geological Survey Professional Paper no. 1360 (Washington, D.C., 1985). General sources discussing applications of seismic mapping include David M. Perkins, "Seismic Risk Maps," *Earthquake Information Bulletin* 6 (November–December 1974): 10–15; Steinbrugge, *Earthquakes, Volcanoes, and Tsunamis*, 39–68; and Joseph I. Ziony and John C. Tinsley, "Mapping the Earthquake Hazards of the Los Angeles Region," *Earthquake Information Bulletin* 15 (July–August 1983): 134–41.

1. Peter L. Ward and Robert A. Page, "The Loma Prieta Earthquake of October 17, 1989," *Earthquakes and Volcanoes* 21 (1989): 215–46.

2. Robert C. Bucknam, Eileen Hemphill-Haley, and Estella B. Leopold, "Abrupt Uplift within the Past 1700 Years at Southern Puget Sound, Washington," *Science* 258 (1992): 1611–14.

3. "Earthquakes and Risk in California," *Earthquake Information Bulletin* 14 (May–June 1982): 98–107.

4. Earl W. Hart, *Fault-Rupture Hazard Zones in California*, Special Publication no. 42 (Sacramento: California Division of Mines and Geology, 1992), 22. In 1994, the legislature officially renamed the special studies zones "earthquake fault zones." See "Name Change for Alquist-Priolo Special Studies Zone Act," *California Geology* 47 (March/April 1994): 62.

5. Hart, *Fault-Rupture Hazard Zones*, 2.

6. Not surprisingly, the number of registered geologists rose significantly after the legislature passed the Alquist-Priolo Act, which cynics labeled the "geologists' retirement act."

7. Berke and Beatley, *Planning for Earthquakes*, 117–9.

8. Blue Print Service Company, 1147 Mission Street, San Francisco, CA 94103. The telephone number is (415) 512-6550.

9. Special Publication no. 42, *Fault-Rupture Hazard Zones in California*, is sold by the California Division of Mines and Geology, 801 K Street, MS 12-30, Sacramento, CA 95814-3532. The telephone number is (916) 445-5716.

10. Hart, *Fault-Rupture Hazard Zones*, 21–2.

11. Jeanne Perkins, Association of Bay Area Governments, interview by author, Oakland, Calif., 31 March 1994; and Earl Brabb, U.S. Geological Survey, interview by author, Menlo Park, Calif., 4 April 1994.

12. Risa Palm, *Real Estate Agents and Special Studies Zone Disclosure: The Response of California Home Buyers to Earthquake Hazards Information*, Program on Technology, Environment and Man Monograph no. 32 (Boulder, Colo.: University of Colorado, Institute of Behavioral Science, 1981).

13. Risa Palm and Michael E. Hodgson, *After a California Earthquake: Attitude and Behavior Change*, Department of Geography Research Paper no. 233 (Chicago: University of Chicago Press, 1992), 90–5; and Michael E. Hodgson and Risa Palm, "Attitude Toward Disaster: A GIS Design for Analyzing Human Response to Earthquake Hazards," *Geo Info Systems* 2 (July/August 1992): 41–51.

14. "World-Wide: Embattled Federal Officials Insisted . . . ," *Wall Street Journal*, 24 January 1994, A-1; and Calvin Sims, "California Cuts Early Estimate of Quake Costs," *New York Times*, 9 February 1994, A18.

15. Mark Petersen, "The January 17, 1994, Northridge Earthquake," *California Geology* 47 (March/April 1994): 40–5; and Richard A. Kerr, "How Many More after Northridge?" *Science* 263 (1994): 460–1.

16. Ross S. Stein and Robert S. Yeats, "Hidden Earthquakes," *Scientific American* 260 (June 1989): 48–57.

17. California Seismic Safety Commission, List of "Significant California Earthquakes," ca. 1993.

18. Kenneth Reich and John Johnson, "Epicenter Moves to Reseda, Name Stays in Northridge," *Los Angeles Times*, 2 February 1994, B1.

19. Enrique E. Bello and Ernest Hardy, "Geographical Information Systems in Natural Hazard Management," in Organization of American States, *Primer on Natural Hazard Management in Integrated Regional Development Planning* (Washington, D.C., 1990), 5-1 to 5-24; "GIS Responds to Los Angeles Basin Earthquake Disaster," *ARC News* (Environmental Systems Research Institute) 16 (winter 1994): 1–2; and Jeffrey Star and John Estes, *Geographic Information Systems: An Introduction* (Englewood Cliffs, N.J.: Prentice Hall, 1990).

20. A. M. Rogers, J. C. Tinsley, and R. D. Borcherdt, "Predicting Relative Ground Response," in K. H. Jacob, ed., *Proceedings from the Symposium on Seismic Hazards, Ground Motions, Soil-Liquefaction and Engineering Practice in Eastern North America, October 20–22, 1987* (Buffalo, N.Y.: National Center for Earthquake Engineering Research, 1987), 403–32; and Stephanie A. King and Anne S. Kiremidjian, "GIS for Regional and Seismic Hazard and Risk Analyses," in J. David Frost and Jean-Lou A. Chameau, eds., *Geographic Information*

Systems and Their Application in Geotechnical Earthquake Engineering (New York: American Society of Civil Engineers, 1993), 56–60.

21. Keiiti Aki and Paul G. Richards, *Quantitative Seismology: Theory and Methods* (San Francisco: W.H. Freeman and Co., 1980), 799–849; and Ralph J. Archuleta, William B. Joyner, and David M. Boore, "A Methodology for Predicting Ground Motion at Specific Sites," in E. E. Brabb, ed., *Progress on Seismic Zonation in the San Francisco Bay Region*, U.S. Geological Survey Circular no. 807 (Washington, D.C., 1979), 26–36.

22. Bernice Bender, "Seismic Hazard Estimation Using a Finite-Fault Rupture Model," *Bulletin of the Seismological Society of America* 74 (1984): 1899–1923; and B. F. Howell, Jr., and T. R. Schultz, "Attenuation of Modified Mercalli Intensity with Distance from the Epicenter," *Bulletin of the Seismological Society of America* 65 (1975): 651–65.

23. Raymond T. Laird and others, *Quantitative Land-Capability Analysis*, U.S. Geological Survey Professional Paper no. 945 (Washington, D.C., 1979), 1–35; and Jeanne B. Perkins, "Regional Planning for Earthquake Hazards in the Eastern Bay Area," in Earl W. Hart, Sue E. Hirschfeld, and Sandra S. Schulz, eds., *Proceedings of the Conference on Earthquake Hazards in the Eastern San Francisco Bay Area*, Special Publication no. 62 (Sacramento: California Division of Mines and Geology, 1982), 405–14 .

24. Jeanne B. Perkins, *The San Francisco Bay Area—On Shaky Ground* (Oakland, Calif.: Association of Bay Area Governments, 1987), 5; also see William J. Kockelman and Earl E. Brabb, "Two Examples of Seismic Zonation in the San Francisco Bay Region," *Earthquake Information Bulletin* 13 (May–June 1981): 80–4.

25. Perkins, interview.

26. H. O. Wood, "Distribution of Apparent Intensity in San Francisco," in Andrew C. Lawson and others, *The California Earthquake of April 18, 1906: Report of the State Earthquake Investigation Commission* (Washington, D.C.: Carnegie Institution of Washington, 1908), vol. I, pt. I, 220–45.

27. Perkins, interview.

28. Perkins, *On Shaky Ground*, 2.

29. Ibid., 10–11.

30. While the six-category maps of shaking intensity employ an unambiguous series of shading symbols, the seven symbols used for the risk maps have two flaws. The symbol for "moderately low" is slightly darker than that for "moderate," and the parallel-line symbols for "moderately high" and "high" differ only in orientation.

31. *On Shaky Ground*, 11.

32. Brabb, interview.

33. Earl E. Brabb, "Analyzing and Portraying Geologic and Cartographic Information for Land-Use Planning, Emergency Response, and Decision-making in San Mateo County, California," *GIS '87—San Francisco: Second Annual International Conference, Exhibits and Workshops on Geographic Information Systems* (Falls Church, Va.: American Society for Photogrammetry and Remote Sensing, and American Congress on Surveying and Mapping, 1987), 362–74.

34. David F. Keefer and others, "Real-Time Landslide Warning during Heavy Rainfall," *Science* 238 (1987): 921–5.

35. E. E. Brabb, "Priorities for Landslide Mapping during the International Decade of Hazard Reduction," *Proceedings of the Seventh International Conference and Field Workshop on Landslides in Czech and Slovak Republics, 28 August–15 September 1993* (Rotterdam: A. A. Balkema, 1993), 7–14.

36. E. E. Brabb, "Proposal for Worldwide Landslide Hazard Maps," *Proceedings of the Seventh International Conference and Field Workshop on Landslides in Czech and Slovak Republics, 28 August–15 September 1993* (Rotterdam: A. A. Balkema, 1993), 15–27.

37. Glenn Borchardt, "Preparation and Use of Earthquake Planning Scenarios," *California Geology* 44 (1991): 195–203.

38. James F. Davis and others, *Earthquake Planning Scenario for a Magnitude 8.3 Earthquake on the San Andreas Fault in the San Francisco Bay Area*, Special Publication no. 61 (Sacramento: California Division of Mines and Geology, 1982).

39. Ward and Page, "The Loma Prieta Earthquake," and U.S. General Accounting Office, *Loma Prieta Earthquake: Collapse of the Bay Bridge and the Cypress Viaduct*, report no. RCED-90-177 (June 1990).

40. Davis, *Earthquake Planning Scenario*, 56.

41. U.S. Geological Survey staff estimate that a magnitude 7.5 earthquake on the Hayward fault, which runs through populated areas, might inflict 4,500 deaths, 135,000 injuries, and over $40 billion in property damage; see U.S. General Accounting Office, *Earthquake Recovery: Staffing and Other Improvements Made Following Loma Prieta Earthquake*, report no. RCED-92-141 (July 1992), 17.

42. These maps are also of interest to the private insurance industry, which considers earthquakes an "uninsurable hazard" because, without subsidies, only those at greatest risk tend to purchase insurance. For discussion of industry efforts to convince Congress to require earthquake insurance for all home owners, see Michael Quint, "Support Grows for Earthquake Insurance Bill," *New York Times*, 21 January 1994, D1–D2.

43. S. T. Algermissen and others, *Probabilistic Estimates of Maximum Acceleration and Velocity in Rock in the Contiguous United States*, U.S. Geological Survey Open File Report no. 82-1033 (Reston, Va., 1982). A second set of maps describe another seismic design factor, peak velocity (measured in cm/sec).

44. D. M. Perkins and S. T. Algermissen, "Seismic Hazards Maps for the U.S.: Present Use and Prospects," in K. H. Jacob, ed., *Proceedings from the Symposium on Seismic Hazards, Ground Motions, Soil-Liquefaction and Engineering Practice in Eastern North America, October 20–22, 1987* (Buffalo, N.Y.: National Center for Earthquake Engineering Research, 1987), 16–25.

45. S. T. Algermissen and others, *Probabilistic Earthquake Acceleration and Velocity Maps for the United States and Puerto Rico*, U.S. Geological Survey Miscellaneous Field Studies Map no. MF-2120 (Reston, Va., 1990). For additional evidence that seismic risk maps are subject to substantial revision, see James F. Dolan and others, "Prospects for Larger or More Frequent Earthquakes in the Los Angeles Metropolitan Region," *Science* 267 (1995): 199–205.

46. Bolt, *Earthquakes*, 82 and 135.

47. S. T. Algermissen and David M. Perkins, "Earthquake-Hazard Map of United States," *Earthquake Information Bulletin* 9 (January–February 1977): 20–21; Dan Morse, "Seismic Codes: Preparing for the Unknown," *Civil Engineering* 59, no. 11 (1989): 70–73; and David B. Rosenbaum, "Pushing for a National Standard," *ENR* 31 (8 November 1993): 8.

48. For example, the Uniform Building Code uses seismic zones slightly less detailed than those delineated for the Standard Building Code, which includes labeled USGS ground-acceleration contour lines on its map; compare International Conference of Building Officials, *Uniform Building Code, 1991 Edition* (Whittier, Calif., 1991), Figure 23-2; and Southern Building Code Congress International, *Standard Building Code, 1991 Edition* (Birmingham, Ala., 1991), 305.

49. Eastern and western areas in zone 2 are labeled 2A and 2B, respectively.

50. Ted Algermissen, telephone interview by author, 10 May 1994; and Building Seismic Safety Council, *Seismic Considerations for Communities at Risk* (Washington, D.C., 1990), 30–31.

51. Algermissen, interview.

52. Memorandum of 2 July 1993 from L. Thomas Tobin, executive director, included in California Seismic Safety Commission, *Hearing on Proposed Maps for NEHRP's Recommended Provisions for the Development of Seismic Regulations for New Buildings*, Report no. 93-02 (Sacramento, Calif., 1993), 32–3.

53. Ibid., 37.

54. Robert B. Olshansky and Paul Hanley, *Reducing Earthquake Hazards in the Central U.S.: Seismic Building Codes* (Memphis, Tenn.: Central U.S. Earthquake Consortium, 1992), 19–22.

55. Comments of Hank Martin, in California Seismic Safety Commission, *Hearing on Proposed Maps*, 24.

56. James Penick, Jr., *The New Madrid Earthquakes of 1811–1812* (Columbia, Mo.: University of Missouri Press, 1976).

57. William Atkinson, *The Next New Madrid Earthquake: A Survival Guide for the Midwest* (Carbondale, Ill.: Southern Illinois University Press, 1989), 9–23.

58. Berke and Beatley, *Planning for Earthquakes*, 83–5; and Edward S. Fratto, John E. Ebel, and Katharine Kadinsky-Cade, "Earthquakes in New England," *Earthquakes and Volcanoes* 22 (1990): 242–9.

59. Richard A. Kerr, "Assessing the Risk of Eastern U.S. Earthquakes," *Science* 214 (1981): 169–71.

60. Edward S. Fratto, "The Earthquake Hazard Program in Massachusetts," *Earthquake Information Bulletin* 17 (September–October 1985): 165–72; and Edward S. Fratto, John E. Ebel, and Katharine Kadinsky-Cade, "Preparing for Earthquakes in Boston," *Earthquakes and Volcanoes* 226 (1990): 250–1.

61. Harold Faber, "Scientists to Test State's Vulnerability to Earthquakes," *New York Times*, 12 April 1994, B6.

62. For an example of CUSEC's educational literature, see Robert B.

Olshansky, *Reducing Earthquake Hazards in the Central U.S.: Seismic Hazard Mapping* (Memphis, Tenn.: CUSEC, 1992).

63. William Spence and others, *Responses to Iben Browning's Prediction of a 1990 New Madrid, Missouri, Earthquake*, U.S. Geological Survey Circular no. 1083 (Washington, D.C., 1993), 1–24. Also see F. A. McKeown and S. F. Diehl, *Evidence of Contemporary and Ancient Excess Fluid Pressure in the New Madrid Seismic Zone of the Reelfoot Rift, Central United States*, U.S. Geological Survey Professional Paper no. 1538-N (Washington, D.C., 1994).

64. Thomas H. Heaton and others, *National Seismic System Science Plan*, U.S. Geological Survey Circular no. 1031 (Washington, D.C., 1989).

65. William L. Ellsworth, "Getting beyond Numerology," *Nature* 363 (1993): 206–7; and Louis C. Pakiser, *Earthquakes* (Reston, Va., 1991), 17–18.

CHAPTER THREE

General references include David Chester, *Volcanoes and Society* (London: Edward Arnold, 1993); Stephen L. Harris, *Agents of Chaos: Earthquakes, Volcanoes, and Other Natural Disasters* (Missoula, Mont.: Mountain Press Publishing Co., 1990); Roger C. Martin and James F. Davis, eds., *Status of Volcanic Prediction and Emergency Response Capabilities in Volcanic Hazard Zones of California*, Special Publication no. 63 (Sacramento: California Division of Mines and Geology, 1982); John H. Latter, ed., *Volcanic Hazards: Assessment and Monitoring*, IAVCEI Proceedings in Volcanology, vol. 1 (Berlin: Springer-Verlag, 1989); Donald W. Peterson, "Volcanic Hazards and Public Response," *Journal of Geophysical Research* 93 (1988): 4161–70; Steinbrugge, *Earthquakes, Volcanoes, and Tsunamis*, 259–91; Haroun Tazieff and Jean-Christophe Sabroux, eds., *Forecasting Volcanic Events* (Amsterdam: Elsevier, 1983); and Robert I. Tilling and Peter W. Lipman, "Lessons in Reducing Volcano Risk," *Nature* 364 (1993): 277–89.

1. Robert J. Tilling, "Introduction and Overview," in Robert J. Tilling, ed., *Volcanic Hazards*, Short Course in Geology, vol. 1 (Washington, D.C.: American Geophysical Union, 1989), 1–8; also see Russell J. Blong, *Volcanic Hazards: A Sourcebook on the Effects of Eruptions* (Orlando, Fla.: Academic Press, 1984), 72.

2. Claude C. Albritton, Jr., *Catastrophic Episodes in Earth History* (London: Chapman and Hall, 1989), 150–6; Walter Alvarez and Frank Asaro, "The Extinction of Dinosaurs," in Janine Bourriau, ed., *Understanding Catastrophe* (Cambridge: Cambridge University Press, 1992), 28–56; and Steinbrugge, *Earthquakes, Volcanoes, and Tsunamis*, 260.

3. Roy A. Bailey, "Other Potential Eruption Centers in California: Long Valley–Mono Lake, Coso, and Clear Lake Volcano Fields," in Martin and Davis, *Status of Volcanic Prediction and Emergency Response Capabilities*, 17–28; and C. Dan Miller and others, *Potential Hazards from Future Volcanic Eruptions in the Long Valley–Mono Lake Area, East-Central California and Southwest Nevada—A Preliminary Assessment*, U.S. Geological Survey Circular no. 877 (Washington, D.C., 1982), 8–9.

4. R. B. Strothers and others, "Volcanic Winter? Climatic Effects of the Largest Volcanic Eruptions," in Latter, *Volcanic Hazards*, 3–9.

5. Roy A. Bailey and others, *The Volcano Hazards Program: Objectives and Long-Range Plans*, U.S. Geological Survey Open File Report no. 83-400 (Reston, Va., 1983), 3–13. For slightly modified descriptions of these three groups, see Chester, *Volcanoes and Society*, 200–1.

6. Federal Emergency Management Agency, *Risks and Hazards*, 10, 24.

7. Thomas L. Wright and Thomas C. Pierson, *Living with Volcanoes: The U.S. Geological Survey's Volcano Hazards Program*, U.S. Geological Survey Circular no. 1073 (Washington, D.C., 1992), vii.

8. Frank Press and Raymond Siever, *Earth*, 4th ed. (New York: W.H. Freeman and Co., 1986), 422.

9. "One of the Pacific Northwest's Volcanic Giants," *Christian Science Monitor*, 10 May 1968, 1. For a more recent assessment of hazards of another active volcano in the region, see National Research Council, Commission on Geosciences, Environment, and Resources, U.S. Geodynamics Committee, *Mount Rainier: Active Cascade Volcano* (Washington, D.C.: National Academy Press, 1994).

10. Dwight R. Crandell, Donal R. Mullineaux, and Meyer Rubin, "Mount St. Helens Volcano: Recent and Future Behavior," *Science* 187 (1975): 438–41; quote on p. 441.

11. Dwight R. Crandell and Donal R. Mullineaux, *Potential Hazards from Future Eruptions of Mount St. Helens Volcano*, U.S. Geological Survey Bulletin no. 1383-C (Washington, D.C., 1978); quote on p. C22.

12. Andrei M. Sarna-Wojcicki and others, "Areal Distribution, Thickness, Mass, Volume, and Grain Size of Air-Fall Ash from the Six Major Eruptions of 1980," in *The 1980 Eruptions of Mount St. Helens, Washington*, U.S. Geological Survey Professional Paper no. 1250 (Washington, D.C., 1981), 577–600.

13. Crandell and Mullineaux, *Potential Hazards from Future Eruptions*, C21.

14. Ibid., C19.

15. In an assessment of the flowage-hazard map, the U.S. Geological Survey used a color overlay showing four types of eruptions effects (in red) and four somewhat different types of potential hazards (in blue); see Wright and Pierson, *Living with Volcanoes*, 22.

16. C. Dan Miller, Donal R. Mullineaux, and Dwight R. Crandell, "Hazards Assessments at Mount St. Helens," in *The 1980 Eruptions of Mount St. Helens, Washington*, U.S. Geological Survey Professional Paper no. 1250 (Washington, D.C., 1981), 789–802.

17. Bailey and others, *The Volcano Hazards Program*, 1; and James Riehle, U.S. Geological Survey, Reston, Va., telephone conversation with author, 2 June 1994.

18. Wright and Pierson, *Living with Volcanoes*, 52–6.

19. Clear Lake, a fourteenth volcano farther south in California's Coast Range, also had no hazard assessment; ibid., 50.

20. Richard P. Hoblitt, C. Dan Miller, and William E. Scott, *Volcanic Hazards with Regards to Siting Nuclear-Power Plants in the Pacific Northwest*, U.S. Geological Survey Open File Report no. 87-297 (Reston, Va., 1987); cited in

C. Dan Miller, "Volcanic Hazards in the Pacific Northwest," *Geoscience Canada* 17 (1990): 183–7. The maps in Figure 2.5 are based on maps Miller generalized from plates 1 and 3 in Hoblitt, Miller, and Scott.

21. Other maps in the 1987 report describe probabilities for 1 cm and 100 cm ashfalls.

22. C. Dan Miller, *Potential Hazards from Future Volcanic Eruptions in California*, U.S. Geological Survey Bulletin no. 1847 (Washington, D.C., 1989).

23. C. Dan Miller, U.S. Geological Survey, Vancouver, Wash., telephone conversation with author, 31 May 1994.

24. C. Dan Miller, *Potential Hazards from Future Eruptions in the Vicinity of Mount Shasta Volcano, Northern California*, U.S. Geological Survey Bulletin no. 1503 (Washington, D.C., 1980).

25. Dwight R. Crandell and Donald R. Nichols, *Volcanic Hazards at Mount Shasta, California*, a U.S. Geological Survey General Interest Publication (Washington, D.C., 1993).

26. Sarna-Wojcicki and others, "Areal Distribution, Thickness, Mass, Volume, and Grain Size."

27. Crandell and Nichols, *Volcanic Hazards at Mount Shasta*, 18.

28. Ibid., 20.

29. Thomas F. Saarinen and James L. Sell, *Warning and Response to the Mount St. Helens Eruption* (Albany: State University of New York Press, 1985), 22–58.

30. Ibid., 55.

31. Miller, *Potential Hazards from Future Eruptions in the Vicinity of Mount Shasta Volcano*, 39.

32. Miller, telephone conversation.

33. Riehle, telephone conversation.

CHAPTER FOUR

General references include Edward Bryant, *Natural Hazards* (Cambridge: Cambridge University Press, 1991); David R. Godschalk, David J. Brower, and Timothy Beatley, *Catastrophic Coastal Storms: Hazard Mitigation and Development Management* (Durham, N.C.: Duke University Press, 1989); National Research Council, Committee on Engineering Implications of Changes in Relative Mean Sea Level, *Responding to Changes in Sea Level* (Washington, D.C.: National Academy Press, 1987); Orrin H. Pilkey and others, *Coastal Land Loss*, Short Course in Geology, vol. 2 (Washington, D.C.: American Geophysical Union, 1989); and Steinbrugge, *Earthquakes, Volcanoes, and Tsunamis*, 233–58.

1. Walter C. Dudley and Min Lee, *Tsunami!* (Honolulu: University of Hawaii Press, 1988), 44–54.

2. *The Everything Pages: Oahu, March 1994–1995* (Honolulu: GTE Hawaiian Telephone Company, 1994), 48–57; and George D. Curtis, *Hawaii Tsunami Inundation/Evacuation Map Project*, Final Report for Contract no. 91-237 (Honolulu: Joint Institute for Marine and Atmospheric Research, 1991), 11.

3. James F. Lander and Patricia A. Lockridge, *United States Tsunamis*,

1690–1988 (Boulder, Colo.: National Geophysical Data Center, 1989), 38–42, 66–7, and 116.

4. Ibid., 46–7.

5. Curtis, *Hawaii Tsunami Inundation/Evacuation Map Project*, 3.

6. Although runup is most often referenced to mean sea level, some reports relate runup to the level of the sea immediately before the tsunami.

7. Anders Wijkman and Lloyd Timberlake, *Natural Disasters: Acts of God or Acts of Man?* (London: International Institute for Environment and Development, 1984), 100.

8. Lander and Lockridge, *United States Tsunamis*, 201–7.

9. Ibid., 102, 149; and Gere and Shah, *Terra Non Firma*, 46–7.

10. Waltraud A. R. Brinkmann, *Hurricane Hazard in the United States: A Research Assessment*, Monograph no. NSF-RA-E-75-007 (Boulder, Colo.: University of Colorado, Institute of Behavioral Science, 1975), 5–7.

11. N. Arthur Pore and Celso S. Barrientos, *Storm Surge*, MESA New York Bight Atlas, monograph no. 6 (Albany, N.Y.: New York Sea Grant Institute, 1976), 7–9.

12. N. Arthur Pore, "The Storm Surge," *Mariners Weather Log* 5 (1961): 151–6.

13. Robert A. Hoover, "Empirical Relationships of the Central Pressures in Hurricanes to the Maximum Surge and Storm Tide," *Monthly Weather Review* 85 (1957): 167–74.

14. Walter R. Davis, "Hurricanes of 1954," *Monthly Weather Review* 82 (1954): 370–3.

15. Federal Emergency Management Agency, *Coastal Construction Manual*, FEMA Publication no. 55 (Washington, D.C., 1986), p. 2-4. FEMA's source for the map is R. H. Simpson and M. B. Lawrence, *Atlantic Hurricane Frequencies along the U.S. Coastline*, NOAA Technical Memorandum no. NWS-SR-58 (Washington, D.C., 1971).

16. For a description of the map more exact than that provided in the *Coastal Construction Manual*, see Brinkmann, *Hurricane Hazard in the United States*, 4–6. Comparison with Brinkmann's version suggests that FEMA exercised significant cartographic license in transcribing the map by Simpson and Lawrence.

17. FEMA, *Coastal Construction Manual*, p. 2-4.

18. Robert E. Davis and Robert Dolan, "Nor'easters," *American Scientist* 81 (1993): 428–39; and Ben Watson, "New Respect for Nor'easters," *Weatherwise* 46 (December 1993–January 1994): 18–23.

19. Pore and Barrientos, *Storm Surge*, 9–21.

20. A. I. Cooperman and H. E. Rosendal, "Great Atlantic Coast Storm, 1962," *Mariners Weather Log* 6 (1962): 79–85.

21. A detailed explanation on the back of the map explains the calculations. "Each category for each coastal factor, except stabilization, was assigned a value from low (1) to very high (4). For each segment of the shoreline the numerical value for each factor was squared and then averaged to determine the overall hazard rating. The overall ratings range from very low (1) to very high (7). Squaring each value gives greater emphasis to extreme events which

generally constitute the greatest hazard. For example, an area might rank low (1) in most hazards but high (3) in shoreline erosion. Unweighted averaging would give this area a very low hazard rating (1), even though the chance of damage to structures by erosion is quite high. By squaring each value first, the overall rating would still be relatively low (2) but the slight increase is more reflective of the actual danger involved in living in that coastal area." See Fred Anders, Suzette Kimball, and Robert Dolan, *Coastal Hazards*, National Atlas 1:7,500,000 map (Reston, Va.: U.S. Geological Survey, 1985).

22. Robert Dolan, University of Virginia, telephone conversation with author, 10 June 1994.

23. For discussion of global warming, rising seas, and geographic variation in relative risk, see Vivien Gornitz, Tammy W. White, and Robert M. Cushman, "Vulnerability of the U.S. to Future Sea Level Rise," in Orville T. Magoon and others, eds., *Coastal Zone '91: Proceedings of the Seventh Symposium on Coastal and Ocean Management* (New York: American Society of Civil Engineers, 1991), 2354–68.

24. Robert Dolan and Harry Lins, *The Outer Banks of North Carolina*, U.S. Geological Survey Professional Paper no. 1177-B (Washington, D.C., 1985), 8–9.

25. Ibid., 13, 26.

26. The map was compiled by S. Riggs. Ibid., 13–14.

27. Dolan, telephone conversation. According to its publications catalogs, the U.S. Geological Survey issued a revised edition in 1986, a year after the original version. Dropping the appendix reduced the report's length from 103 to 47 pages.

28. Federal Emergency Management Agency and Office of Ocean and Coastal Resource Management, NOAA, *Preparing for Hurricanes and Coastal Flooding: A Handbook for Local Officials*, FEMA publication no. 50 (Washington, D.C., 1983), 105–6; and Robert M. White, "The National Hurricane Warning Program," *Bulletin of the American Meteorological Society* 53 (1972): 631–3.

29. A. Todd Davison, "The National Flood Insurance Program and Coastal Hazards," in Orville T. Magoon and others, eds., *Coastal Zone '93: Proceedings of the Eighth Symposium on Coastal and Ocean Management* (New York: American Society of Civil Engineers, 1993), 1377–91.

30. Godschalk, Brower, and Beatley, *Catastrophic Coastal Storms*, 103–5.

31. William L. Waugh, Jr., "Hurricanes," in William L. Waugh, Jr., and Ronald John Hy, eds., *Handbook of Emergency Management: Programs and Policies Dealing with Major Hazards and Disasters* (Westport, Conn.: Greenwood Press, 1990), 61–79.

32. Warren Horst, Nassau County Emergency Management Office, telephone conversation with author, 16 June 1994.

33. Allan McDuffie, U.S. Army Corps of Engineers, Wilmington District, telephone conversation with author, 16 June 1994.

34. Brian R. Jarvinen and Miles B. Lawrence, "An Evaluation of the SLOSH Storm-Surge Model," *Bulletin of the American Meteorological Society* 66 (1985): 1408–13. For an application of the SLOSH model to hazard mitigation

through land-use planning see Philip Berke and Carlton Ruch, "Application of a Computer System for Hurricane Emergency Response and Land Use Planning," *Journal of Environmental Management* 21 (1985):117–34.

35. *New York State Hurricane Evacuation Study, 1993: Technical Data Report* (Wilmington, N.C.: U.S. Army Corps of Engineers, 1993), 17–21.

36. Local officials also use HURREVAC, a computerized decision-support tool provided by the National Hurricane Center to help local officials adjust evacuation planning to tides and current forecasts. John Dinizzo, Nassau County Emergency Management Office, telephone conversation with author, 24 June 1994.

37. *New York State Hurricane Evacuation Study, 1993*, 14, 39, 97–106.

38. Ibid., 33.

39. For three to six reference points per sheet, the inundation maps also report maximum surge heights for the four different hurricane categories. Based on mean sea level, these surge heights help local officials assess the severity of inundation for significant landmarks. Warren Horst, telephone conversation with author, 24 June 1994.

40. McDuffie, telephone conversation.

41. Waugh, "Hurricanes," 72–3.

42. For a brief critique of FEMA's difficulties in providing proactive leadership, see William L. Waugh, Jr., "Current Policy and Implementation Issues in Disaster Preparedness," in Louise K. Comfort, ed., *Managing Disaster: Strategies and Policy Perspectives* (Durham, N.C.: Duke University Press, 1988), 111–25; see especially 120–4. For a news report on a recent study that suggests the Northeast is more vulnerable than previously believed to severe coastal storms, see William K. Stevens, "Historic Hurricane Could Catch Northeast with Its Guard Down," *New York Times*, 23 August 1994, C4.

43. For example, see Steve Kemper, "This Beach Boy Sings a Song Developers Don't Want to Hear," *Smithsonian* 23 (October 1992): 72–86; and Orrin H. Pilkey and William J. Neal, "Save Beaches, Not Buildings," *Issues in Science and Technology* 8 (spring 1992): 36–41.

44. Orrin Pilkey, Jr., Duke University, telephone interview by author, 8 November 1989.

45. Orrin H. Pilkey, Jr., Orrin H. Pilkey, Sr., and Robb Turney, *How to Live with an Island: A Handbook to Bogue Banks, North Carolina* (Raleigh: North Carolina Department of Natural and Economic Resources, 1975) is now out of print. An expanded treatment is available in Orrin H. Pilkey, Jr., and others, *From Currituck to Calabash: Living with North Carolina's Barrier Islands* (Durham, N.C.: Duke University Press, 1982). For an overview of coastal hazards, see Wallace Kaufman and Orrin Pilkey, Jr., *The Beaches Are Moving: The Drowning of America's Shoreline* (Durham, N.C.: Duke University Press, 1983).

46. Pilkey, interview.

CHAPTER FIVE

General references include Christopher R. Church and others, eds., *The Tornado: Its Structure, Dynamics, Prediction, and Hazards*, Geophysical Monograph no. 79 (Washington, D.C.: American Geophysical Union, 1993); Joe R. Eagleman, *Severe and Unusual Weather* (New York: Van Nostrand Reinhold, 1983); Snowden D. Flora, *Tornadoes of the United States* (Norman, Okla.: University of Oklahoma Press, 1953); Loran B. Smith and David T. Jervis, "Tornadoes," in Waugh and Hy, *Handbook of Emergency Management*, 106–28; and John T. Snow, "The Tornado," *Scientific American* 250 (April 1984): 86–96.

1. National Oceanic and Atmospheric Administration, *Tornado Safety: Surviving Nature's Most Violent Storms* (Washington, D.C., 1982), 6; and U.S. Bureau of the Census, *Statistical Abstract of the U.S.*, 113th ed. (Washington, D.C., 1993), 230.

2. National Weather Service, *Tornado* (Washington, D.C., 1978 [government document no. C 55.102: T 63/978]), 3.

3. NOAA, *Tornado Safety*, 4.

4. Eagleman, *Severe and Unusual Weather*, 102–6.

5. National Oceanic and Atmospheric Administration, *Tornado Preparedness Planning* (Washington, D.C., 1978 [government document no. C 55.102: T 63.2/978]), 25.

6. This is my interpretation of the index. The calculations are not fully explained, nor are the data described in detail sufficient to replicate their results precisely.

7. Thomas P. Grazulis, *Significant Tornadoes, 1880–1989*, 2 vols. (St. Johnsbury, Vt.: Tornado Project, 1991); and Thomas P. Grazulis, Joseph T. Schaefer, and Robert F. Abbey, Jr., "Advances in Tornado Climatology, Hazards, and Risk Assessment Since Tornado Symposium II," in Church and others, *The Tornado*, 409–26.

8. Joseph T. Schaefer and others, "The Stability of Climatological Tornado Data," in Church and others, *The Tornado*, 459–66.

9. T. Theodore Fujita, "Tornadoes around the World," *Weatherwise* 26 (April 1973): 56–62, 79–83.

10. For examples of the three categories of intensity, see Robert W. Christopherson, *Geosystems: An Introduction to Physical Geography* (New York: Macmillan, 1994), 230; and Joseph T. Schaefer and others, "Tornadoes: When, Where, How Often," *Weatherwise* 33 (April 1980): 52–9. The six categories of damage are from Fujita, "Tornadoes around the World," 59; for an example of their use, see Frederick K. Lutgens and Edward J. Tarbuck, *The Atmosphere: An Introduction to Meteorology*, 4th ed. (Englewood Cliffs, N.J.: Prentice-Hall, 1989), 280.

11. Category descriptions for the Fujita intensity scale are from *Natural Disaster Survey Report: Tampa Bay Area Tornadoes, October 3, 1992* (Silver Spring, Md.: National Weather Service), B-1.

12. T. Theodore Fujita quoted in "Tornadoes: An American Experience," *Journal of American Insurance* 60, no. 2 (1984): 15–20; quote on p. 19.

13. *Natural Disaster Survey Report: Tampa Bay Area Tornadoes.*

14. Rodger Brown, National Severe Storms Laboratory, Norman, Okla., telephone conversation with author, 8 July 1994.

15. Charles A. Doswell III and Donald W. Burgess, "On Some Issues of United States Tornado Climatology," *Monthly Weather Review* 116 (1988): 495–501.

16. Ibid., 497.

17. Thomas P. Grazulis, "A 110-Year Perspective of Significant Tornadoes," in Church and others, *The Tornado*, 467–74.

18. Schaefer and others, "The Stability of Climatological Tornado Data," 465.

19. Schaefer and others, "Tornadoes: When, Where, How Often," 56.

20. Schaefer and others, "The Stability of Climatological Tornado Data," 465.

21. Dennis McNeese, State Farm Insurance, letter to author, 13 July 1994.

22. Dennis McNeese, telephone conversation with author, 15 July 1994.

23. Eagleman, *Severe and Unusual Weather*, 89.

24. Fujita, "Tornadoes around the World," 80–3.

25. Thomas Schlatter, "Tornado Belt, Halo Arcs," *Weatherwise* 39 (April 1986): 110–11.

26. Flora, *Tornadoes of the United States*, 13.

27. Annual estimate of 50,000 to 100,000 thunderstorms is reported in Mark Roman, "Tornado Tracker," *Discover* 10 (June 1989): 50–6. My "fewer than 3,000 tornadoes" reflects approximately 700 to 800 officially recognized tornadoes and Grazulis's estimate of 2,000 unreported tornadoes.

28. *Natural Disaster Survey Report: Tampa Bay Area Tornadoes*, viii–xi.

29. Smith and Jervis, "Tornadoes," 113.

30. Sue Bowler, "Radar Network Watches Where the Wind Blows," *New Scientist* 125 (24 March 1990): 30–1; Donald W. Burgess, Ralph J. Donaldson, Jr., and Paul R. Desrochers, "Tornado Detection and Warning by Radar," in Church and others, *The Tornado*, 203–31; and Samuel Milner, "NEXRAD: The Coming Revolution in Radar Storm Detection and Warning," *Weatherwise* 39 (April 1986): 72–85.

31. See, for example, H. Richard Crane, "Doppler Radar: The Speed of Air in a Tornado," *Physics Teacher* 27 (1989): 212–13.

32. Charles A. Doswell III, Steven J. Weiss, and Robert H. Jones, "Tornado Forecasting: A Review," in Church and others, *The Tornado*, 557–71.

33. Committee on National Weather Service Modernization, of the Commission on Engineering and Technical Systems, National Research Council, *Toward a New National Weather Service—Second Report* (Washington, D.C.: National Academy Press, 1992), 5.

34. Burgess, Donaldson, and Desrochers, "Tornado Detection and Warning by Radar," 217–19.

35. *Strategic Plan for the Modernization and Associated Restructuring of the National Weather Service*, dated March 1989, is reproduced in U.S. Congress, House Committee on Science, Space, and Technology, *Tornado Warnings and Weather Service Modernization: Hearing before the Subcommittee on Natural Resources, Agricultural Research and Environment*, 101st Cong., 1st sess., 1989, 68–90. Although the National Weather Service is the lead agency, the NEXRAD network includes several stations operated by the Air Force and the Federal Aviation Administration.

36. National Weather Service Employees Organization, *Response to the Strategic Plan for the Modernization and Associated Restructuring of the National*

Weather Service and Recommendations, dated 1989, is reproduced in House Committee, *Tornado Warnings and Weather Service Modernization,* 91–115.

37. Brown, telephone conversations, 8 and 11 July 1994.

38. Burgess, Donaldson, and Desrochers, "Tornado Detection and Warning by Radar," 206.

39. *Natural Disaster Survey Report: Tampa Bay Area Tornadoes,* ix.

40. Cost reported in Bowler, "Radar Network Watches Where the Wind Blows."

CHAPTER SIX

General references include Federal Emergency Management Agency, *Answers to Questions about the National Flood Insurance Program,* FEMA publication no. FIA-2 (Washington, D.C., 1992); Federal Interagency Floodplain Management Task Force, *Floodplain Management in the United States: An Assessment Report,* vol. 2: Full Report, FEMA publication no. FIA-18 (Washington, D.C., 1992); FEMA, *Guide to Flood Insurance Rate Maps,* FEMA publication no. FIA-14 (Washington, D.C., 1988); Interagency Floodplain Management Review Committee, *Sharing the Challenge: Floodplain Management into the 21st Century* (Washington, D.C., 1994); John H. McShane, *Floodplain Management and the Administration of the National Flood Insurance Program: A Handbook for New York State Community Officials* (Cornell University, project report, Master of Professional Studies, 1989); Beth Millemann, "The National Flood Insurance Program," *Oceanus* 36 (spring 1993): 6–8; and *Natural Disaster Survey Report: The Great Flood of 1993* (Silver Spring, Md.: National Weather Service, 1994).

1. Daniel Cotter, FEMA, telephone conversation with author, 9 August 1994.

2. Interagency Floodplain Management Review Committee, *Sharing the Challenge,* 100.

3. The address is Flood Map Distribution Center, 6930 (A–F) San Tomas Road, Baltimore, MD 21227-6227, and the toll-free telephone number is 1-800-358-9616.

4. D. Earl Jones, Jr., and Jonathan E. Jones, "Floodway Delineation and Management," *Journal of Water Resources Planning and Management* 113 (1987): 228–42.

5. Sample rates are reported in U.S. General Accounting Office, *Flood Insurance: Financial Resources May Not Be Sufficient to Meet Future Expected Losses,* report no. RCED-94-80 (March 1994), 31–2.

6. Don M. Corbett and others, *Stream-Gaging Procedure: A Manual Describing Methods and Practices of the Geological Survey,* U.S. Geological Survey Water-Supply Paper no. 888 (Washington, D.C., 1943), 13–108; and Reginald W. Herschy, *Streamflow Measurement* (London: Elsevier Applied Science Publishers, 1985), 93–163.

7. E. J. Kennedy, "Discharge Ratings at Gaging Stations," *Techniques of Water-Resources Investigations of the United States Geological Survey,* chap. A10 of bk. 3, *Applications of Hydraulics* (Washington, D.C., 1984).

8. Tate Dalrymple, *Flood-Frequency Analyses* [Manual of Hydrology: Part 3, Flood-Flow Techniques], U.S. Geological Survey Water-Supply Paper no. 1543-A (Washington, D.C., 1960), 7–8.

9. W. B. Langbein, "Plotting Positions in Frequency Analysis," in Dalrymple, *Flood-Frequency Analyses*, 48–9.

10. H. C. Riggs, "Frequency Curves," *Techniques of Water-Resources Investigations of the United States Geological Survey*, chap. A2 of bk. 4, *Hydrologic Analysis and Interpretation* (Washington, D.C., 1968), 7–9.

11. Ven Te Chow, *Handbook of Applied Hydrology* (New York: McGraw-Hill Book Co., 1964), 8-1 to 8-37.

12. Charles T. Haan, *Statistical Methods in Hydrology* (Ames: Iowa State University Press, 1977), 140–4; and Roy F. Powell, L. Douglas James, and D. Earl Jones, Jr., "Approximate Method for Quick Flood Plain Mapping," *Journal of Water Resources Planning and Management* 106 (1980): 103–22.

13. Wilbert O. Thomas, Jr., "Techniques Used by the U.S. Geological Survey in Estimating the Magnitude and Frequency of Floods," in L. Mayer and D. Nash, eds., *Catastrophic Flooding* (Boston: Allen and Unwin, 1987), 267–88.

14. H. C. Riggs, *Streamflow Characteristics* (Amsterdam: Elsevier Science Publishers, 1985), 161–2, 233–4.

15. FEMA's highly quantitative, more or less standardized technique for mapping flood risk is not the only approach. For example, see M. Gordon Wolman, "Evaluating Alternative Techniques of Floodplain Mapping," *Water Resources Research* 7 (1971): 1383–92.

16. FEMA, *Answers to Questions about the National Flood Insurance Program*, 2.

17. Brian R. Mrazik, "Flood Risk Analysis for the National Flood Insurance Program," in Vijay P. Singh, ed., *Application of Frequency and Risk in Water Resources* (Dordrecht, The Netherlands: D. Reidel Publishing Co., 1987), 443–53. Also see Richard M. Vogel and Jery R. Stedinger, "Flood-Plain Delineations in Ice Jam Prone Regions," *Journal of Water Resources Planning and Management* 110 (1984): 206–19.

18. U.S. General Accounting Office, *Flood Insurance: Financial Resources May Not Be Sufficient*, 31–2.

19. Federal Emergency Management Agency, *FIRM, Town of Nags Head, Dare County, North Carolina*, Community-Panel no. 375356 0016 C, map revised 2 April 1993.

20. Federal Emergency Management Agency, *Appeals, Revisions, and Amendments to National Flood Insurance Program Maps: A Guide for Community Officials*, FEMA publication no. FIA-12 (Washington, D.C., 1993).

21. D. E. Burkham, "Accuracy of Flood Mapping," *Journal of Research of the U.S. Geological Survey* 6 (1978): 515–27. For a similar study, see Stephen J. Burges, "Analysis of Uncertainty in Flood Plain Mapping," *Water Resources Bulletin* 15 (1979): 227–43.

22. Interagency Floodplain Management Review Committee, *Sharing the Challenge*, 60.

23. For an anecdotal yet daunting comparison of a low-risk structure inside

a 100-year flood zone with a high-risk structure outside the 100-year zone of another river, see L. Douglas James and Brad Hall, "Risk Information for Floodplain Management," *Journal of Water Resources Planning and Management* 112 (1986): 485–99.

24. Cotter, telephone conversation.

25. Gene Stakhiv, U.S. Army Corps of Engineers, telephone conversation with author, 28 July 1994.

26. Frank Thomas, FEMA, telephone conversation with author, 27 July 1994.

27. Emmett M. Laursen, "The 100-Year Flood," in Singh, *Application of Frequency and Risk in Water Resources*, 299–308; quote on p. 304.

28. "A Primer on Flood Insurance," *Journal of American Insurance* 62, no. 3 (1986): 22–4.

29. Cotter, telephone conversation.

30. U.S. General Accounting Office, *National Flood Insurance: Marginal Impact on Flood Plain Development, Administrative Improvements Needed*, report no. CED-82-105 (16 August 1982), 39–45.

31. Ibid., 45.

32. Interagency Floodplain Management Review Committee, *Sharing the Challenge*, x, xii, 100; and U.S. General Accounting Office, *Flood Insurance: Financial Resources May Not Be Sufficient*, 11–12.

33. Cotter, telephone conversation.

34. U.S. Congress, House Committee on Banking, Finance, and Urban Affairs, *Flood Disaster Protection Act of 1973: Hearing before the Subcommittee on Policy Research and Insurance*, 101st Cong., 2d sess., 1990, 32.

35. Robert T. Aangeenbrug, "The Map Information Facility—A Cooperative Federal and Private Venture in Geocoding," *Proceedings of the 1980 Annual Conference of the Association for Computing Machinery*, Nashville, Tenn., 27–29 October 1980, 19–28; and Robert T. Aangeenbrug, "The National Flood Insurance Program Mapping Project," in Patricia A. Moore, ed., *Computer Mapping of Natural Resources and the Environment Plus Satellite Derived Data Applications*, Harvard Library of Computer Graphics, vol. 15 (Cambridge, Mass.: Harvard University Laboratory for Computer Graphics and Spatial Analysis, 1981), 5–8.

36. U.S. General Accounting Office, *Memo to Hon. Jake Garn and Hon. Howard M. Metzenbaum, United States Senate, Subject: Termination of the Map Information Facility Contract by the Federal Emergency Management Agency*, report no. CED-81-99 (12 May 1981), 5.

37. Robert T. Aangeenbrug, letter to author, 26 July 1994.

38. Interagency Floodplain Management Review Committee, *Sharing the Challenge*, 100–1.

39. Ibid., 100, 158–9.

40. A comprehensive study of the 1993 Midwest flooding recommended increasing the waiting period from five to fifteen days. Ibid., xii.

41. Jane Birnbaum, "Flood Insurance: Many Homeowners Are Facing a Costly Oversight," *New York Times*, 7 August 1993, 35.

CHAPTER SEVEN

General references include Center of Environmental Research Information, U.S. Environmental Protection Agency, *Protection of Public Water Supplies from Ground-Water Contamination*, publication no. 625/4-85/016 (Washington, D.C., 1985); Committee on Ground Water Modeling Assessment, Water Science and Technology Board, Commission on Physical Sciences, Mathematics, and Resources, National Research Council, *Ground Water Models: Scientific and Regulatory Applications* (Washington, D.C.: National Academy Press, 1990); Committee on Techniques for Assessing Ground Water Vulnerability, Water Science and Technology Board, Commission on Geosciences, Environment, and Resources, National Research Council, *Ground Water Vulnerability Assessment: Predicting Relative Contamination Potential under Conditions of Uncertainty* (Washington, D.C.: National Academy Press, 1990); M. Erdélyi and J. Gálfi, *Surface and Subsurface Mapping in Hydrogeology* (Chichester, U.K.: John Wiley and Sons, 1988); Wendy Gordon, *A Citizen's Handbook on Groundwater Protection* (New York: Natural Resources Defense Council, 1984); Eric P. Jorgensen, ed., *The Poisoned Well: New Strategies for Groundwater Protection* (Washington, D.C.: Island Press, 1989); William F. McTernan and Edward Kaplan, eds., *Risk Assessment for Groundwater Pollution Control* (New York: American Society of Civil Engineers, 1990); G. William Page, ed., *Planning for Groundwater Protection* (Orlando, Fla.: Academic Press, 1987); and Ruth Patrick, Emily Ford, and John Quarles, *Groundwater Contamination in the United States*, 2d ed. (Philadelphia: University of Pennsylvania Press, 1983, 1987).

1. Jack T. Dugan and Donald E. Schild, *Water-Level Changes in the High Plains Aquifer—Predevelopment to 1990*, U.S. Geological Survey Water-Resources Investigations Report no. 91-4165 (Washington, D.C., 1992), 12; and Jack Lewis, "The Ogallala Aquifer," *EPA Journal* 16 (November–December 1990): 42–4. The rate of decline has decreased since 1980 because of reduced pumping and increased precipitation, and in some places the water table is now rising.

2. As suggested by a 1991 General Accounting Office study, a national map at the comparatively detailed county-unit level would have limited value. For a sample of counties, the GAO found "as much variability in hydrogeologic vulnerability within counties as there is between counties." See U.S. General Accounting Office, *Groundwater Protection: Measurement of Relative Vulnerability to Pesticide Contamination*, report no. PEMD-92-8 (October 1991), 7.

3. Gordon, *A Citizen's Handbook*, 9.

4. Jorgensen, *Poisoned Well*, 41–4.

5. R. J. Perkins, "Septic Tanks, Lot Size, and the Pollution of Water Table Aquifers," *Journal of Environmental Health* 46 (1984): 298–301.

6. O. D. von Engeln, *Geomorphology Systematic and Regional* (New York: Macmillan, 1948), 563. For an examination of hydrologic problems in karst areas, see William B. White, *Geomorphology and Hydrology of Karst Terrains* (New York: Oxford University Press, 1988).

7. For a philosophical overview of intergenerational equity, see K. S. Shrad-

er-Frechette, *Burying Uncertainty: Risk and the Case against Geological Disposal of Nuclear Waste* (Berkeley: University of California Press, 1993), 189–207.

8. See, for example, Robert H. Boyle, "The Killing Fields," *Sports Illustrated* 78 (22 March 1993): 62–9; Jorgensen, *Poisoned Well*, 278 90; and Peeyush Varshney, U. Sunday Tim, and Carl E. Anderson, "Risk-Based Evaluation of Ground-Water Contamination by Agricultural Pesticides," *Ground Water* 31 (1993): 356–62. Groundwater is more susceptible to pesticide contamination when the water table is near the surface. See L. H. Tornes and M. E. Brigham, *Nutrients, Suspended Sediment, and Pesticides in Waters of the Red River of the North Basin, Minnesota, North Dakota, and South Dakota, 1970–90*, U.S. Geological Survey Water-Resources Investigations Report no. 93-4231 (Washington, D.C., 1994), 32–5, 59.

9. R. J. Dingman, H. F. Ferguson, and Robert O. R. Martin, *The Water Resources of Baltimore and Harford Counties*, Bulletin no. 17 (Baltimore, Md.: Maryland Department of Geology, Mines and Water Resources, 1956).

10. Morris M. Thompson, *Maps for America: Cartographic Products of the U.S. Geological Survey and Others*, 3d ed. (Washington, D.C.: U.S. Geological Survey, 1987), 157.

11. Ute J. Dymon, "The Communication Structure Surrounding the Groundwater Maps within the USGS Hydrological Atlas Series," *The American Cartographer* 15 (1988): 387–98.

12. Ute J. Dymon, "Groundwater Mapping: An Analysis of the United States Hydrological Investigation Atlas Series," *Environmental Management* 11 (1987): 775–92.

13. Kevin J. Breen and Denise H. Dumouchelle, *Geohydrology and Quality of Water in Aquifers in Lucas, Sandusky, and Wood Counties, Northwestern Ohio*, U.S. Geological Survey Water-Resources Investigations Report no. 91-4024 (Columbus, Ohio: U.S. Geological Survey, 1991), 98–9.

14. G. L. Barr, *Ground-Water Contamination Potential and Quality in Polk County, Florida*, U.S. Geological Survey Water-Resources Investigations Report no. 92-4086 (Tallahassee, Fla.: U.S. Geological Survey, 1992).

15. Committee on Techniques for Assessing Ground Water Vulnerability, 139–44.

16. Julio C. Olimpio and others, *Use of a Geographic Information System to Assess Risk to Ground-Water Quality at Public-Supply Wells, Cape Cod, Massachusetts*, U.S. Geological Survey Water-Resources Investigations Report no. 90-4140 (Boston, Mass.: U.S. Geological Survey, 1991), 18–22.

17. Ibid., 26–31.

18. J. Boonstra and N. A. de Ridder, *Numerical Modelling of Groundwater Basins: A User-Oriented Manual* (Wageningen, The Netherlands: International Institute for Land Reclamation and Improvement, 1981); and Bruce Hunt, *Mathematical Analysis of Groundwater Resources* (London: Butterworths, 1983).

19. See, for example, Mary P. Anderson and William W. Woessner, *Applied Groundwater Modeling: Simulation of Flow and Advective Transport* (San Diego, Calif.: Academic Press, 1992); and Geneviève Ségol, *Classic Groundwater Simulations: Proving and Improving Numerical Models* (Englewood Cliffs, N.J.: PTR Prentice-Hall, 1994).

20. James W. Mercer and Charles R. Faust, "Ground-Water Modeling: An Overview," *Ground Water* 18 (1980): 108–15.

21. This approach, called the *finite-difference* method, includes models in which the spacing of grid lines varies to provide greater detail in selected parts of the study area. Another approach, the *finite-element* method, is based on an irregular network of triangles. See Herbert F. Wang and Mary P. Anderson, *Introduction to Groundwater Modeling: Finite Difference and Finite Element Methods* (San Francisco: W.H. Freeman, 1982).

22. Karen L. Vogel and Andrew G. Reif, *Geohydrology and Simulation of Ground-Water Flow in the Red Clay Creek Basin, Chester County, Pennsylvania, and New Castle County, Delaware*, U.S. Geological Survey Water-Resources Investigations Report no. 93-4055 (Lemoyne, Pa.: U.S. Geological Survey, 1993).

23. Chin-Fu Tsang, "The Modeling Process and Model Validation," *Ground Water* 29 (1991): 825–31.

24. Thomas C. Beard and others, "Iterative Approach to Groundwater Flow Modeling of the Martinsville Alternative Site, under Consideration for Low-level Radioactive Waste Storage in Clark County, Illinois, U.S.A.," *Environmental Geology and Water Science* 18 (1991): 195–207.

25. Vogel and Reif, *Geohydrology and Simulation of Ground-Water Flow in the Red Clay Creek Basin*, 30–44.

26. Charles R. Wilson and others, "Design of Ground-Water Monitoring Networks Using the Monitoring Efficiency Model (MEMO)," *Ground Water* 30 (1992): 965–70.

27. For a concise comparison of remediation strategies, see Jorgensen, *Poisoned Well*, 181–7.

28. E. Scott Bair, Abraham E. Springer, and George S. Roadcap, "Delineation of Traveltime-Related Capture Areas of Wells Using Analytical Flow Models and Particle-Tracking Analysis," *Ground Water* 29 (1991): 387–97.

29. Ralph L. Nichols, Brian B. Looney, and Jonathan E. Huddleston, "3-D Digital Imaging," *Environmental Science and Technology* 26 (1992): 642–9; and Kristine Uhlman and Mark E. Portman, "Ground-Water Modeling without Fear," *Civil Engineering* 61 (September 1991): 64–5.

30. Abraham E. Springer and E. Scott Bair, "Comparison of Methods Used to Delineate Capture Zones of Wells: 2. Stratified-Drift Buried-Valley Aquifer," *Ground Water* 30 (1992): 908–17.

31. R. A. Hodge and Andrew J. Roman, "Ground-Water Protection Policies: Myths and Alternatives," *Ground Water* 28 (1990): 498–504.

CHAPTER EIGHT

General references include Michael H. Brown, *The Toxic Cloud* (New York: Harper and Row, 1987); Committee on Advances in Assessing Exposure to Airborne Pollutants, Board on Environmental Studies and Toxicology, Commission on Geosciences, Environment, and Resources, National Research Council, *Human Exposure Assessment for Airborne Pollutants: Advances and Opportunities* (Washington, D.C.: National Academy Press, 1991); Derek

Elsom, *Atmospheric Pollution: Causes, Effects, and Control Policies* (Oxford: Basil Blackwell, 1987); E. Willard Miller and Ruby Miller, *Environmental Hazards: Air Pollution: A Reference Handbook* (Santa Barbara, Calif.: ABC-CLIO, 1989); David R. Patrick, ed., *Toxic Air Pollution Handbook* (New York: Van Nostrand Reinhold, 1994); U.S. Environmental Protection Agency, *The Plain English Guide to the Clean Air Act*, EPA publication no. 400-K-93-001 (Washington, D.C., 1993); and *What Price Clean Air? A Market Approach to Energy and Environmental Policy* (New York: Committee for Economic Development, 1993).

1. See, for example, Douglas W. Dockery and others, "An Association between Air Pollution and Mortality in Six U.S. Cities," *New England Journal of Medicine* 329 (1993): 1753–9.

2. For an insightful examination of air pollution in the L.A. basin, see James M. Lents and William J. Kelly, "Clearing the Air in Los Angeles," *Scientific American* 269 (October 1993): 32–9.

3. Jack R. Farmer, "Technology Standards," in Patrick, *Toxic Air Pollution Handbook*, 325–40; and David R. Patrick, "Other Programs That Control Toxic Air Pollutants," in Patrick, *Toxic Air Pollution Handbook*, 489–504.

4. See, for example, Carol M. Browner, "Environmental Tobacco Smoke: EPA's Report," *EPA Journal* 19 (October–December 1993): 18–19; Ken Sexton, "An Inside Look at Air Pollution," *EPA Journal* 19 (October–December 1993): 9–12; and D. R. Wernette and L. A. Nieves, "Breathing Polluted Air: Minorities Are Disproportionately Exposed," *EPA Journal* 18 (March–April 1992): 16–17.

5. U.S. Environmental Protection Agency, Office of Air Quality Planning and Standards Research, *National Air Quality and Emissions Trends Report, 1992*, EPA publication no. 454/R-93-031 (Research Triangle Park, N.C., 1993), 2-1 to 2-2.

6. The standard for ozone is based on daily estimates of the highest hourly concentration. The expected number of days per year with ozone levels above 0.12 parts per million should not exceed one. Exceedance criteria also consider the second highest daily maximum as well as the number of daily exceedances. Ibid., p. 3-21.

7. Robert J. Brian, "Ambient Air Surveillance," in Arthur C. Stern, ed., *Air Pollution*, 3d ed., vol. 7, *Supplement to Measurements, Monitoring, Surveillance, and Engineering Control* (Orlando, Fla.: Academic Press, 1986), 143–62; Paul N. Cheremisinoff and Richard A. Young, *Pollution Engineering Practice Handbook* (Ann Arbor, Mich.: Ann Arbor Science Publishers, 1976), 53–7; and Robin R. Segall and Peter R. Westlin, "Source Sampling and Analysis," in Patrick, *Toxic Air Pollution Handbook*, 166–216.

8. U.S. Environmental Protection Agency, Environmental Monitoring Systems Laboratory, *Test Method, Section 2.6, Reference Method for Determination of Carbon Monoxide in the Atmosphere (Nondispersive Infrared Photometry)*, EPA publication no. 600/4-77-027a (Research Triangle Park, N.C., 1983).

9. Gordon Clickman, New York State Department of Environmental Protection, telephone conversation with author, 14 November 1994.

10. For detailed specifications on siting air monitors and "probes," see Appendices D and E of U.S. Environmental Protection Agency, "Part 58—

Ambient Air Quality Surveillance," *Code of Federal Regulations,* Title 40, 1 July 1994 edition (Washington, D.C., 1994), 177–210. Corrections or amendments appear in the *Federal Register.*

11. Brian W. Swinn, "How the 1990 Clean Air Act Amendments Affect New York State," *The Conservationist* 47 (April 1993): 46–7. For a concise examination of the benefits of oxygenated gasoline, see U.S. General Accounting Office, *Air Pollution: Oxygenated Fuels Help Reduce Carbon Monoxide,* GAO report no. RCED-91-176 (August 1991).

12. Clickman, telephone conversation.

13. Although football draws larger crowds, games are scheduled on Saturday afternoons in the autumn, when the weather is less hostile and other traffic is comparatively light.

14. David Coburn, telephone conversation with author, 16 November 1994.

15. Robert W. Andrews, "Dome Drivers Take Detours for Clean Air," *Syracuse Post-Standard,* 26 October 1993, D-2.

16. Several years after its initiation, the traffic rerouting plan was the focus of a protest led by African American politicians, who objected to increased traffic and bad air. In response, local officials modified the route and reopened East Adams Street. See Cindy E. Rodriguez, "3 Arrested in Protest of Dome Traffic Route," *Syracuse Post-Standard,* 19 January 1995, B1; and Jacqueline Arnold, "Detour to Dome Is Down," *Syracuse Post-Standard,* 28 January 1995, A1.

17. Computed from table 5-3 in EPA, *National Air Quality and Emissions Trends Report, 1992,* 5-9 to 5-10.

18. Ibid., 5-1; and *Code of Federal Regulations,* 40 CFR Part 81.

19. One of the *Trends Report*'s more obvious cartographic flaws is that its choropleth maps often use hue (rather than value) as the principal visual variable or display count data rather than intensity data. Some maps sport aesthetically harsh colors while others lack needed contrast. Whatever the motivation, this unrestrained use of color extends to relatively detailed maps that suggest nonattainment throughout an entire state because of a single nonattainment county. For discussion of misuses of cartographic color and choropleth maps, see Mark Monmonier, *How to Lie with Maps* (Chicago: University of Chicago Press, 1991), 20–4 and 147–56.

20. U.S. Environmental Protection Agency, Office of Air Quality Planning and Standards Research, *National Air Quality and Emissions Trends Report, 1990,* EPA publication 490/4-91-023 (Research Triangle Park, N.C., 1991), p. 1-2.

21. Jerry J. Berman, "The Right to Know: Public Access to Electronic Public Information," *Software Law Journal* 3 (1989): 491–530, especially 507–16; and Patrick, "Other Programs That Control Toxic Air Pollutants," 493–4.

22. For example, air emissions accounted for 58.0 percent of the 3,181,646,757 pounds of total releases reported in 1992. These included 549,351,729 pounds of fugitive or nonpoint air emissions and 1,295,606,607 pounds of stack or point air emissions. See U.S. Environmental Protection Agency, *1992 Toxics Release Inventory: Public Data Release,* EPA publication no. 745-R-94-001 (Washington, D.C., 1994), 35.

23. The Clean Air Act, among other laws, requires air-emissions reporting by municipal waste incinerators, commercial waste-treatment firms, and electrical generating plants, the latter including some huge producers of sulfur dioxide and nitrogen oxides. Emissions data for incinerators and power plants are not directly compatible with the TRI data, which focus on specific toxic compounds. For a concise discussion of the variety of air emissions as well as estimation techniques and data bases, see David R. Patrick, "Emissions Estimation," in Patrick, *Toxic Air Pollution Handbook*, 217–25. The EPA provides public access to AIRS (Aerometric Information Retrieval System), an on-line system that integrates several air-pollution data bases. In addition to direct access, the EPA also publishes AIRS data on CD-ROMs for use with personal computers.

24. U.S. Environmental Protection Agency, *State Fact Sheets: 1992 Toxics Release Inventory: Public Data Release*, EPA publication no. 745-F-94-001 (Washington, D.C., 1994).

25. Not identified as the "national summary" in its title, this report is usually published simultaneously with the *State Fact Sheets* report, with which it shares the same title and subtitle. For an example, see EPA, *1992 Toxics Release Inventory: Public Data Release*.

26. Had the EPA's national summary reported carcinogenic air emissions by state, I would have demonstrated a cartographically correct alternative, probably with the dot-array technique that worked well in describing the state-level pattern of failed savings and loan associations. See Mark Monmonier, *Mapping It Out: Expository Cartography for the Humanities and Social Sciences* (Chicago: University of Chicago Press, 1993), 86.

27. Paolo Zannetti, *Air Pollution Models: Theories, Computational Methods, and Available Software* (Southampton, U.K.: Computational Mechanics Publications; New York: Van Nostrand Reinhold, 1990), 263–7 and 273–5.

28. U.S. General Accounting Office, *Air Pollution: Reliability and Adequacy of Air Quality Dispersion Models*, GAO report no. RCED-88-192 (August 1988), 11.

29. See, for example, Perry W. Fisher, John A. Foster, and James W. Sumner, "Comparison of the ISCST Model with Two Alternative U.S. EPA Models in Complex Terrain in Hamilton County, Ohio," *Journal of the Air and Waste Management Association* 44 (1994): 418–27.

30. Zannetti, *Air Pollution Models*, 361–2.

31. William F. Cosulich Associates, *Onondaga County Solid Waste Management Program. Appendix J: Air Quality Impact Analysis* (Woodbury, N.Y., June 1988).

32. The variable contour interval on the hourly map is inherently confusing. I cannot fully explain or justify the distinctly uneven upward progression (0.5-1-3-6-8) of the labeled isolines and their variable interval. Although a 0.5 interval between concentrations of 0.5 and 1 provides useful detail at the low end of the range and an increment of 2 avoids the graphic clutter of extra isolines at the upper end, the jump from 3 to 6 is puzzling. My hunch is that between concentrations of 2 and 8 the map (compiled, no doubt, from a more detailed computer plot) includes numerous distracting details not relevant to

either regulators or environmental scientists. Page-size isoline maps in an environmental impact statement do not require the fixed contour interval useful on topographic maps.

33. William F. Cosulich Associates, *Onondaga County Solid Waste Management Program. Waste-to-Energy Facility: Draft Supplemental Environmental Impact Statement.* (Woodbury, N.Y., June 1988), p. 6-38.

34. For discussion of controversies in siting incinerators, see E. Malone Steverson, "Provoking a Firestorm: Waste Incineration," *Environmental Science and Technology* 25 (1991): 1808–14. For evidence that skeptics' fears are sometimes justified, see U.S. General Accounting Office, *Hazardous Waste: A North Carolina Incinerator's Noncompliance with EPA and OSHA Requirements,* GAO report no. RCED-92-78 (June 1992).

35. Dan Wartenberg and Caron Chess, "Risky Business: The Inexact Art of Hazard Assessment," *The Sciences* 32 (March/April 1992): 17–21.

36. See, for example, B. Drummond Ayres, Jr., "Pollution Shrouds Shenandoah Park," *New York Times,* 2 May 1991, A20.

37. For discussion of acid precipitation and a series of relevant maps, see Robert J. Mason and Mark T. Mattson, *Atlas of United States Environmental Issues* (New York: Macmillan, 1990), 81–7.

38. National Acid Precipitation Assessment Program, *1992 Report to Congress* (Washington, D.C., 1993), 41.

39. Climatologist Douglas Carter called the Tug Hill "the snowiest place in the United States east of the Rocky Mountains." See Douglas Carter, "Climate," in John H. Thompson, ed., *Geography of New York State* (Syracuse, N.Y.: Syracuse University Press, 1966), 54–78; quote on p. 69.

40. For examples of the effects of sample points and interpolation method on isolines, see J. C. Simpson and A. R. Olsen, *Uncertainty in North American Wet Deposition Isopleth Maps: Effect of Site Selection and Valid Sample Criteria,* EPA publication no. 600/4-90/005 (Washington, D.C., 1990).

41. See, for example, comparative statistics reported in U.S. Environmental Protection Agency, Office of Air and Radiation, *Acid Rain Program: Overview,* EPA publication no. 430/F-92/019 (Washington, D.C., 1992), 2.

42. For a spirited and insightful discussion of the correlation between emissions and deposition, see Archie M. Kahan, *Acid Rain: Reign of Controversy* (Golden, Colo.: Fulcrum, 1986), 75–89.

43. For discussion of the goals and operation of emissions trading, see Joe DiLeo, "The Ecological Economist," *Environmental Forum* 10 (March/April 1993): 12–16; Carol B. Goldburg and Lester B. Lave, "Progress on Market Incentives for Abating Pollution: Trading Sulfur Dioxide Allowances," *Environment, Science and Technology* 26 (1992): 2076–8; Joseph I. Lieberman, "To Market, to Market," *Issues in Science and Technology* 8 (summer 1992): 25–9; and Robert B. Stavins and Bradley W. Whitehead, "Dealing with Pollution," *Environment* 34 (September 1992): 6–11, 29–42.

44. Adam J. Rosenberg, "Emissions Credit Futures Contracts on the Chicago Board of Trade: Regional and Rational Challenges to the Right to Pollute," *Virginia Environmental Law Journal* 13 (1994): 501–36.

45. For discussion of pros and cons, see Joe Alper, "Protecting the Envi-

ronment with the Power of the Market," *Science* 260 (1993): 1884–5; William F. Pedersen, Jr., "The Limits of Market-Based Approaches to Environmental Protection," *Environmental Law Reporter* 24 (1994): 10173–6; and Thomas Michael Power and Paul Rauber, "The Price of Everything," *Sierra* 78 (November/December 1993): 87–96. For discussion of a proposal by Northeast Utilities, operating in Connecticut, to buy five hundred tons of emissions credits from the New Jersey Public Service Electric and Gas Company, see Matthew L. Wald, "2 Utilities Intend Interstate Trade of Smog Pollution," *New York Times*, 16 March 1994, A1, B2.

46. Eugene J. McCarthy, "Pollution Absolution," *The New Republic* 203 (29 October 1990): 9.

47. Pedersen, "The Limits of Market-Based Approaches," 10174.

48. For a synopsis, see Carl Jensen, *Censored! The News That Didn't Make the News—and Why* (Chapel Hill, N.C.: Shelburne Press, 1993), 76–7.

49. Luther F. Bliven, "EPA Agreement Reduces Acid Rain," *Syracuse Post-Standard*, 10 August 1994, C1; and Rosenberg, "Emissions Credit Futures Contracts," 522–5.

50. Paul Rauber, "Schemes That Go Clunk," *Sierra* 77 (July/August 1992): 38–40.

51. Paul Carrier, "Governor Steadfast in His Plan to Give 'Pollution Credits,'" *Portland Press Herald*, 20 July 1994, 1A, 6A.

52. I converted the EPA display to my Visibility Base Map, which shows Delaware, Rhode Island, and other small states clearly with appropriately abstract area symbols that deemphasize the intricate state boundaries.

53. Fran Du Melle, "Laws Protecting Nonsmokers," *EPA Journal* 19 (October–December 1993): 21–2.

CHAPTER NINE

General references for indoor radon include David Bodansky, Maurice A. Robkin, and David R. Stadler, eds., *Indoor Radon and Its Hazards* (Seattle, Wash.: University of Washington Press, 1987); Douglas G. Brookins, *The Indoor Radon Problem* (New York: Columbia University Press, 1990); Leonard A. Cole, *Element of Risk: The Politics of Radon* (Washington, D.C.: AAAS Press, 1993); John W. Gofman, *Radiation and Human Health* (San Francisco: Sierra Club Books, 1981); Shyamal K. Majumdar, Robert F. Schmalz, and E. Willard Miller, eds., *Environmental Radon: Occurrence, Control, and Health Hazards* (Easton, Pa.: Pennsylvania Academy of Science, 1990); and Steve Page, "EPA's Strategy to Reduce Risk of Radon," *Journal of Environmental Health* 56 (December 1993): 27–36. General references for electromagnetic fields include William Ralph Bennett, Jr., *Health and Low-Frequency Electromagnetic Fields* (New Haven, Conn.: Yale University Press, 1994); Jack M. Lee, Jr., and others, *Electrical and Biological Effects of Transmission Lines: A Review* (Portland, Oreg.: U.S. Department of Energy, Bonneville Power Administration, 1985); and Stephen Prata, *EMF Handbook: Understanding and Controlling Electromagnetic Fields in Your Life* (Corte Madera, Calif.: Waite Group Press, 1993).

1. John Hersey, *Hiroshima* (New York: Alfred A. Knopf, 1946). For casualty figures, see Samuel Glasstone and Philip J. Dolan, comps., *The Effects of Nuclear Weapons*, 3d ed. (Washington, D.C.: U.S. Department of Defense and U.S. Department of Energy, 1977), 544.

2. Douglas Brode, *The Films of the Fifties* (Secaucus, N.J.: Citadel Press, 1976), 106–7; and Spencer R. Weart, *Nuclear Fear: A History of Images* (Cambridge, Mass.: Harvard University Press, 1988), 191–5.

3. Brookins, *The Indoor Radon Problem*, 7; and Margaret A. Reilly, "The Index House: Pennsylvania Radon Research and Demonstration Project, Pottstown, Pennsylvania, 1986–1988," in Majumdar, Schmalz, and Miller, *Environmental Radon*, 26–38.

4. Michael Lafayore, "Warning! This House Contains Radon," *Readers Digest* 128 (June 1986): 110–14.

5. In contrast, cumulative exposure is measured in rems or millirems (thousandths of a rem), and absorbed doses are measured in rads or millirads (thousandths of rads).

6. David Bodansky, "Overview of the Indoor Radon Problem," in Bodansky, Robkin, and Stadler, *Indoor Radon and Its Hazards*, 3–16.

7. Maurice A. Robkin, "Dosimetry Models," in Bodansky, Robkin, and Stadler, *Indoor Radon and Its Hazards*, 76–90.

8. Edward A. Martel, "Critique of Current Lung Dosimetry Models for Radon Progeny Exposure," in Philip K. Hopke, ed., *Radon and Its Decay Products: Occurrence, Properties, and Health Effects* (Washington, D.C.: American Chemical Society, 1987), 444–61, especially 454; and Maurice A. Robkin, "Terminology for Describing Radon Concentrations and Exposures," in Bodansky, Robkin, and Stadler, *Indoor Radon and Its Hazards*, 17–29.

9. Because of uncertainty, the EPA reports that the actual number might range from 7,000 to 30,000 deaths per year. See U.S. Environmental Protection Agency, *A Citizen's Guide to Radon*, 2d ed., EPA publication no. 402-K92-001 (Washington, D.C., 1992), 2.

10. Committee on the Biological Effects of Ionizing Radiations, Board of Radiation Effects Research, Commission on Life Sciences, National Research Council, *Health Effects of Exposure to Low Levels of Ionizing Radiation: BEIR V* (Washington, D.C.: National Academy Press, 1990), 270–2; and N. H. Harley and J. H. Harley, "Indoor Radon: A Natural Risk," *Nuclear Safety* 32 (1991), 537–43.

11. Eric J. Hall, *Radiation and Life*, 2d ed. (New York: Pergamon Press, 1984), 31–3.

12. For a concise review, see Kenneth L. Jackson, Joseph P. Geraci, and David Bodansky, "Observations of Lung Cancer: Evidence Relating Lung Cancer to Radon Exposure," in Bodansky, Robkin, and Stadler, *Indoor Radon and Its Hazards*, 91–111, especially 100–2. Also see Shrader-Frechette, *Burying Uncertainty*, 134–6.

13. See, for example, Bernard L. Cohen, "Experimental Tests of the Linear-No Threshold Theory of Radiation Carcinogenesis," in Majumdar, Schmalz, and Miller, *Environmental Radon*, 319–30; Gofman, *Radiation and Human Health*, 467–8; and John W. Gofman, *Radiation-Induced Cancer from*

Low-Dose Exposure: An Independent Analysis (San Francisco: Committee for Nuclear Responsibility, 1990), chaps. 13 and 14.

14. Philip H. Abelson, "Mineral Dusts and Radon in Uranium Mines," *Science* 254 (1991): 777.

15. Ibid.

16. Because only a small fraction of homes have been tested, estimates vary. See, for example, Leonard A. Cole, "Radon: The Silent Killer?" *Garbage* 6 (spring 1994): 22–8; and Warren E. Leary, "Studies Raise Doubts about Need to Lower Home Radon Levels," *New York Times*, 6 September 1994, C4.

17. Anthony V. Nero, Jr., "A National Strategy for Indoor Radon," *Issues in Science and Technology* 9 (fall 1992): 33–40.

18. For discussion of Abelson's paper, presented at the 1991 Science Writer's Workshop, see Cole, *Element of Risk*, 76–7; quote on p. 77.

19. Abelson, "Mineral Dusts and Radon."

20. See Nero, "A National Strategy for Indoor Radon," as well as U.S. Congress, House Committee on Energy and Commerce, *Radon Awareness and Disclosure: Joint Hearing before the Subcommittee on Health and the Environment and the Subcommittee on Transportation and Hazardous Materials*, 103d Cong., 1st sess., 14 July 1993, 184–202.

21. Cole, "Radon: The Silent Killer?" 28.

22. See "EPA's Map of Radon Zones," *Radon Bulletin* [Conference of Radiation Control Program Directors] 4 (winter 1993): 2–3. Published without apparent fanfare in 1993, the EPA map illustrated articles in *Scientific American* and the *New York Times*, among other publications, in 1994. See, for example, John Horgan, "Radon's Risks," *Scientific American* 271 (August 1994): 14–16; Tammy Kutzmark and Donald Geis, "The Radon Problem: A Solution Is Easier Than You Think," *Public Management* 76 (May 1994): 6–15; and Leary, "Studies Raise Doubts."

23. The map's legend box identifies the categories only as "Zone 1," "Zone 2," and "Zone 3." A one-page fact sheet, provided separately, describes the categories, their development, and the map's purpose. The National Safety Council, which operates the EPA's 1-800-SOS-RADON hotline, distributes the map and fact sheet.

24. In its eagerness to compile a radon-hazard map from available data, the EPA relied on geologic information and apparently produced more than one version of the map within a year. Generally similar maps with the same title ("Areas with Potentially High Radon Levels") but noteworthy differences appeared in Michael Lafavore, *Radon, the Invisible Threat: What It Is, Where It Is, How to Keep Your House Safe* (Emmaus, Pa.: Rodale Press, 1987), 30; and Mason and Mattson, *Atlas of United States Environmental Issues*, 148.

25. Lafavore, *Radon, the Invisible Threat*, 31; and Thomas M. Gerusky, "Protecting the Homefront," *Environment* 29 (January/February 1987): 12–37.

26. Arthur W. Rose, John W. Washington, and Daniel J. Greeman, "Geology and Geochemistry of Radon Occurrence," in Majumdar, Schmalz, and Miller, *Environmental Radon*, 64–77; especially map on p. 92. In a revealing yet frustrating search of geographic trends, a colleague and I mapped indoor radon-test data for Pennsylvania counties and Zip Code areas. See

George A. Schnell and Mark Monmonier, "Radon in Residences: a Carto-graphic Analysis," in Majumdar, Schmalz, and Miller, *Environmental Radon*, 39–53.

27. Vernon N. Houk, Daniel A. Hoffman, and Christie Eheman, "Public Health Implications of Radon Exposures in the United States," in Majumdar, Schmalz, and Miller, *Environmental Radon*, 216–22.

28. See, for example, Charles Laymon, Charles Kunz, and Lawrence Keefe, *Indoor Radon in New York State: Distribution, Sources, and Controls* (Albany: State of New York Department of Health, 1990), 24–5.

29. Houk, Hoffman, and Eheman, "Public Health Implications of Radon Exposures in the United States."

30. Linda C. S. Gundersen and others, "Geology of Radon in the United States," in Alexander E. Gates and Linda C. S. Gundersen, eds., *Geologic Controls on Radon*, Geological Society of America Special Paper no. 271 (Boulder, Colo.: Geological Society of America, 1992), 1–16.

31. R. Randall Schumann, ed., *Geological Radon Potential of EPA Region 5*, U.S. Geological Survey Open File Report no. 93-292-E (Denver, Colo., 1994).

32. Linda C. S. Gundersen, R. Randall Schumann, and Sharon W. White, "The USGS/EPA State Radon Potential Assessments: An Introduction," in Schumann, *Geological Radon Potential of EPA Region 5*, 1–35.

33. See, for example, R. Randall Schumann and Kevin M. Schmidt, "Preliminary Geological Radon Potential Assessment of Minnesota," in Schumann, *Geological Radon Potential of EPA Region 5*, 122–47; caveats on pp. 122 and 142.

34. James K. Otton, *The Geology of Radon*, a U.S. Geological Survey general interest publication (Washington, D.C., 1992).

35. Colored red, orange, and blue, these categories are based on the 4 pCi/L threshold, with "low" meaning "the majority of homes contain less" than the EPA action level, "high" meaning "the majority of homes contain more," and "moderate" indicating somewhat confusingly that "one third to one half of the homes have more than 4 pCi/L." Ibid., 25.

36. L. C. S. Gundersen and others, *Map Showing Radon Potential of Rocks and Soils in Montgomery County, Maryland*, 1:62,500, U.S. Geological Survey Miscellaneous Field Studies Map no. 2043 (Reston, Va., 1988).

37. Ibid.

38. Anthony Nero, Lawrence Berkeley National Laboratory, telephone conversation with author, 27 January 1995.

39. Still under way, the project is mentioned briefly in Linda Gundersen, "DOE, EPA Begin Search for High-Radon Homes," *Radon Research Notes* [Oak Ridge National Laboratory] no. 9 (December 1992), 1–3, 11; and Susan L. Rose, "Current and Future Perspectives," in Niren L. Nagada, ed., *Radon: Prevalence, Measurements, Health Risks and Control* (Philadelphia: American Society for Testing Materials, 1994), 148–58, especially 155. For a concise introduction to the methodology see Anthony V. Nero, Jr., "Methodologies for Identifying High-Radon Areas: A Brief Review," in House Committee on Energy and Commerce, *Radon Awareness and Disclosure*, 191–6.

40. U.S. General Accounting Office, *Air Pollution: Uncertainty Exists in Radon Measurements*, GAO report no. RCED-90-25 (October 1989), 5.

41. Nero, telephone conversation.

42. "Lung Cancer's Gassy Ally," *U.S. News and World Report* 104 (18 January 1988): 13.

43. Louis Slesin, "Power Lines and Cancer: the Evidence Grows," *Technology Review* 90 (October 1987): 53–9.

44. David Noland, "Power Play," *Discover* 10 (December 1989): 62–8; quote on p. 64.

45. Nancy Wertheimer and Ed Leeper, "Electrical Wiring Configurations and Childhood Cancer," *American Journal of Epidemiology* 109 (1979): 273–84.

46. For a concise graphic description of the Wertheimer-Leeper wire codes, see "Sharpening the Focus in EMF Research," *EPRI Journal* 17 (March 1992): 4–13.

47. A more recent study that reinforced Wertheimer and Leeper's findings reported "an association between childhood leukemia risk and wiring configuration but not direct measurements of electric and magnetic fields." See Stephanie J. London and others, "Exposure to Residential Electric and Magnetic Fields and Risk of Childhood Leukemia," *American Journal of Epidemiology* 134 (1991): 923–37. But physicists and engineers remain skeptical of a causal connection; see, for example, William R. Bennett, Jr., "Cancer and Power Lines," *Physics Today* 47 (April 1994): 23–9.

48. Prata, *EMF Handbook*, 6.

49. See, for example, M. Granger Morgan and Indira Nair, "Electromagnetic Fields: The Jury's Still Out. Part 3: Managing the Risks," *IEEE Spectrum* 29 (August 1990): 32–5; and Leonard A. Sagan, "Epidemiological and Laboratory Studies of Power Frequency Electric and Magnetic Fields," *JAMA* [Journal of the American Medical Association] 268 (1992): 625–9. A panel of the federal Committee on Interagency Radiation Research and Policy Coordination concluded that ELF fields are not a research priority; see "Health Effects of Low-Frequency Electric and Magnetic Fields," *Environment, Science and Technology* 27 (January 1993): 42–51. But the same issue carried a less dismissive rebuttal from a respected epidemiologist; see David A. Savitz, "Health Effects of Low-Frequency Electric and Magnetic Fields," *Environment, Science and Technology* 27 (January 1993): 52–4.

50. See Paul Brodeur, *Currents of Death: Power Lines, Computer Terminals, and the Attempt to Cover Up Their Threat to Your Health* (New York: Simon and Schuster, 1989); Paul Brodeur, *The Great Power-Line Cover-Up: How the Utilities and the Government Are Trying to Hide the Cancer Hazard Posed by Electromagnetic Fields* (Boston: Little, Brown and Co., 1993); and Paul Brodeur, *The Zapping of America: Microwaves, Their Deadly Risk, and the Cover-Up* (New York: W. W. Norton, 1977).

51. Brodeur, *The Great Power-Line Cover-Up*, 90–101.

52. Ibid., 139–49.

53. Ibid., 144–5.

54. Gary Taubes, "Fields of Fear," *Atlantic Monthly* 274 (November 1994): 94–108; quote on p. 107.

55. Michael Fumento, *Science under Seige: Balancing Technology and the Environment* (New York: William Morrow and Co., 1993), 218–55; quote on p. 255.

56. H. Keith Florig, "Containing the Costs of the EMF Problem," *Science* 257 (24 July 1992): 468–9, 488–92.

57. See, for example, James Barron, "Court Allows Damages for Fear of High-Voltage Power Lines," *New York Times*, 13 November 1993, B5; Peter Marks, "Electric Fields Create Nebulous Peril but Real Fear on L.I.," *New York Times*, 6 January 1994, B1, B7; and Lisa N. Mulhall, "The Transmission Line Siting Act—Balancing Power and People," *Stetson Law Review* 20 (1991): 909–27.

58. For example, a recent EPRI-sponsored study of electrical workers that failed to find a correlation between magnetic fields and leukemia nonetheless suggested a link with brain cancer; see David A. Savitz and Dana P. Loomis, "Magnetic Field Exposure in Relation to Leukemia and Brain Cancer Mortality among Electric Utility Workers," *American Journal of Epidemiology* 141 (1995): 123–34.

59. Maria Feychting and Anders Ahlbom, "Magnetic Fields and Cancer in Children Residing Near Swedish High-voltage Power Lines," *American Journal of Epidemiology* 138 (1993): 467–81.

60. Daniel Wartenberg, conversation with author, 13 February 1995. Also see D. Wartenberg and others, "Using a Geographic Information System to Identify Populations Living Near High Voltage Electric Power Transmission Lines in New York State," *Abstracts of Papers at the International Symposium on Computer Mapping in Epidemiology and Environmental Health, Tampa, Florida, 12–15 February 1995*, 70–1.

CHAPTER TEN

General references include William R. Freudenburg and Eugene A. Rosa, eds., *Public Reactions to Nuclear Power: Are There Critical Masses?* (Boulder, Colo.: Westview Press, 1984); Gordon T. Goodman and William D. Rowe, eds., *Energy Risk Management* (London: Academic Press, 1979); Peter Gould, *Fire in the Rain: The Democratic Consequences of Chernobyl* (Baltimore, Md.: Johns Hopkins University Press, 1990); League of Women Voters Education Fund, *The Nuclear Waste Primer*, rev. ed. (New York: Lyons and Burford, 1993); and Richard E. Webb, *The Accident Hazards of Nuclear Power Plants* (Amherst: University of Massachusetts Press, 1976).

1. See Mark Monmonier, *Drawing the Line: Tales of Maps and Cartocontroversy* (New York: Henry Holt, 1995), 220–55.

2. Paul Slovic, "Perception of Risk," *Science* 236 (1987): 280–5.

3. Although correlations of radioactive releases at Three Mile Island with increased mortality suggest that the TMI-2 accident might have produced "excess deaths" in Pennsylvania and nearby states, no single death has ever been directly attributed to these releases. See Jay M. Gould and Benjamin A. Goldman, *Deadly Deceit: Low-Level Radiation, High-Level Cover-up* (New York: Four Walls Eight Windows, 1990), 57–70.

4. Weart, *Nuclear Fear*, 421.

5. Ibid., 422.

6. See, for example, Paul Slovic, Sarah Lichtenstein, and Baruch Fischhoff,

"Images of Disaster: Perception and Acceptance of Risks from Nuclear Power," in Goodman and Rowe, *Energy Risk Management*, 223–45.

7. Paul Slovic, James H. Flynn, and Mark Layman, "Perceived Risk, Trust, and the Politics of Nuclear Waste," *Science* 254 (1991): 1603–7.

8. Gerard H. Clarfield and William M. Wiecek, *Nuclear America: Military and Civilian Nuclear Power in the United States, 1940–1980* (New York: Harper and Row, 1984), 349–50. For contemporary media accounts, which omitted the more gruesome details, see "And Three Were Dead," *Newsweek* 57 (16 January 1961): 74; "Runaway Reactor," *Time* 77 (13 January 1961): 18–19; and "What Happened in Reactor Blast," *Business Week* no. 1636 (7 January 1961): 28. Also see Walter C. Patterson, "Chernobyl: Worst but Not First," *Bulletin of the Atomic Scientists* 42 (August/September 1986): 43–5.

9. Nuclear reactors generate energy by driving turbines with superheated steam. The heat comes from a controlled chain reaction initiated by positioning fuel rods of uranium or plutonium pellets sufficiently close so that neutrons released by radioactive decay in one fuel rod can bombard fissionable atoms in a neighboring fuel rod. Between the fuel rods are control rods of cadmium, which absorbs neutrons. By automatically raising or lowering the control rods, the reactor allows fission to occur without bursting the steam lines or melting the "core" (reactor vessel).

10. Zhores Medvedev, "Two Decades of Dissidence," *New Scientist* 72 (1976): 264–7.

11. Ibid., 265.

12. For discussion of Hill's attacks, see Zhores Medvedev, "Facts behind the Soviet Nuclear Disaster," *New Scientist* 74 (1977): 761–4; Zhores A. Medvedev, *Nuclear Disaster in the Urals*, trans. George Saunders (New York: W. W. Norton and Co., 1979), 13; and Sarah White, "Medvedev Repeats Accident Claim," *New Scientist* 72 (1976): 315.

13. In November 1977 CIA officials confirmed the accident in responding to an information request from R. B. Pollock, representing the Citizen's Movement for Safe and Efficient Energy. See John R. Trabalka, L. Dean Eyman, and Stanley I. Auerbach, "Analysis of the 1957–1958 Soviet Nuclear Accident," *Science* 209 (1980): 345–53; especially note 4 on p. 351.

14. For the full text of Tumerman's letter, see Medvedev, *Nuclear Disaster in the Urals*, 11–12.

15. Medvedev, "Facts behind the Soviet Nuclear Disaster."

16. J. R. Trabalka, L. D. Eyman, and S. I. Auerbach, *Analysis of the 1957–1958 Soviet Nuclear Accident*, ORNL report no. 5613 [Environmental Sciences Division publication no. 1445] (Oak Ridge, Tenn.: Oak Ridge National Laboratory, 1979); and J. R. Trabalka, L. D. Eyman, F. L. Parker, E. G. Struxness, and S. I. Auerbach, "Another Perspective of the 1958 Soviet Nuclear Accident," *Nuclear Safety* 20 (1979): 206–10. Trabalka and his two coauthors also published the former report in the July 18, 1980, issue of *Science*.

17. In a later effort to reconstruct the disaster, Medvedev advanced several explanations. See Medvedev, *Nuclear Disaster in the Urals*, 155–64.

18. Trabalka, Eyman, and Auerbach, "Analysis of the 1957–1958 Soviet Nuclear Accident," 351.

19. For descriptions of the Chernobyl disaster, see Peter Gould, *Fire in the Rain*, especially pp. 41–2; Eliot Marshall, "Reactor Explodes Amid Soviet Silence," *Science* 232 (1986): 814–15; and Gordon Thompson, "What Happened at Reactor Four," *Bulletin of the Atomic Scientists* 42 (August/September 1986): 26–31.

20. Serge Schmemann, "Soviet Announces Nuclear Accident at Electric Plant," *New York Times*, 29 April 1986, A1, A11.

21. Alexander Amerisov, "A Chronology of Soviet Media Coverage," *Bulletin of the Atomic Scientists* 42 (August/September 1986): 38–40.

22. Herbert L. Abrams, "How Radiation Victims Suffer," *Bulletin of the Atomic Scientists* 43 (August/September 1986): 13–17; Ragnar E. Löfstedt, "Chernobyl: Four Years Later, the Repercussions Continue," *Environment* 32 (April 1990): 2–5; David Marples, "Chernobyl's Lengthening Shadow," *Bulletin of the Atomic Scientists* 49 (September 1993): 38–43; and "Soviet Union 'Showed the World How to Evacuate,'" *New Scientist* 111 (4 September 1986): 20.

23. Grigori Medvedev, *The Truth about Chernobyl*, trans. Evelyn Rossiter (New York: Basic Books, 1991), 78–9.

24. Carroll Bogert, "Chernobyl's Legacy," *Newsweek* 115 (7 May 1990): 30–1.

25. Cristina J. Battista, "Chernobyl: GIS Model Aids Nuclear Disaster Relief," *GIS World* 7 (March 1994): 32–5; Marco Bojcun, "The Legacy of Chernobyl," *New Scientist* 130 (20 April 1991): 30–5; Jay M. Gould, "Chernobyl—the Hidden Tragedy," *Nation* 256 (15 March 1993): 331–4; and Frank von Hippel and Thomas B. Cochran, "Chernobyl: Estimating Long-Term Health Effects," *Bulletin of the Atomic Scientists* 42 (August/September 1986): 18–24.

26. For a study that concluded the effects of Chernobyl were largely social and economic, see Lynn R. Anspaugh, Robert J. Catlin, and Marvin Goldman, "The Global Impact of the Chernobyl Reactor Accident," *Science* 242 (1988): 1513–19. For a far less optimistic assessment, see Robert Peter Gale, "Chernobyl: Answers Slipping Away," *Bulletin of the Atomic Scientists* 42 (September 1990): 19–23.

27. See, for example, Matthew L. Wald, "Study Finds Gap in Security at Nuclear Plants," *New York Times*, 8 April 1993, A7; Matthew L. Wald, "Doubts Raised about Rules to Protect Nuclear Reactors," *New York Times*, 11 February 1994, A20; and Matthew L. Wald, "Regulator Says Economics Could Lead Nuclear Plants to Scrimp," *New York Times*, 9 September 1994, A19.

28. See Daniel F. Ford, *Three Mile Island: Thirty Minutes to Meltdown* (New York: Viking Press, 1982); Cynthia B. Flynn, "The Local Impacts of the Accident at Three Mile Island," in Freudenburg and Rosa, *Public Reactions to Nuclear Power*, 205–32; and John G. Kemeny and others, *Report of the President's Commission on the Accident at Three Mile Island: The Need for Change—the Legacy of TMI* (Washington, D.C., 1979).

29. For concise accounts of the five-day incident, see Kent Barnes, James Brosius, Susan L. Cutter, and James K. Mitchell, *Responses of Impacted Populations to the Three Mile Island Nuclear Reactor Accident: An Initial Assessment,*

Department of Geography Discussion Paper no. 13 (New Brunswick, N.J.: Rutgers University, 1979); Ford, *Three Mile Island: Thirty Minutes to Meltdown*, 16–22; Peggy M. Hassler, *Three Mile Island: A Reader's Guide to Selected Government Publications and Government-Sponsored Research Publications* (Metuchen, N.J.: Scarecrow Press, 1988), 14–20; Howard Morland, "The Meltdown That Didn't Happen," *Harpers* 259 (October 1979): 16–23; and Philip Starr and William Pearman, *Three Mile Island Sourcebook: Annotations of a Disaster* (New York: Garland Publishing, 1983), 3–7.

30. Cynthia Bullock Flynn, "Reactions of Local Residents to the Accident at Three Mile Island," in David L. Sills, C. P. Wolf, and Vivien B. Shelanski, eds., *Accident at Three Mile Island: The Human Dimensions* (Boulder, Colo.: Westview Press, 1982), 49–63; especially 52.

31. David Burnham, "Report on Nuclear Accident Holds Agency Is Unable to Insure Safety," *New York Times*, 25 January 1980, A13; and Kemeny and others, *Report of the President's Commission on the Accident at Three Mile Island*, 30–31.

32. Evacuation plans included with the plant's license application covered only a 2.2-mile radius around the plant. As required by state regulations, plans prepared by police, fire, and civil defense officials in the three closest counties extended the planning radius to 5 miles, but only one plan was complete. Townships and other local jurisdictions lacked their own plans, and the NRC had yet to approve two plans filed previously by the state. See Donald J. Ziegler, Stanley D. Brunn, and James H. Johnson, Jr., "Evacuation from a Nuclear Technological Disaster," *Geographical Review* 71 (1981): 1–16, especially 15.

33. Kemeny and others, *Report of the President's Commission on the Accident at Three Mile Island*, 40.

34. Ibid., 15.

35. Ibid., 38.

36. NRC officials considered evacuation zones extending as far as 50 miles from the plant. See Russell R. Dynes, "The Accident at Three Mile Island: The Contribution of the Social Sciences to the Evaluation of Emergency Preparedness and Response," in Sills, Wolf, and Shelanski, *Accident at Three Mile Island: The Human Dimensions*, 119–29.

37. See Morland, "The Meltdown That Didn't Happen," 18.

38. Webb, *Accident Hazards of Nuclear Power Plants*, 4.

39. H. E. Collins, B. K. Grimes, and F. Galpin, *Planning Basis for the Development of State and Local Government Radiological Emergency Response Plans in Support of Light Water Nuclear Power Plants: A Report Prepared by a U.S. Nuclear Regulatory Commission and U.S. Environmental Protection Agency Task Force on Emergency Planning* (Washington, D.C., 1978), especially pp. 15–17. The conclusions of the task force report appear in U.S. Congress, House Committee on Government Operations, *Emergency Planning around U.S. Nuclear Powerplants: Nuclear Regulatory Commission Oversight*, 96th Cong., 1st sess. (7, 10, and 14 May 1979), 553–6. Also see Webb, *Accident Hazards of Nuclear Power Plants*, 45; and U.S. General Accounting Office, *Areas around Nuclear Facilities Should Be Better Prepared for Radiological Emergencies*, report no. EMD-78-110 (30 March 1979).

The terms are defined in the *Code of Federal Regulations*, in section 10 CFR 50.47: "*Emergency planning zone* is a generic area around a commercial nuclear facility used to assist in offsite emergency planning and the development of a significant response base. For commercial nuclear power plants, EPZs of about 10 and 50 miles are delineated for the plume and ingestion exposure pathways; *Plume Exposure Pathway* refers to whole body external exposure to gamma radiation from the plume and from deposited materials and inhalation exposure from the passing radioactive plume. The duration of primary exposures could range in length from hours to days; *Ingestion Exposure Pathway* refers to exposure primarily from the ingestion of water or foods such as milk or fresh vegetables that have been contaminated with radiation. The duration of primary exposure could range from hours to months."

40. Ibid., 47.

41. Ibid.

42. William B. Cottrell, "General Administrative Activities," *Nuclear Safety* 21 (1980): 665–71.

43. M. Silberberg and others, *Reassessment of the Technical Bases for Estimating Source Terms*, U.S. Nuclear Regulatory Commission report no. 956 (Washington, D.C., 1985), xvii.

44. Webb, *Accident Hazards of Nuclear Power Plants*, 85–6.

45. For descriptions of probabilistic risk assessment for nuclear plant accidents, see Ralph R. Fullwood and Robert E. Hall, *Probabilistic Risk Assessment in the Nuclear Power Industry: Fundamentals and Applications* (Oxford: Pergamon Press, 1988); and Webb, *Accident Hazards of Nuclear Power Plants*, especially pp. 1–6, 74–9.

46. J. J. DiNunno and others, *Calculation of Distance Factors for Power and Test Reactor Sites*, U.S. Atomic Energy Commission report no. TID-14844 (Washington, D.C., 1962).

47. N. C. Rasmussen and others, *Reactor Safety Study: An Assessment of Accident Risks in U.S. Commercial Nuclear Power Plants*, U.S. Nuclear Regulatory Commission report no. 75/014 (WASH-1400) (Washington, D.C., 1975). Also see "Emergency Plans Approved by Federal Nuclear Agency," *New York Times*, 6 December 1979, A26. For discussion of comparative mortality estimates, see Webb, *Accident Hazards of Nuclear Power Plants*, 96–8.

48. Webb, *Accident Hazards of Nuclear Power Plants*, 86.

49. Steven C. Sholly and Gordon Thompson, principal authors, *The Source Term Debate: A Report by the Union of Concerned Scientists: A Review of the Current Basis for Predicting Severe Accident Source Terms with Special Emphasis on the NRC Source Term Reassessment Program (NUREG-0956)* (Cambridge, Mass.: Union of Concerned Scientists, 1986).

50. Ibid., 5-7 to 5-8; and Matthew L. Wald, "A-plants Warned to Be Wary of Truck Bombs," *New York Times*, 1 July 1993, A15. For additional discussion of nuclear plant sabotage, see Webb, *Accident Hazards of Nuclear Power Plants*, 150–3.

51. U.S. Nuclear Regulatory Commission, *Implications of the Accident at Chernobyl for Safety Regulation of Commercial Nuclear Power Plants in the United States*, NRC report no. 1251 (Washington, D.C., 1989); for discussion of the size of emergency planning zones, see pp. 4-1 to 4-4.

52. Ibid., 4-4.

53. Ibid.

54. The *Code of Federal Regulations* defines FEMA's role and responsibilities in 44 CFR 350; for criteria for the review and approval of state and local emergency plans see §350.5.

55. See U.S. Federal Emergency Management Agency, *A Guide to Preparing Emergency Public Information Materials*, FEMA publication no. REP-11 (Washington, D.C., 1985).

56. For discussion of these guidelines and the design of off-site emergency response maps, see Ute J. Dymon and Nancy L. Winter, "Evacuation Mapping: The Utility of Guidelines," *Disasters* 17 (1993): 12–24. By contrast, more specific standards for detailed off-site emergency maps require state and local governments to show population distribution, evacuation areas, evacuation routes, relocation centers, and radiological sampling points and monitoring stations; see, for example, U.S. Nuclear Regulatory Commission and U.S. Federal Emergency Management Agency, *Criteria for Preparation and Evaluation of Radiological Emergency Response Plans and Preparedness in Support of Nuclear Power Plants, Criteria for Utility Offsite Planning and Preparedness, Final Report*, NRC publication no. 654/FEMA publication no. FEMA-REP-1, Rev. 1 (Washington, D.C., 1980), 61.

57. Mary M. Schneider, Vermont Yankee Nuclear Power Corp., letter to author, 18 April 1995. To reduce possible confusion, three separate calendars provide relocation center directions, telephone numbers, and other information customized for residents of Vermont, New Hampshire, and Massachusetts. The three calendars use the same EPZ map but different detail maps for their respective relocation centers.

58. For cartographic scholar Ute Dymon's critique of the design and content of emergency maps distributed by several nuclear plants in the Northeast, see Dymon and Winter, "Evacuation Mapping: The Utility of Guidelines."

59. For FEMA guidelines on adjustment, see *Criteria for Preparation and Evaluation of Radiological Emergency Response Plans*, FEMA publication no. REP-1, Rev. 1, 11; and 10 CFR 50.47(c)(2).

60. See "Reactor Goes Critical Inadvertently," *Nuclear Safety* 15 (March–April 1974): 210–12; the AEC description is quoted on p. 211. Also see John Kifner, "Shutdown of a Nuclear Plant, 17th in 19 Months, Spurs U.S. Debate," *New York Times*, 31 March 1974, 43; and Webb, *Accident Hazards of Nuclear Power Plants*, 75–6, 195–6.

61. "NRC Plans to Fine Vermont Yankee, Cites Violations at A Plant," *Wall Street Journal*, 4 August 1993, C16.

62. Ibid., 49–51; and U.S. Nuclear Regulatory Commission and U.S. Federal Emergency Management Agency, *Criteria for Preparation and Evaluation of Radiological Emergency Response Plans and Preparedness in Support of Nuclear Power Plants, Criteria for Utility Offsite Planning and Preparedness, Final Report*, NRC publication no. 654/FEMA publication no. REP-1, Rev. 1, Supp. 1 (Washington, D.C., 1988), 14–15.

63. U.S. General Accounting Office, *Nuclear Regulation: Public Knowledge of Radiological Emergency Procedures*, GAO report no. RCED-87-122 (June 1987), 6–8.

64. Susan L. Cutter, "Emergency Preparedness and Planning for Nuclear Power Plant Accidents," *Applied Geography* 4 (1984): 235–45; Barbara M. Vogt and John Sorensen, *Evacuation in Emergencies: An Annotated Guide to Research* (Oak Ridge, Tenn.: Oak Ridge National Laboratory, 1987), 107–29; and Donald J. Zeigler and James H. Johnson, Jr., "Evacuation Behavior in Response to Nuclear Power Plant Accidents," *Professional Geographer* 36 (1984): 207–15.

65. Figure 10.7 includes eastern Tennessee's Watts Bar nuclear plant, with EPZs delineated for two reactors under construction, but omits the Yankee Rowe nuclear station, in western Massachusetts, the Shoreham power plant, on Long Island, and other commercial reactors once licensed to operate but now in various stages of decommissioning, entombment, or monitored storage. See Taylor Moore, "Decommissioning Nuclear Plants," *EPRI Journal* 10 (July/August 1985): 14–21. Unless the site contains a reactor currently licensed for operation, the map also omits locations at which a reactor was canceled during planning or construction.

66. See, for example, William J. Broad, "Theory on Thread of Blast at Nuclear Waste Site Gains Support," *New York Times*, 23 March 1995, A18.

67. See, for example, Scott Allen, "Moving New England's Nuclear Waste," *Boston Globe*, 18 January 1995, 15, 16.

68. In the United Kingdom, which has an extensive railway network, nearly all nuclear sites are on a railroad, some of which have been kept open specifically for transporting nuclear waste. This strategy avoids the safety hazard— and the political controversy—of shipping radioactive materials along public roads.

69. Chet Gray, telephone communication, 19 March 1995. For an annual report on the cleanup effort, see *Environmental Management 1995: Progress and Plans of the Environmental Management Program*, DOE publication no. EM-0228 (Washington, D.C., 1995). Also see U.S. Congress, Office of Technology Assessment, *Complex Cleanup: The Environmental Legacy of Nuclear Weapons Production*, publication no. OTA-O-484 (Washington, D.C., 1991).

70. Seth Shulman, *The Threat at Home: Confronting the Toxic Legacy of the U.S. Military* (Boston: Beacon Press, 1992).

71. U.S. General Accounting Office, *Environmental Cleanup: Better Data Needed for Radioactively Contaminated Defense Sites*, GAO report no. NSIAD-94-168 (24 August 1994); quote on p. 1.

72. Keith Schneider, "By Default and without Debate, Utilities Ready Long-Term Storage of Nuclear Waste," *New York Times*, 15 February 1994, A19; Matthew L. Wald, "Nuclear Waste, with Nowhere to Go," *New York Times*, 28 March 1994, A10; and Keith Schneider, "Nuclear Plants to Become De Facto Radioactive Dumps," *New York Times*, 15 February 1995, A19. Also see "Companies Face Space Shortage to Dispose of Nuclear Waste," *New York Times*, 11 June 1994, 26; and William J. Broad, "Deadly Nuclear Waste Piles Up with No Clear Solution at Hand," *New York Times*, 14 March 1995, C1, C11.

73. For a case study of an abortive, ill-conceived effort to site a low-level radioactive waste dump in New York State, see Monmonier, *Drawing the Line*, 220–55. For insights on legal and political aspects of NIMBY (not in my backyard) controversies, see Michael B. Gerrard, *Whose Backyard, Whose Risk: Fear*

and Fairness in Toxic and Nuclear Waste Siting (Cambridge, Mass.: MIT Press, 1994); and G. Tomas Murauskas and Fred M. Shelley, "Local Political Responses to Nuclear Waste Disposal," *Cities* 3 (1986): 157–62. Also see Robert Reinhold, "A Test Case for Nuclear Disposal," *New York Times*, 24 January 1994, A8.

74. For an optimistic view of high-level radioactive waste disposal at Yucca Mountain, see Taylor Moore, "The Hard Road to Nuclear Waste Disposal," *EPRI Journal* 15 (July/August 1990): 4–17. For a critique of the Yucca Mountain site, see Dan Grossman and Seth Shulman, "Verdict at Yucca Mountain: Will America's Nuclear Waste Dump Be Safe?" *Earth* 3 (March 1994): 54–63. For discussion of New Mexico opposition to the Mescalero site, see Matthew L. Wald, "Tribe on Path to Nuclear Waste Site," *New York Times*, 6 August 1993, A12. Not all Mescalero Apaches were eager to have a nuclear waste storage site; see George Johnson, "Apache Tribe Rejects Move to Store Nuclear Waste on Reservation," *New York Times*, 2 February 1995, A16; and George Johnson, "Nuclear Waste Dump Gets Tribe's Approval in Re-vote," *New York Times*, 11 March 1995, 6.

75. Despite opposition from persons living near nuclear power plants, centralized at-reactor storage of nonutility waste may prove more politically expedient than dispersed on-site storage at hospitals, universities, and other generators. Another option is a centralized, state-run low-level radioactive waste storage site. New York State, in which the New York Power Authority owns a nuclear plant, might well elect this interim solution. For a concise discussion of the issue, see Mary R. English, *Siting Low-Level Radioactive Waste Disposal Facilities: The Public Policy Dilemma* (New York: Quorum Books, 1992), 22–4.

76. Using the 1980 census, FEMA estimated that 3.4 million people resided within ten miles of a nuclear power plant. See U.S. General Accounting Office, *Nuclear Regulation: Public Knowledge of Radiological Emergency Procedures*, GAO report no. RCED-87-122 (June 1987), 4. If they grew at the same rate as the country as a whole, the EPZs would have a total 1995 population of over 3.95 million.

CHAPTER ELEVEN

General references include Michael T. Charles and John Choon K. Kim, eds., *Crisis Management: A Casebook* (Springfield, Ill.: Charles C. Thomas, 1990); Louise K. Comfort, ed., *Managing Disaster: Strategies and Policy Perspectives* (Durham, N.C.: Duke University Press, 1988); Susan L. Cutter, *Living with Risk: The Geography of Technological Hazards* (London: Edward Arnold, 1993); Richard J. Healy, *Emergency and Disaster Planning* (New York: John Wiley and Sons, 1969); J. M. Holmes and C. H. Byers, *Countermeasures to Airborne Hazardous Chemicals* (Park Ridge, N.J.: Noyes Data Corporation, 1990); Richard T. Sylves and William L. Waugh, Jr., eds., *Cities and Disaster: North American Studies in Emergency Management* (Springfield, Ill.: Charles C. Thomas, 1990); Waugh and Hy, *Handbook of Emergency Management*; and Donald J. Zeigler, James H. Johnson, Jr., and Stanley D. Brunn, *Technological Hazards* (Washington, D.C.: Association of American Geographers, 1983).

1. Alton Hall Blackington, "The Molasses Flood," in Benjamin Watson, ed., *New England's Disastrous Weather* (Camden, Maine: Yankee Books, 1990), 151–5.

2. James Dodson, "The Wind That Shook the World," in Watson, *New England's Disastrous Weather*, 3–13; and "Storm Batters All New England; Providence Hit by Tidal Wave," *New York Times*, 22 September 1938, 1, 18.

3. The catastrophe occurred on the morning of 3 December 1984. See, for example, Sanjoy Hazarika, "Gas Leak in India Kills at Least 410 in City of Bhopal," *New York Times*, 4 December 1984, A1; "Poison Gas Leak at Union Carbide Plant in India Kills Hundreds, Injures 10,000," *Wall Street Journal*, 4 December 1984, 3; Sanjoy Hazarika, "India Police Seize Factory Records of Union Carbide," *New York Times*, 7 December 1984, A1, A10. Also see William Bogard, *The Bhopal Tragedy: Language, Logic, and Politics in the Production of a Hazard* (Boulder, Colo.: Westview Press, 1989).

4. Dan Kurzman, *A Killing Wind: Inside Union Carbide and the Bhopal Catastrophe* (New York: McGraw-Hill, 1987), 20–4.

5. See ibid., 37–98; and Ward Morehouse and M. Arun Subramanian, *The Bhopal Tragedy: What Really Happened and What It Means for American Workers and Communities at Risk* (New York: Council on International and Public Affairs, 1986), 1–22.

6. Kurzman, *Killing Wind*, 206–7. For news accounts of the 11 August 1985 accident, see Associated Press, "Chemical Leak Warning System Failed; Hospitals Treat 131 near W.Va. Plant," *Syracuse Herald-Journal*, 12 August 1985, A1, A6; "Union Carbide Plant Leaks Gas; at Least 100 Hurt," *Wall Street Journal*, 12 August 1985, 5; and Barry Meier and Terence Roth, "Union Carbide Says Site Lacked New Safety Gear; Concern Waited 19 Minutes to Tell Officials of Leak of Gas at Institute Plant," *Wall Street Journal*, 13 August 1985, 3, 20. Because it manufactured MIC, Union Carbide's Institute plant aroused public concern immediately after the Bhopal disaster; see Carol Hymowitz and Terence Roth, "In West Virginia's 'Chemical Valley,' India's Toxic Gas Disaster Stirs Fears," *Wall Street Journal*, 5 December 1984, 4.

7. EPCRA was part of a larger legislative action, the Superfund Amendment and Reauthorization Act (SARA), which revised and extended the Comprehensive Environmental Response, Compensation, and Liability Act (CERCLA), whereby Congress established the Superfund program in 1980. Because EPCRA is the third principal section, or title, of SARA, writers often refer to it as SARA Title III, or merely Title III. For a concise introduction to EPCRA and its requirements, see U.S. Environmental Protection Agency, *Chemicals in Your Community: A Guide to the Emergency Planning and Community Right-to-Know Act*, EPA publication no. 550-K-93-003 (Washington, D.C., 1988). Also see Susan G. Hadden, *A Citizen's Right to Know: Risk Communication and Public Policy* (Boulder, Colo.: Westview Press, 1989).

8. Evelina R. Moulder, *Local Emergency Response Plans*, Special Data Issue (Washington, D.C.: International City/County Management Association, 1993), 1. For an examination of the operations and effectiveness of LEPCs—but not their use of maps or modeling—see Rosemary O'Leary, *Emergency*

Planning: Local Government and the Community Right-to-Know Act (Washington, D.C.: International City/County Management Association, 1993).

9. For discussion of philosophical differences between public administration experts and civil defense personnel, see Loran B. Smith, "Civil Defense," in Waugh and Hy, *Handbook of Emergency Management*, 271–92.

10. For a critique of LEPCs, see Susan G. Hadden, "Citizen Participation in Environmental Policy Making," in Sheila Jasanoff, ed., *Learning from Disaster: Risk Management after Bhopal* (Philadelphia: University of Pennsylvania Press, 1994), 91–112.

11. Ibid., 98.

12. Ute J. Dymon, "Mapping—The Missing Link in Reducing Risk under SARA III," *Risk: Health, Safety and Environment* 5 (1994): 337–49.

13. For guidelines concerning map content, see Federal Emergency Management Agency, *A Guide for the Review of State and Local Emergency Operations Plans*, FEMA publication no. CPG 1-8A (Washington, D.C., 1992), A-3.

14. See Dymon, "Mapping—The Missing Link"; and Aaron Zitner, "500 Marlborough Residents Return Home after Explosions," *Boston Globe*, 7 January 1991, 18.

15. U.S. Defense Civil Preparedness Agency, *Disaster Operations: A Handbook for Local Government* (Washington, D.C., 1972), 13.

16. Dymon, "Mapping—The Missing Link."

17. U.S. Environmental Protection Agency, Office of Solid Waste and Emergency Response, *Opportunities and Challenges for Local Emergency Planning Committees*, Technical Assistance Bulletin vol. 10 (November 1991). CAMEO was widely promoted by the EPA, first in a Macintosh version and later in a DOS version. Several short reports from LEPCs mention CAMEO; see, for example, U.S. Environmental Protection Agency, Office of Solid Waste and Emergency Response, *Successful Practices in Title III Implementation*, EPA publication no. OSWER-90-006.2, series 6, no. 5 (June 1990), 7, 9, 16–18.

18. Hadden, *Citizen's Right to Know*, 104.

19. Joseph Falge, interviews by author, 2 March 1994 and 17 May 1995.

20. See, for example, Patrick Lakamp and Brian Carr, "Critics Fault Emergency Office," *Syracuse Herald-American*, 3 October 1993, D1, D4; and "Where Is Falge? Courting Disaster in Flood Battle" [editorial], *Syracuse Herald-American*, 27 March 1994, C4. A story following the bombing of the Federal Building in Oklahoma City was more kind; see Dan McGuire, "Onondaga County Is Prepared for the Worst," *Syracuse Herald-Journal*, 10 May 1995, A1, A6.

21. The highway map, with a 1986 copyright date, was produced for Onondaga County by the National Survey, a Chester, Vermont, firm providing highway maps for many counties in the Northeast.

22. The town maps were cut (and joined where necessary) into town-unit portions by the Onondaga Environmental Management Council. The street atlas, produced by a regional street map company with headquarters in Syracuse, is *Central New York Atlas, Metro Syracuse, Utica & Ithaca*, new ed., rev. and updated (Syracuse, N.Y.: Marshall Penn-York, 1992).

23. For example, a list of thirteen "elements in emergency response plans"

contains no explicit mention of maps; see Moulder, *Local Emergency Response Plans*, 2.

24. Dick Friess, New York State Emergency Management Office, telephone conversation with author, 16 May 1995. As of May 1995, only fourteen of the fifty-seven counties outside New York City did not have EIS. But not all of the forty-three offices had effectively implemented the software.

25. Friess, telephone conversation; and Dan O'Brien, New York State Emergency Management Office, telephone conversation with author, 16 May 1995.

26. Len Egol, "Hazardous Materials Trigger a Software Explosion," *Chemical Engineering* 96 (April 1989): 179–82; "Florida County Uses ARC/INFO and EGRESS for Toxic Emergencies," *ARC News* 16 (winter 1994): 14; Lawrence G. Mondschein, "The Role of Spatial Information Systems in Environmental Emergency Management," *Journal of the American Society for Information Science* 45 (1994): 678–85; and J. W. Morentz, D. Griffith, and J. Applebaum, "SARA Title III Software: A Review of What's Available and What It Can Do for You," *Environmental Management Review* 26 (1992): 36–46.

27. See Constance Wench, "Software Review: Emergency Information System/Chemical Emergencies (EIS/C)," *American Industrial Hygiene Association Journal* 54 (June 1993): 335–9.

28. Richard Cobb, interview by author, 22 June 1995.

29. See John H. Sorensen and Sam A. Carnes, "An Approach for Deriving Emergency Planning Zones for Chemical Munitions Emergencies," *Journal of Hazardous Materials* 30 (1992): 223–42. For discussion of the use of vulnerability zones in siting hazardous facilities, see Michael K. Lindell, "Assessing Emergency Preparedness in Support of Hazardous Facility Risk Analysis: Application to Siting a US Hazardous Waste Incinerator," *Journal of Hazardous Materials* 40 (1995): 299–319.

30. For discussion of the use of wind frequency data to refine vulnerability zones, see M. P. Singh and others, "Estimation of Vulnerability Zones Due to Accidental Release of Toxic Materials Resulting in Dense Gas Clouds," *Risk Analysis* 11 (1991): 425–40.

31. J. Andrew Walker, George E. Ruberg, and John J. O'Dell, "Simulation for Emergency Management: Taking Advantage of Automation in Emergency Preparedness," *Simulation* 53 (1989): 95–100. In addition to plume modeling, expert systems software can help emergency response officials address complex issues arising during an emergency; see Judith M. Hushon, "Response to Chemical Emergencies," *Environmental Science and Technology* 20 (1986): 118–21.

32. Thomas Bowman, interview by author, 23 May 1995.

33. Thomas Bowman, telephone conversation with author, 29 June 1995.

34. For descriptions of TIGER/Line files and the TIGER concept, see Frederick R. Broome and David B. Meixler, "The TIGER Data Base Structure," *Cartography and Geographic Information Systems* 17 (1990): 39–47; U.S. Bureau of the Census, *1990 Census of Population and Housing: Guide, Part A. Text*, Census publication 1990 CPH-R-1A (Washington, D.C., 1992), 63–7; U.S. Bureau of the Census, "1992 TIGER/Line Files: Helping You Map

Things Out," *Product Profile* no. 6 (November 1993); and J. Paul Wyatt, *TIGER, the Coast-to-Coast Digital Map Data Base* (Washington, D.C.: U.S. Department of Commerce, Bureau of the Census, 1990).

35. Richard Cobb, interview by author, 23 May 1995.

36. Address matching works only when the data cooperate. A "hit" is not guaranteed unless the address is "well-formed" and the TIGER file up-to-date and error-free. For discussion of the potential and pitfalls of address matching, see Donald F. Cooke, "Spatial Data for Business," in Gilbert H. Castle III, ed., *Profiting from a Geographic Information System* (Fort Collins, Colo.: GIS World, 1993), 211–30.

37. The house at 415 is half a street width S and 7.5 lot widths L of the distance up the street from the low-address intersection. Because the total distance between the intersections is a whole street width (assuming streets are equal in width) plus 12 lot widths, the northing at 415 can be calculated as

$$N(415) = N(low) + [(0.5S + 7.5L)/(S + 12L)](N(high) - N(low)),$$

where N(low) and N(high) are the northings of the low-address and high-address intersections, respectively. This process is called *linear interpolation*. A similar formula estimates the easting at 415 as

$$E(415) = E(low) + [(0.5S + 7.5L)/(S + 12L)](E(high) - E(low)).$$

38. Robert Laurini and Derek Thompson, *Fundamentals of Spatial Information Systems* (London: Academic Press, 1992), 226–44, 314–24; Donna J. Peuquet, "An Examination of Techniques for Reformatting Digital Cartographic Data, Part 1: The Raster-to-Vector Process," *Cartographica* 18 (spring 1981): 34–48; and Jeffrey Star and John Estes, *Geographic Information Systems: An Introduction* (Englewood Cliffs, N.J.: Prentice-Hall, 1990), 33–48.

39. C. Dana Tomlin, *Geographic Information Systems and Cartographic Modeling* (Englewood Cliffs, N.J.: Prentice-Hall, 1990), 134–49.

40. For example, see Mark Abkowitz, Paul Der-Ming Cheng, and Mark Lepofsky, "Ship It by GIS," *Civil Engineering* 60 (April 1990): 64–6; Anthony C. Gatrell and Peter Vincent, "Managing Natural and Technological Hazards," in Ian Masser and Michael Blakemore, eds., *Handling Geographical Information: Methodology and Potential Applications* (London: Longman Scientific and Technical, 1991), 148–80; Sylvain Lassarre, Kurt Fedra, and Elisabeth Weigkricht, "Computer-Assisted Routing of Dangerous Goods for Haute-Normandie," *Journal of Transportation Engineering* 119 (1993): 200–10; and Mark Lepofsky, Mark Abkowitz, and Paul Cheng, "Transportation Hazard Analysis in Integrated GIS Environment," *Journal of Transportation Engineering* 119 (1993): 239–54. Also see Douglas W. Harwood, John G. Viner, and Eugene R. Russell, "Procedure for Developing Truck Accident and Release Rates for Hazmat Routing," *Journal of Transportation Engineering* 119 (1993): 189–99. In semiarid regions, GIS is also useful for identifying areas vulnerable to wildfire as well as managing mitigation and forecasting the spread of an actual fire. For discussion of the need for intergovernmental coordination, see Lisa Warnecke, "GI–GIS Projects in the Field: Overcoming the Government Maze," *Geo Info Systems* 5 (October 1995): 34–43.

41. For discussion of community vulnerability modeling with geographic information systems, see George F. Hepner and Harvey J. Miller, "Using GIS

for Environmental Assessment," *ACSM Bulletin* no. 146 (November/December 1993): 42–6; and Robert B. McMaster, "Modeling Community Vulnerability to Hazardous Materials Using Geographic Information Systems," *Proceedings of the Third International Symposium on Spatial Data Handling*, 17–19 August 1988, Sydney, Australia, 143–56.

42. For discussion of hazard geometries, see James H. Johnson, Jr., and Donald J. Zeigler, "Evacuation Planning for Technological Hazards: An Emerging Imperative," *Cities* 3 (1986): 148–56.

43. Identification of threatened subpopulations is one of many key questions in Susan Cutter's list "What You Can Do in Your Community," included as a sidebar in Susan L. Cutter, "Airborne Toxic Releases," *Environment* 29 (July/August 1987): 12–17, 28–31.

44. Vulnerability can be defined as the discrepancy between hazardousness and preparedness; see Susan L. Cutter and William D. Solecki, "The National Pattern of Airborne Toxic Releases," *Professional Geographer* 41 (1989): 149–61; and Susan L. Cutter and John Tiefenbacher, "Chemical Hazards in Urban America," *Urban Geography* 12 (1991): 417–30. Also see William Duncan Solecki, "Acute Chemical Disasters and Rural United States Hazardousness" (Ph.D. diss., Rutgers University, 1990), abstract in *Dissertation Abstracts International* 52 (1991): 1035-A.

45. For an example of a local library providing GIS access to an environmental and emergency data base, see Kevin Corbley, "Citizens Enjoy GIS Access," *GIS World* 27 (March 1995): 66–8.

46. The models were ALOHA, ARCHIE, and "The Green Book," and vulnerability zones were based on a level of concern (LOC) concentration of 3 ppm. Letter from William C. Stone, chairperson of the Local Emergency Planning Committee of Schenectady County, to Donald DeVito, chairperson of the State Emergency Response Commission Working Group, 17 February 1995.

47. *LEPC Hazard Analysis: Potential Chlorine Release Accident, Town of Rotterdam Sewage Treatment Plant*, report approved by the Schenectady County Local Emergency Planning Committee, 20 April 1995.

48. Bradford C. Mank, "Preventing Bhopal: 'Dead Zones' and Toxic Death Risk Index Taxes," *Ohio State Law Journal* 53 (1992): 761–804.

49. Keith Schneider, "Chemical Plants Buy Up Neighbors for Safety Zone," *New York Times*, 28 November 1990, A1, B8.

CHAPTER TWELVE

General references include Carolyn Rebecca Block and Margaret Dabdoub, eds., *Workshop on Crime Analysis through Computer Mapping Proceedings: 1993* (Chicago: Illinois Criminal Justice Information Authority, 1993); Paul J. Brantingham and Patricia L. Brantingham, eds., *Environmental Criminology* (Beverly Hills, Calif.: Sage Publications, 1981); Robert J. Bursik, Jr., and Harold G. Grasmick, *Neighborhoods and Crime: The Dimensions of Effective Community Control* (New York: Lexington Books, 1993); David J. Evans and David T. Herbert, eds., *The Geography of Crime* (London: Routledge, 1989);

Robert M. Figlio, Simon Hakim, and George F. Rengert, eds., *Metropolitan Crime Patterns* (Monsey, N.Y.: Willow Tree Press, 1986); Daniel E. Georges-Abeyie and Keith D. Harries, eds., *Crime: A Spatial Perspective* (New York: Columbia University Press, 1980); Keith D. Harries, *The Geography of Crime and Justice* (New York: McGraw-Hill, 1974); Michael D. Maltz, Andrew C. Gordon, and Warren Friedman, *Mapping Crime in Its Community Setting* (New York: Springer-Verlag, 1990); and Susan J. Smith, *Crime, Space, and Society* (Cambridge: Cambridge University Press, 1986).

1. For a concise history of Uniform Crime Reporting, see U.S. Department of Justice, Federal Bureau of Investigation, *Crime in the United States, 1993* (Washington, D.C., 1994), 1–3. For a brief evaluation, see Victoria W. Schneider and Brian Wiersema, "Limits and Use of the Uniform Crime Reports," in Doris Layton MacKenzie, Phyllis Jo Baunach, and Roy R. Roberg, eds. *Measuring Crime: Large-Scale, Long-Range Efforts* (Albany: State University of New York Press, 1990), 21–48. Since the late 1980s, the Department of Justice has been converting to a more detailed, more fully computerized data collection process called the National Incident-Based Reporting System (NIBRS), which will provide data for twenty-two kinds of offenses. For examples of the uses of the new data, see Brian A. Reaves, *Using NIBRS Data to Analyze Violent Crime* (Washington, D.C.: U.S. Department of Justice, Bureau of Justice Statistics, 1993).

2. U.S. Department of Justice, Federal Bureau of Investigation, *Uniform Crime Reporting Handbook* (Washington, D.C., 1984).

3. Although some information is provided earlier, in press releases, the FBI publishes *Crime in the United States* toward the end of the year following the calendar year covered by the report. Another annual report is U.S. Department of Justice, Federal Bureau of Investigation, *Law Enforcement Officers Killed and Assaulted* (Washington, D.C., 1982–). For an example of a special report using UCR data, see Association of State Uniform Crime Reporting Programs and the Center for Applied Social Research, Northwestern University, *Hate Crime Statistics, 1990: A Resource Book* (Washington, D.C.: Federal Bureau of Investigation, 1993).

4. *Uniform Crime Reporting Handbook*, 16.

5. *Crime in the United States, 1993*, 57. But some tabulations include a so-called Modified Crime Index, obtained by adding incidents of arson to totals for the other seven index crimes; see, for example, table 8 in ibid., 109–57.

6. *Uniform Crime Reporting Handbook*, 16, 28.

7. See, for example, Schneider and Wiersema, "Limits and Use of the Uniform Crime Reports."

8. U.S. Department of Justice, Bureau of Justice Statistics, *Criminal Victimization in the United States, 1992* (Washington, D.C., 1994), 102.

9. Ibid., 108.

10. Ibid., 107.

11. Ibid., 103.

12. R. I. Mawby, "Policing and the Criminal Area," in Evans and Herbert, *Geography of Crime*, 260–81.

13. James Q. Wilson, *Varieties of Police Behavior: The Management of Law and*

Order in Eight Communities (Cambridge, Mass.: Harvard University Press, 1968), 140–99.

14. For a facsimile of the questionnaire, see *Criminal Victimization in the United States, 1992*, 120–40.

15. Ibid., 52.

16. *Crime in the United States, 1993*, 77.

17. See, for example, ibid., 9

18. A quintile classification is a five-category quantile classification, obtained by ranking the data values from lowest to highest and assigning (as much as possible) an equal number of areas to each category—easy to do when dividing fifty states among five categories of ten each. Interval ranges for each category are then found by inspecting the list of ranked data values. For a study that demonstrates the relative efficacy of quantile classification for intermap comparisons, see Judy M. Olson, "The Effects of Class Interval Systems on Choropleth Map Correlation," *Proceedings of the Association of American Geographers* 3 (1971): 127–30.

19. See, for example, Harries, *Geography of Crime and Justice*, 16–37; Keith D. Harries and Stephen J. Stadler, "Assault and Heat Stress: Dallas As a Case Study," in Evans and Herbert, *Geography of Crime*, 38–58; and Savageau and Boyer, *Places Rated Almanac*, 212–15.

20. Barbara Gimla Shortridge, *Atlas of American Women* (New York: Macmillan, 1987), 125–7. For general discussion of other factors underlying spatial variations in rape, see Larry Baron and Murray A. Straus, *Four Theories of Rape in American Society: A State-Level Analysis* (New Haven, Conn.: Yale University Press, 1989).

21. I say "suggests" because the NCVS reports results for households, not for individuals. Reporting rates cited here are for females. See *Criminal Victimization in the United States, 1992*, 103.

22. See *Crime in the United States, 1993*, 32–3, 39–40. Although burglaries are least common in February and most frequent in August, because of a pre-Christmas rush, the seasonal trend is less pronounced than for aggravated assault.

23. Savageau and Boyer, *Places Rated Almanac*, 215–16.

24. Ibid., 214.

25. For maps describing variations in crime among metropolitan areas, see Roger Doyle, *Atlas of Contemporary America: Portrait of a Nation—Politics, Economy, Environment, Ethnic and Religious Diversity, Health Issues, Demographic Patterns, Quality of Life, Crime, Personal Freedoms* (New York: Facts on File, 1994), 166–7. Using data for a single year, 1990, Doyle provides separate summary maps for violent and property crimes.

26. The tables of 343 metro areas include 25 in Canada, but no Canadian city placed in either the top or the bottom group.

27. The Punta Gorda metropolitan area is Charlotte County, which includes the heavily promoted retirement community of Port Charlotte. For further information, see Edward A. Fernald and others, *Atlas of Florida* (Gainesville: University Press of Florida, 1992), 132, 154, 254, 268.

28. For criteria used to identify metropolitan areas, see U.S. Bureau of the Census, *State and Metropolitan Area Data Book, 1991* (Washington, D.C., 1991), 353–9. For a critique of the application of these criteria by the Office of Management and Budget, see Jeremy Schlosberg, "The MSA Mess: Business Depends on Metropolitan Rankings That Are Meaning Less," *American Demographics* 11 (January 1989): 53–8. For discussion of new metropolitan areas designated in late 1992, see Judith Waldrop, "Why Metro Numbers Are Meaning Less," *American Demographics* 15 (March 1993): 9–11.

29. Savageau and Boyer, *Places Rated Almanac*, 230.

30. Phyllis Kaniss, *Making Local News* (Chicago: University of Chicago Press, 1991), 58, 76, 78.

31. Susan J. Smith, "News and the Dissemination of Fear," in Jacquelin Burgess and John R. Gold, eds., *Geography, The Media, and Popular Culture* (London: Croom Helm, 1985), 229–53. Although Smith's study focused on the British press, her criticism was broad and she referred frequently to the writings of sociologist Robert Park, who examined the press in the U.S., particularly in Chicago.

32. Roy Edward Lotz, *Crime and the American Press* (New York: Praeger Publishers, 1991), 61–2.

33. See, for example, Mitchell Stephens, *A History of News: From the Drum to the Satellite* (New York: Viking, 1988), 263–4.

34. Ibid., 113–17, 239–40, 285.

35. Kai Erikson, "Notes on the Sociology of Deviance," in Howard Becker, ed., *The Other Side: Perspectives on Deviance* (New York: Free Press, 1964), 9–21; quote on p. 14.

36. See Lotz, *Crime and the American Press*, 45–8.

37. Sue Ann Wood, "How 'Where' Gets Handled," *St. Louis Post-Dispatch*, 4 February 1990, D1.

38. For discussion of the map's dual role in newspaper design and news reporting, see Mark Monmonier, *Maps with the News: The Development of American Journalistic Cartography* (Chicago: University of Chicago Press, 1989), 19–24 , 125–94.

39. A good example is the map "A Five-Day Rampage," which describes the "trail of violence" that ended when police shot the perpetrator during a car crash. See N. R. Kleinfeld, "Ex-Convict's Murder Odyssey: Seven People Die in Two States," *New York Times*, 22 June 1995, A1, B4.

40. For the story accompanying the map, see Lori Duffy, "Couple's Fight Ends in Fatal Stabbing," *Syracuse Post-Standard*, 2 June 1995, A1.

41. See, for example, Matthew Purdy, "1993 Homicides Fewer and More Clustered in New York City," *New York Times*, 10 January 1994, B1, B4.

42. For the principal story, see Pamela Sampson and Theresa Brady, "Carjacking: When Thieves Become Bold," *Greensburg Tribune-Review*, 27 June 1993, A1, A10.

43. See Mitchell R. Joelson and Glenn M. Fishbine, "The Display of Geographic Information in Crime Analysis," in Georges-Abeyie and Harries, *Crime: A Spatial Perspective*, 249–63; Tod Newcombe, "High-Tech Tools for the

War on Crime," *Governing* 6 (August 1993): 20–1; and Sandra Wendelken, "GIS Enhances Preventive Law Enforcement," *GIS World* 8 (January 1995): 58–61.

44. See, for example, Ralph A. Sanders, "Theory and Practice in Urban Police Response: A Case Study of Syracuse, New York," in Georges-Abeyie and Harries, *Crime: A Spatial Perspective*, 276–89.

45. For discussion of geographic profiling, see D. Kim Rossmo, "Overview: Multivariate Spatial Profiles as a Tool in Crime Investigation," in Block and Dabdoub, *Workshop on Crime Analysis through Computer Mapping*, 89–126.

46. Federal Bureau of Investigation, *Manual of Law Enforcement Records* (Washington, D.C., 1984), 38–40; quote on p. 38.

47. For a history and concise description of STAC, see Carolyn Rebecca Block, "STAC Hot Spot Areas: A Statistical Tool for Law Enforcement Decisions," in Block and Dabdoub, *Workshop on Crime Analysis through Computer Mapping*, 33–53.

48. Mapping packages used to display STAC analyses include ArcInfo/ArcView, ATLAS, MapInfo, and Streets-on-a-Disk. Daniel Higgins, letter to author, 19 June 1995.

49. For discussion of the nearest-neighbor statistic used to confirm that a pattern of points is more clustered than random, see Barry N. Boots and Arthur Getis, *Point Pattern Analysis*, Scientific Geography Series vol. 8 (Newbury Park, Calif.: Sage Publications, 1988).

50. For examples and discussion of the effects of using different radii, see George F. Rengert, "Comparing Cognitive Hot Spots to Crime Hot Spots," in Block and Dabdoub, *Workshop on Crime Analysis through Computer Mapping*, 55–67.

51. For information on the calculation and use of dispersion ellipses (also known as standard deviational ellipses), see D. W. Lefever, "Measuring Geographical Concentration by Means of a Standard Deviational Ellipse," *American Journal of Sociology* 32 (1962): 89–94; and Larry K. Stephenson, "Centrographic Analysis of Crime," in Georges-Abeyie and Harries, *Crime: A Spatial Perspective*, 146–55.

52. See Carolyn Rebecca Block, "Automated Spatial Analysis as a Tool in Violence Reduction," *CJ the Americas* 6 (February–March 1993): 7–8, 10; and Carolyn Rebecca Block and Richard Block, *Street Gang Crime in Chicago*, NIJ Research in Brief series (Washington, D.C.: National Institute of Justice, December 1993).

53. Carolyn Rebecca Block, telephone conversation with author, 15 June 1995.

54. Ibid.

55. See Colin McMahon, "Crimefighting Tool Puts Rogers Park on the Map," *Chicago Tribune*, 31 May 1993, sec. 2C, 1–2.

56. For discussion of the use of maps to communicate to legislative bodies and the general public, see Douglas Hicks and Jim Wilson, "Geographic Analysis to Support Community Policing: An Information Foundation for the Future," in Block and Dabdoub, *Workshop on Crime Analysis through Computer Mapping*, 295–308.

57. For discussion of crime prevention through urban design, see Cortus T. Koehler, *Urban Design and Crime: A Partially Annotated Bibliography*, CPL bibliography no. 218 (Chicago: Council of Planning Librarians, 1988); and Oscar Newman, *Defensible Space: Crime Prevention through Urban Design* (New York: Macmillan, 1972).

58. In an impressive example of cooperation, New York City shared crime, prosecution, and corrections data with the Vera Institute of Justice, which produced and published an innovative graphic summary of crime and justice. See Lola E. Odubekun and others, *The Vera Institute Atlas of Crime and Justice in New York City* (New York: Vera Institute of Justice, 1993).

CHAPTER THIRTEEN

General references include Melinda S. Meade, John W. Florin, and Wilbert M. Gesler, *Medical Geography* (New York: Guilford Press, 1988); Richard F. Mould, *Cancer Statistics* (Bristol, U.K.: Adam Hilger, 1983); Gerald F. Pyle, *Applied Medical Geography* (Washington, D.C.: V.H. Winston and Sons, 1979); Gary W. Shannon and Gerald F. Pyle, *Disease and Medical Care in the United States: A Medical Atlas of the Twentieth Century* (New York: Macmillan, 1993); R. W. Thomas, ed., *Spatial Epidemiology*, London Papers in Regional Science no. 21 (London: Pion, 1990); and J. A. M. van Oers, *A Geographic Information System for Local Public Health Policy* (Assen, The Netherlands: Van Gorcum, 1993).

1. See E. W. Gilbert, "Pioneer Maps of Health and Disease in England," *Geographical Journal* 124 (June 1958): 172–83.

2. Snow's fame might well exceed his impact. By the time he disabled the pump, the epidemic had already begun to subside.

3. For a concise, well-illustrated examination of map use in epidemiology and medical geography, see Pyle, *Applied Medical Geography*, 37–79.

4. For a description of death registration, see *Physician's Handbook on Medical Certification of Death*, Department of Health and Human Services publication no. (PHS) 87-1108 (Hyattsville, Md.: National Center for Health Statistics, 1987); and the technical appendix in *Vital Statistics of the United States, 1990*, vol. 2, *Mortality*, pt. A (Washington, D.C.: National Center for Health Statistics, 1994).

5. The Center typically issues a "final advance report" two years after end of the calendar year covered and a final hardbound report two years later. Separate volumes of *Vital Statistics* contain counts for births, marriages, and divorces.

6. In addition to the NCHS death tallies, the Centers for Disease Control (CDC), in Atlanta, counts cases of HIV, measles, and other highly contagious diseases. Physicians who diagnose ailments on the CDC's official list of "notifiable" diseases are supposed to report them immediately, but many don't, largely to shield patients from nosy public health investigators intent on tracking down other persons who might have been exposed. Other cases are missed because victims don't seek treatment or are misdiagnosed. For more information on the data network for contagious diseases, see Elizabeth W. Etheridge,

Sentinel for Health: a History of the Centers for Disease Control (Berkeley: University of California Press, 1992).

7. *Vital Statistics of the United States, 1990,* vol. 2, *Mortality,* pt. A, technical appendix, 18.

8. Malcolm A. Murray, "Geography of Death in the United States and the United Kingdom," *Annals of the Association of American Geographers* 57 (1967): 301–14; especially 311.

9. In 1989 a revision of the standard death certificate added items on educational attainment and Hispanic origin.

10. Two methods are available for age adjustment: direct and indirect. The indirect method, described here, is especially useful for small areas, for which age-specific death rates are not available. For a fuller description of age adjustment of death rates, see J. H. Abramson, *Making Sense of Data: A Self-Instruction Manual on the Interpretation of Epidemiological Data* (New York: Oxford University Press, 1988), 136–45; and Donald J. Bogue, *Principles of Demography* (New York: John Wiley and Sons, 1969), 122–4.

Direct adjustment provides more reliable comparisons but requires age-specific mortality data for each areal unit, a requirement that accounts for the wider use of the indirect method. See S. D. Walter and S. E. Birnie, "Mapping Mortality and Morbidity Patterns: An International Comparison," *International Journal of Epidemiology* 20 (1991): 678–89.

11. Mortality counts and age-adjusted death rates are from *Monthly Vital Statistics Report* 43 (24 October 1994): 8–9.

12. As with the crime rates in chapter 12, I used the Visibility Base Map, but because age-adjusted rates are not readily merged, I omitted the District of Columbia, a non-state, rather than combine it with Maryland.

13. For discussion of heart disease among African Americans, see John Z. Ayanian, "Heart Disease in Black and White," *New England Journal of Medicine* 329 (26 August 1993): 656–8; Sybil L. Crawford and others, "Do Blacks and Whites Differ in Their Use of Health Care for Symptoms of Coronary Heart Disease?" *American Journal of Public Health* 84 (June 1994): 957–64; and Julian E. Keil and others, "Mortality Rates and Risk Factors for Coronary Disease in Black As Compared with White Men and Women," *New England Journal of Medicine* 329 (8 July 1993): 73–8.

14. See, for example, Morton H. Shaevitz, *Lean and Mean: The No Hassle, Life-Extending Weight Loss Program for Men* (New York: G.P. Putnam's Sons, 1993), 17–19; and Redford Williams and Virginia Williams, *Anger Kills* (New York: Times Books, 1993), 40–5.

15. See the chapter on coronary heart disease in Shannon and Pyle, *Disease and Medical Care in the United States,* 44–54.

16. Ibid., 49–50.

17. This section is a generalized, nontechnical explanation of statistical significance. Standard texts on biostatistics offer a more detailed discussion, complete with mathematical formulas. For a lucid comparative treatment, see R. F. Mould, *Introductory Medical Statistics* (Tunbridge Wells, Kent, U.K.: Pitman Medical Publishing, 1976), 31–9.

18. A probability model generates theoretical distribution curves tailored to

the number of deaths recorded for an area. Epidemiologists typically assess the significance of nonzero age-adjusted death rates using the estimated standard error and confidence limits described in Chin Long Chiang, "Standard Error of the Age-Adjusted Death Rate," *Vital Statistics Selected Reports* 47, no. 9 (1961). Assessments of the significance of zero rates based on no reported deaths are based on the exact Poisson test, described in George W. Snedecor and William G. Cochran, *Statistical Methods*, 6th ed. (Ames: Iowa State University Press, 1967), 223–7.

19. For examples of other disease atlases, see Owen J. Devine and others, *Injury Mortality Atlas of the United States, 1979–1987* (Atlanta, Ga.: U.S. Public Health Service, 1991); and Thomas J. Mason and others, *An Atlas of Mortality from Selected Diseases*, National Institutes of Health publication no. 81-2397 (Washington, D.C., 1981).

20. See, for example, Thomas J. Mason and others, *Atlas of Cancer Mortality for U.S. Counties: 1950–1969*, Department of Health, Education and Welfare publication no. (NIH) 75-780 (Washington, D.C., 1975).

21. Wilson B. Riggan and others, *U.S. Cancer Mortality Rates and Trends, 1950–1979*, vol. 4, *Maps*, Environmental Protection Agency publication no. 600/1-83/015e (Washington, D.C., 1987), iii.

22. Linda Williams Pickle and others, *Atlas of U.S. Cancer Mortality among Whites: 1950–1980*, Department of Health and Human Services publication no. (NIH) 87-2900 (Washington, D.C., 1987); and Linda Williams Pickle and others, *Atlas of U.S. Cancer Mortality among Nonwhites: 1950–1980*, Department of Health and Human Services publication no. (NIH) 90-1582 (Washington, D.C., 1990).

23. For discussion of the trend maps, see Pickle and others, *Atlas of U.S. Cancer Mortality among Whites: 1950–1980*, 6–8.

24. Linda Williams Pickle, telephone conversation with author, 13 July 1995. For the maps of female lung cancer, see Pickle and others, *Atlas of U.S. Cancer Mortality among Whites: 1950–1980*, 78–9.

25. For maps of lung and related cancers among white males, see ibid., 76–7.

26. For the maps of female lung cancer, see Pickle and others, *Atlas of U.S. Cancer Mortality among Nonwhites: 1950–1980*, 92–3.

27. For maps of prostate cancer among white males, see Pickle and others, *Atlas of U.S. Cancer Mortality among Whites: 1950–1980*, 90–1.

28. Pickle and others, *Atlas of U.S. Cancer Mortality among Nonwhites: 1950–1980*, 16.

29. See Aaron Blair and Joseph F. Fraumeni, Jr., "Geographic Patterns of Prostate Cancer in the United States," *Journal of the National Cancer Institute* 61 (1978): 1379–84.

30. For a collection of reports on projects supported by the center, see Linda Williams Pickle and Douglas J. Hermann, eds., *Cognitive Aspects of Statistical Mapping*, Cognitive Methods Staff Working Paper Series no. 18 (Hyattsville, Md.: National Center for Health Statistics, 1995).

31. Linda Pickle and others, *Atlas of United States Mortality* (Hyattsville, Md.: National Center for Health Statistics, 1996). Health service areas are

groups of one or more counties that are comparatively self-contained market areas for ordinary hospital services. See Diane M. Makuc and others, "Health Service Areas for the United States," *Vital and Health Statistics—Series 2: Data Evaluation and Methods Research* no. 112 (November 1991): 1–102.

32. The design also reflects the semiotic theory expounded by the French cartographer Jacques Bertin. See Linda Williams Pickle and others, "Designing the New U.S. Mortality Atlas," in Robert T. Aangeenbrug, ed., *Mapping in Epidemiology and Environmental Health* (Tampa, Fla.: World Computer Graphics Foundation, in press).

33. For a report on cognitive cartographic research sponsored by the center, see Stephan Lewandowsky and others, "Perception of Clusters in Statistical Maps," *Applied Cognitive Psychology* 7 (1993): 533–51.

34. In addition to consulting on map design, the cartographic research group in the Department of Geography at the Pennsylvania State University tested and refined the color sequences. For discussion, see Alan M. MacEachren, Cynthia A. Brewer, and Linda W. Pickle, "Mapping Health Statistics: Representing Data Reliability," *Proceedings of the 17th International Cartographic Conference*, Barcelona, Spain, 3–9 September 1995, 311–19.

35. For a comparison of statistical and visual notions of risk and error, see Alan M. MacEachren, *How Maps Work: Representation, Visualization, and Design* (New York: Guilford Press, 1995), 444–8.

36. For examples of this strategy in other contexts, see Peter Lloyd, ed., *Groupware in the 21st Century: Computer Supported Cooperative Working toward the Millennium* (Westport, Conn.: Praeger, 1994); and Mike Sharples, ed., *Computer Supported Collaborative Writing* (New York: Springer-Verlag, 1993).

37. See Mark Monmonier, *Atlas Touring: Concepts and Development Strategies for a Geographic Visualization Support System*, CASE Center Technical Report no. 9011 (Syracuse, N.Y.: New York State Center for Advanced Technology in Computer Applications and Software Engineering, 1990); and Mark Monmonier, "Pattern Templates as a Geography-Side Approach to Epidemiological Visualization," *Statistical Computing and Statistical Graphics Newsletter* [American Statistical Association] 6 (April 1995): 8–10.

38. For a concise review of the scope and goals of the National Spatial Data Infrastructure, see Mapping Science Committee, Board on Earth Sciences and Resources, Commission on Geosciences, Environment and Resources, National Research Council, *Toward a Coordinated Spatial Data Infrastructure for the Nation* (Washington, D.C.: National Academy Press, 1993).

39. For a concise description of Epi Map, see "New Computer Program Presents Public Health Data in Map Form," *Public Health Reports* 108 (1993): 795.

40. New York was one of the first states to establish a tumor registry. For a concise history of tumor registries, see G. Wagner, "History of Cancer Registration," in O. M. Jensen and others, eds., *Cancer Registration: Principles and Methods*, IARC Scientific Publication no. 95 (Lyon, France: International Agency for Research on Cancer, 1991), 3–6. According to a 1993 survey by the Centers for Disease Control, the states varied considerably in their laws and regulations for cancer registries. Nine states lacked laws authorizing a registry

as well as all eight regulations the CDC deemed essential, and four states had either no authorizing legislation or none of the eight regulations. See "State Cancer Registries: Status of Authorizing Legislation and Enabling Regulations—United States, October 1993," *Morbidity and Mortality Weekly Report* 43 (4 February 1994): 71–5.

41. The State Health Department covers upstate New York and the two outer counties of Long Island. Although New York City has its own health department and cancer surveillance program, the state cancer registry has covered the entire state since 1973. See New York State Department of Health, *Cancer Surveillance Program—Operations Manual* (Albany, N.Y., April 1991), 1–6.

42. Aura Weinstein, interview by author, Albany, N.Y., 10 July 1995.

43. New York State Department of Health, *Cancer Surveillance Program—Operations Manual*, 21.

44. Cancer Surveillance Program, Bureau of Cancer Epidemiology, New York State Department of Health, *Cancer Incidence in Zip Code Area 10504 (Armonk) Westchester County, New York* (Albany, N.Y., March 1992).

45. Steve Forand and Peter Lauridsen, interview by author, Albany, N.Y., 10 July 1995.

46. James M. Melius and others, "Residence Near Industries and High Traffic Areas and the Risk of Breast Cancer on Long Island" (New York State Department of Health, April 1994, photocopy).

47. More-affluent women would tend to live in more-affluent neighborhoods, in areas farther removed from chemical plants.

48. For other examples of the use of GIS in exploratory epidemiological research, see Mary Braddock and others, "Using a Geographic Information System to Understand Child Pedestrian Injury," *American Journal of Public Health* 84 (1994): 1158–61; and van Oers, *A Geographic Information System for Local Public Health Policy*, 88–108. High-interaction computer mapping systems that lack overlay and other integrative functions are also useful, especially for experimenting with race and sex categories, the beginning and end of the time period, the range of ages of patients or deceased, and disease categories (which can be broad or narrow); for an example, see Forrest A. Pommerenke and others, "Targeting Cancer Control: The State Cancer Control Map and Data Program," *American Journal of Public Health* 84 (1994): 1479–82.

49. For examples of promising developments in geographic analysis applied to epidemiology, see S. Openshaw, "Automating the Search for Cancer Clusters: A Review of Problems, Progress, and Opportunities," in Thomas, *Spatial Epidemiology*, 48–78; D. Wartenberg and M. Greenberg, "Space–Time Models for the Detection of Clusters of Disease," in Thomas, *Spatial Epidemiology*, 17–34; and S. D. Walter, "The Analysis of Regional Patterns in Health Data: II. The Power to Detect Environmental Effects," *American Journal of Epidemiology* 136 (1992): 742–59.

50. Gould discusses his narrative maps and their use in Peter Gould, "Source Error in a Map Series, or Science as a Socially Negotiated Enterprise," *Cartographic Perspectives* no. 21 (spring 1995): 30–6.

51. "Lyme Disease—United States, 1994," *Morbidity and Mortality Weekly Report* 43 (4 February 1994): 459–62.

52. See Lawrence K. Altman, "U.S. Agency Reports Lyme Disease Cases Up by 58% in '94," *New York Times*, 23 June 1995, A16.

CHAPTER FOURTEEN

1. See, for example, Arthur H. Robinson, *Early Thematic Mapping in the History of Cartography* (Chicago: University of Chicago Press, 1982). Robinson's temporal focus is the eighteenth and nineteenth centuries, and hazards maps are conspicuously absent from his discussion of maps of physical phenomena. Moreover, the catalog of the Geography and Map Division at the Library of Congress yielded a mere handful of pre-1950 hazard maps.

2. For a concise survey of the terrestrial impact hypothesis and other explanations, such as massive volcanic eruptions, see Antony Milne, *The Fate of the Dinosaurs: New Perspectives in Evolution and Extinction* (Bridport, Dorset, U.K.: Prism Press, 1991), especially pp. 166–92.

3. See David Morrison, ed., *The Spaceguard Survey: Report of the NASA International Near-Earth-Orbit Detection Workshop* (Pasadena, Calif.: Jet Propulsion Laboratory, 1992). The seventy-two-page Spaceguard report is included in U.S. Congress, House Committee on Science, Space, and Technology, *The Threat of Large Earth-Orbit Crossing Asteroids: Hearing before the Subcommittee on Space*, 103d Cong., 1st sess., 1993, 112–76. For additional discussion, see William J. Broad, "When Worlds Collide: A Threat to the Earth Is a Joke No Longer," *New York Times*, 1 August 1994, A1, A12; Michael Szpir, "Close Encounters of the Bad Kind," *American Scientist* 82 (May–June 1994): 220–1; and Gerrit L. Verschuur, "The End of Civilization," *Astronomy* 19 (September 1991): 51–5.

4. House Committee, *Threat of Large Earth-Orbit Crossing Asteroids*, 121.

5. See Thomas J. Ahrens and Alan W. Harris, "Deflection and Fragmentation of Near-Earth Asteroids," *Nature* 360 (1992): 429–33; and John D. H. Rather and others, *Summary Report of the Near-Earth-Object Interception Workshop* (Pasadena, Calif.: Jet Propulsion Laboratory, 1992). The sixty-five-page report is included in House Committee, *Threat of Large Earth-Orbit Crossing Asteroids*, 44–111.

6. See H. J. Melosh and I. V. Nemchinov, "Solar Asteroid Diversion," *Nature* 366 (1993): 21–2.

7. NASA's risk analysis relied on work of astrogeologist Eugene M. Shoemaker, who integrated astronomical data for smaller, more frequent impactors with paleontological evidence for the larger, less frequent objects, like the NEO believed to have exterminated the dinosaurs sixty-five million years ago. See, for example, Eugene M. Shoemaker, Ruth F. Wolfe, and Carolyn S. Shoemaker, "Asteroid and Comet Flux in the Neighborhood of Earth," in Virgil L. Sharpton and Peter D. Ward, eds., *Global Catastrophes in Earth History: An Interdisciplinary Conference on Impacts, Volcanism, and Mass Mortality*, Special Paper no. 247 (Boulder, Colo.: Geological Society of America, 1990), 155–70. For further discussion of the risk of colliding objects, see Clark R. Chapman

and David Morrison, "Impacts on the Earth by Asteroids and Comets: Assessing the Hazard," *Nature* 367 (1994): 3–40.

8. House Committee, *Threat of Large Earth-Orbit Crossing Asteroids*, 128–9.

9. For discussion of the difficulty of detecting small asteroids and infrequent comets, see House Committee, *Threat of Large Earth-Orbit Crossing Asteroids*, 145–8, 161 2.

10. Carl Sagan and Steven J. Ostro, "Dangers of Asteroid Deflection," *Nature* 368 (1994): 501.

11. For example, see Carl Sagan, "Nuclear War and Climatic Catastrophe: Some Policy Implications," in Lester Grinspoon, ed., *The Long Darkness: Psychological and Moral Perspectives on Nuclear Winter* (New Haven, Conn.: Yale University Press, 1986), 7–62; S. L. Thompson and others, "Global Climatic Consequences of Nuclear War: Simulations with Three-Dimensional Models," *Ambio* 13 (1984): 236–43; and U.S. Congress, Senate Committee on Armed Services, *Nuclear Winter and Its Implications*, 99th Cong., 1st sess., 1993, especially 38–40.

Index

Abelson, Philip, 178–79
Acid rain, 167
Address matching, 232–33
African Americans, mortality of, 268–70
Age-adjusted death rates, 11, 266–68
Age-specific incidence, in case-cluster analysis, 279
AIDS, geographic pattern, 284–85
Air pollution: allowances, 169–70; criteria pollutants, 150; dispersion models, 161–67; emissions credits, 169–70; geographic patterns, 155–56, 159–60, 171; health hazards, 148–50; measurement of, 151; meteorological conditions favoring, 149; monitoring of, 150–54; from motor vehicles, 152–54, 157–58; National Ambient Air Quality Standards, 155; non-attainment counties, 154–55; regional transfer of, 169; releases of, 158–59; from smoking, 171; wind dispersion of, 161, 163, 169
Air-pollution mapping: count data incorrectly portrayed, 161; dispersion models, 161–67; interpolation models, 167; national maps, 155–56; nitrate wet-deposition, 167–68; probability statements in, 166; and public policy, 172; reliability, 162; of smoking restrictions, 171; standardization, 155; symbols for, 156–58, 159, 163, 167
Air quality. *See* Air pollution
Alaska, tsunami hazard in, 67–69, 74–75
Algermissen, Ted, 41–44
ALOHA (Areal Locations of Hazardous Atmospheres), dispersion model for emergency management, 225, 227–29, 236–37
Alpha particles, from decay of radon gas, 174, 175–76
Alquist-Priolo Special Studies Zones Act, 21–25
Altoona, Pennsylvania, crime reporting in, 254
American Journal of Epidemiology, 187
Anders, Fred, 72
Anhydrous ammonia, as a chemical hazard, 225–26
Anti-smoking laws, geographic variations in, 171
Appalachia, low crime rates in, 250–51
Aquicludes, 129–30
Aquifers: and aquicludes, 129–30; confined, 130, 137–39; defined, 127; hydrologic role of, 129–32; zones of aeration and saturation, 129
Armonk (Westchester County), New York, cancer study of, 280–81